Intimate Facts about Wheat Grain Storage Proteins, Gliadins and Glutenins

Intimate Facts about Wheat Grain Storage Proteins, Gliadins and Glutenins

By

Eugene Metakovsky

Cambridge Scholars Publishing

Intimate Facts about Wheat Grain Storage Proteins, Gliadins and Glutenins

By Eugene Metakovsky

This book first published 2022

Cambridge Scholars Publishing

Lady Stephenson Library, Newcastle upon Tyne, NE6 2PA, UK

British Library Cataloguing in Publication Data
A catalogue record for this book is available from the British Library

ISBN (10): 1-5275-8837-8
ISBN (13): 978-1-5275-8837-0

CONTENTS

PREFACE BY THE AUTHOR

This book is based on my 40 years of experience in the field of storage protein, mainly gliadin, research. It is a complete review of more than 740 publications, which describe methods of wheat genotype analysis and identification, polymorphism and genetics (genes, loci, inheritance) of seed storage proteins, primary structure of gliadin and glutenin polypeptides and their encoding genes, the structure of the *Gli* and *Glu* loci, polypeptides having attributes of both gliadins and glutenins, variability at the *Gli* loci and types of spontaneous mutations, families of gliadin alleles, homoalleles and the origin of heteroalleles at the *Gli* loci, worldwide highly structured gliadin polymorphism of wheat germplasm (gene geography), the genetic diversity of common wheat germplasm during the 20th century and the problem of its genetic erosion, the origin of polyploid wheat and relationships among wheat species, the polymorphism of local varieties and landraces and problems of conservation and maintaining of wheat genetic resources in collections, and the relation of polymorphism at the *Glu* and *Gli* loci to the end-use quality… A special emphasis is made on inconsistencies and discrepancies of experimental data and on the discussion of some doubts and confusions.

My first steps in the field of wheat storage proteins were in Ukraine in contacts with Drs M.M. Kopus, A.A. Pomortsev, F.A. Poperelya, A.I. Rybalka, T.A. Sobko, A.A. Sozinov, I.A. Sozinov, as well as colleagues from other countries, Drs G. Boggini, B. Javornik, D. Lafiandra, A.Yu. Novoselskaya, P.I. Payne, R. Redaelli, J. Sutka, P. Vaccino, …

I am very much indebted to all my colleagues and assistants from different countries and laboratories, friends that I still know and have known. In some cases, we have known one another for many years and we have performed joint research. Our free discussions and analyses with Drs C.W. Wrigley, G. Branlard, N.E. Pogna, J.M. Carrillo, V.M. Chernakov, Z.A. Iakobashvili, S.F. Koval, V.A. Melnik, L. Pascual, S.V. Chebotar, G.A Romanov… and their help have encouraged me to write this book. I remember and appreciate the contributions of my dear colleagues and friends who have already passed away Drs E.V. Ananiev, W. Bushuk, S.F. Koval, N.E. Pogna, F.A. Poperelya, M. Rodriguez-Quijano, A.A. Sozinov....

I especially thank very much Dr C.W. Wrigley for his proposals, ideas and advices, always excellent, and who tried to improve my English a little in the final version of this book.

I would also like to thank my sweet wife who always supports me.

INTRODUCTION

The history of wheat domestication and use is closely linked to the efforts of humans to protect themselves from hunger and to gain control over their food supply [Igrejas & Branlard 2020]. Human agricultural activity began about ten thousand years ago in some small region of the Near East [Ozkan et al 2005]. The lifestyle transition from that based on hunting and gathering to farming and agriculture was a key event for the subsequent evolution of Western civilization [Dvorak et al 2012]. Now, up to 38% of the Earth's surface (excluding oceans, rivers and Antarctica) is sown to various agricultural plants [Matsuoka 2011].

The birth of agriculture is associated with the beginning of wheat cultivation. Throughout centuries, wheat always was and currently remains one of the most important crops. For example, in 2019, wheat was grown worldwide on 214 million hectares and its production exceeded 730 million tonnes. Common (bread) wheat represents 90-95% of the total wheat production [FAOSTAT, 2021].

In addition to the quantity of wheat grown (the yield), its quality is of great importance. The history of wheat improvement is a story of progressive advances, especially during the 20th century, from manual selection of the stronger plants being selected for future propagation (often through the individual efforts of farmers who did not use genetics but their art and experience [Duvick 1996]), ranging from cross-pollination of plants that together are expected to produce improved progeny, to more sophisticated breeding technologies. In parallel, laboratory methods have been developed for the biochemical study of proteins responsible for wheat quality and for elucidating the genetics of the controlling loci. Nevertheless, even in the 1970s, the discovery or creation by selection of a new valuable wheat genotype was considered as an accidental success ("un heureux hasard") [Branlard 2009].

Wheat-grain storage proteins, gliadins and glutenins created by nature [Wrigley 1996; Wrigley et al 2006b], make up the basis of the unique suitability of wheat for making bread and other bakery products, which represent a considerable part of the human diet. Wheat provides about 20% of daily protein and food calorie intake worldwide. Hence, any advances related to our knowledge or breeding in this species will have a major worldwide impact [Ribeiro et al 2013].

In recent decades, there have been unprecedented increases in our knowledge of the chemistry, genetics and functionality of gliadin and glutenin. This knowledge helps in the production of new genotypes with improved characteristics including field, dough-quality and medical aspects (see **4.**). Nevertheless, the immense complexity of the hexaploid common-wheat genome [Walkowiak et al 2020], and of the genetic loci encoding the synthesis of seed storage proteins, leaves many questions unanswered.

This book is a review. The original research data (Figures, Tables, etc.) can be found in the cited publications. For the spelling of names of cultivars and other information about them, the internet site wheatpedigree.net is used throughout this book.

1. GLIADIN AND THE DETECTION OF ITS POLYMORPHISM

1.1. Gliadin, an extremely polymorphic alcohol-soluble protein, and glutenin, a giant macromolecule

The endosperm of a wheat grain (70% dry weight) is mainly composed of starch, while protein amounts to 10-15%. About 70% of all flour proteins [Altenbach et al 2020], or up to 80% of the grain protein [MacRitchie et al 1990; Branlard et al 2003] consists of storage proteins. The full spectrum of wheat endosperm proteins contains at least 1300 individual polypeptides [Gianibelli et al 2001].

Dough made from wheat flour consists of starch and a mixture of the various polypeptides called "gluten". Gluten is easily separated from the starch and is almost entirely composed of two fractions, gliadin (40% of the gluten), which is extractable with 70% ethanol, and glutenin, which remains in the sediment after removal of the gliadin. These two proteins obtained their names, gliadin and glutenin, about 200 years ago [Wrigley et al 2006b]. Gliadin and glutenin are also called "prolamins". This book is about the prolamins of wheat grain.

Gliadin is one of the most unusual proteins synthesized by living organisms.

It is well known that almost any protein can be precipitated with alcohol (ethanol), which causes the protein to coagulate and precipitate. On the contrary, being extremely hydrophobic due to its high content of the amino acids glutamine and proline, gliadin is readily soluble in 70% aqueous ethanol. Therefore, for analysis, gliadin can be easily extracted from flour by adding 70% aqueous alcohol solution to it. Recently, gliadins have been shown to be soluble in pure water [Urade et al 2018].

Another prominent characteristic of gliadin is that it is extremely polymorphic and, being extracted from one wheat seed, represents a highly heterogeneous mixture of similar but well-distinguishable polypeptides (gliadins). The polymorphism of gliadin may be called "multi-level". Firstly, there are three main types of gliadin polypeptides, the α-, γ-, and ω-gliadins, are very different in their domain structure (see **3.1.**). Secondly, there are families of gliadin polypeptides of the same type, so molecules belonging to the same family are more similar to each other than gliadins from different family types. Thirdly, there is considerable variation of gliadins within each type and within each family. Individual gliadins (gliadin polypeptides) of the same type may differ from each other in their amino acid sequence, molecular weight (length), charge and genetic control (see **2.5.**). Fourthly, any common wheat genotype produces tens of distinct gliadin molecules, and there is a strong difference between genotypes in terms of the sets of gliadins synthesized (see **1.2.**). Theoretically, two common wheat genotypes may be found (or made up) that produce completely different sets of gliadin polypeptides. Fifthly, wheat species differ in the sets of their gliadins produced (see **8.4.**). The inheritance of gliadin polypeptides (see **2.2**) and the structure of gliadin-encoding (*Gli*) loci (see **5.**) are unusually complex. There are major (see **2.1.**) and minor (see **6.**) *Gli* loci on different chromosomes of wheat.

Thus, gliadins, due to their enormous intraspecific and interspecific polymorphism, offer a great opportunity for genetic studies. Also, the difference of common wheat genotypes can be used to confirm, in a controversial situation, the origin of a given wheat grain (see **9.8.**).

Many approaches have been proposed for the analysis of gliadin polymorphism (see **1.2.**). The ability of this protein to be easily extracted from flour in a relatively pure form facilitates its analysis.

The amino-acid composition of gliadin polypeptides is very peculiar being rich in proline and glutamine and poor in charged amino acids, especially lysine. Some gliadins lack cysteine. Those gliadins that have cysteine form only intra-molecular disulphide bonds [Shewry & Tatham 1997]. Therefore. gliadins are monomeric molecules and have a characteristic three-dimensional structure [Müller & Weiser 1995; 1997]. Generally, the molecular weight (MW) of different gliadin polypeptides varies from about 20 kDa up to more than 65 kDa (up to 80 kDa [Branlad 2009]), although for a vast majority of gliadins, their MW is 30-45 kDa [Payne et al 1982; Metakovsky et al 1984b; Shewry & Halford 2002] (see **2.5.**). The vast majority of gliadin polypeptides are not glycosylated [Ewart 1977; Bunce et al 1985], but some of them may undergo post-translational processing so that one gliadin-encoding gene produces two polypeptides of slightly different length [DuPont et al 2004].

Molecules of glutenin, commonly called "glutenin subunits" are similar to gliadins in some characteristics, in other respects, the two protein classes differ significantly [reviewed in Payne 1987; Shewry & Tatham 1997; Gianibelli et al 2001; Shewry et al 2003; Li et al 2021]. The similarity of the two protein classes is in their amino-acid composition: both have high glutamine and proline contents and low lysine content. The genes encoding each of these proteins lack introns, have a domain structures and typical nucleotide sequences that regulate transcription [Cassidy et al 1998; Shewry & Halford 2002]. The fundamental difference is that gliadin is extracted as individual polypeptides, while glutenin polypeptides are linked by covalent disulphide bonds through cysteines [Shewry & Tatham 1997], thereby creating a network (gluten matrix) [Kasarda 1989; Tao et al 1992]. (The term "subunit" emphasizes the fact that this molecule originates from the gluten network. In this book, the term "subunit" is used interchangeably with the term "polypeptide". Besides, individual molecules of glutenin we call "glutenins") Non-covalent hydrogen bonds, which are formed between the amino acid residues of glutamine in the polypeptide chain, also participate in the stabilization of the matrix structure [Feeney et al 2003]. Therefore, the gluten matrix is, in fact, one giant macromolecule, the biggest present in nature [D'Ovidio & Masci 2004]. The MW of this macromolecule may reach several thousand kDa [Wrigley 1996; etc.]. The size of the glutenin macromolecule is highly dependent on environmental factors, primarily the temperature at which plants are grown [Branlard et al 2020].

The polymorphism of glutenin is usually investigated using SDS-electrophoresis (SDS-EP) although many other techniques have been proposed (see **1.4.**, **3.5.**, **4.1.**). Before the use of SDS-EP, the gluten matrix must be separated into its individual polypeptides by a reducing agent, for example, β-mercaptoethanol, to break the disulphide bonds between the polypeptide chains.

In the SDS-electrophoregram of glutenin, there are two main groups of glutenin polypeptides (glutenins), the high-molecular-weight (HMW) subunits (see **4.1.**), and low molecular weight (LMW) subunits (see **3.2.**), distinct in their electrophoretic (SDS-EP) mobility (EM), biochemical and structural characteristics [Shewry et al 1992; Margiotta et al 1993]. In the formation of the gluten matrix, HMW and LMW glutenin subunits play equally important roles [Shewry & Tatham 1997; D'Ovidio & Masci 2004; Juhász & Gianibelli 2006; Shewry et al 2006]. Due to its ability to form a matrix, glutenin determines the dough quality, influencing parameters such as dough strength and elasticity [Payne 1987; etc.] (see **4.3.**, **4.5.**). Some glutenins may undergo, like gliadins, post-translational processing [Nunes-Miranda et al 2017].

For genetic studies on wheat, in addition to specific gliadins and glutenins, several highly polymorphic molecular (DNA) genetic markers have been developed and are widely used [Sharp et al 1989; He et al 1992; Liu JG et al 1992; Vaccino et al 1993; Talbert et al 1994; Bryan et al 1999; Somers et al 2003; Landjeva et al 2007; Ravel et al 2007; etc.].

1.2. Methods proposed for detecting gliadin polymorphism

Many techniques and their modifications have been proposed and used to detect and analyse gliadin polymorphism, including

- many different procedures of one-dimensional electrophoresis (EP) [Shewry et al 1978; Autran et al 1979a; Du Cros & Wrigley 1979; Konarev et al 1979; Sozinov & Poperelya 1980; Lookhart et al 1982; Wrigley et al 1982a; Pogna et al 1986; Werner et al 1994; Bietz & Schmalzried 1995; Lookhart & Bean 1995; Poperelya 1996; Bean et al 1998; Bean & Lookhart 2001; Uthayakumaran et al 2005; Kozub et al 2009; etc],
- of two-dimensional EP [Wrigley 1970; Mecham et al 1978; Payne et al 1982; 1988b; Lafiandra et al 1984; Metakovsky et al 1984b; Howes 1986; Dumur et al 2004; Hurkman & Tanaka 2004; Gobaa et al 2008; Dupont et al 2011; etc.],
- electrofocusing [Wrigley & Shepherd 1973],
- chromatography [Bietz & Burnouf 1985; Lookhart et al 1986; DuPont et al 2000; Gianibelli et al 2002a; Uthayakumaran et al 2006; Tanaka et al 2011, Pronin et al 2020; etc.],
- immunological approaches [Konarev 1980; Skerritt 1985; Skerritt et al 1991],
- analysis of gliadin-coding DNA [Scheets & Hedgcoth 1988; Vaccino & Metakovsky 1995; Anderson & Greene 1997; Zhang et al 2003; etc.].

It was established that the set of gliadins is specific for a given wheat genotype and does not depend on the zone or specific eco-climatic conditions of growing, on the level of fertilizer applied, the grain protein content or on incomplete germination [Autran & Bourdet 1975; Ellis & Beminster 1977; Du Cros & Wrigley 1979; Zillman & Bushuk 1979a; Lookhart & Finney 1984]. Cultivars differed in sets of encoded gliadins [Kasarda et al 1976; etc.].

Any technique used for the analysis of gliadin polymorphism has its own advantages and disadvantages. For example, neither chromatography nor immunological approach provides better resolution (detection of a larger number of different gliadin polypeptides synthesized by a given genotype) compared with one-dimensional EP. These approaches may be used to reveal the polymorphism of gliadins present in a given genotype, but few attempts have been made to carry out genetic research using these methods, for example, to establish the inheritance of individual gliadins

[Bietz et al 1975; Bietz & Burnouf 1985; Skerritt et al 1991]. Chromatography has been successfully used to analyse the biochemical properties of individual gliadin polypeptides or their groups [DuPont et al 2000], but not their genetics. Similar to some other approaches, various modifications of two-dimensional EP can be successfully used to reveal and analyse the gliadin polypeptides present in a few genotypes studied [Wrigley 1970; Mecham et al 1978; Payne et al 1982; Lafiandra et al 1984; Dumur et al 2004; Gobaa et al 2008].

One-dimensional EP is more suitable than any other approach for the routine inspection of a large number of wheat genotypes when there is a need, for example, to routinely identify and distinguish many cultivars or examine progeny from crosses. The procedure permits the analysis of gliadin extracted from one seed (seed-by-seed analysis) and even from a half-seed, if is necessary to grow up the germ end of the analysed seed. Since gliadin polypeptides have a very low content of charged amino acids, an acidic buffer (pH 3.0-3.2) provides the best resolution for their one-dimensional separation. The electrophoregram of the gliadin composition of any wheat genotype obtained by EP in acid gel provides a complex pattern (about 20-25 EP bands of different staining intensities) which is traditionally divided into four zones, depending on the electrophoretic mobility (EM) of the bands, α-gliadins (the fastest group of bands), β-, γ-, and ω-gliadins (the slowest) [Jones et al 1959; Woychik et al 1961]. The earliest EP procedures, which could resolve gliadins of one common wheat genotype into 20-25 EP bands, were proposed in Australia [Wrigley & Baxter 1974], France [Autran & Bourdet 1975], Russia [Gubareva et al 1975] and Ukraine [Sozinov et al 1975a]. When several procedures of one-dimensional EP in acid buffer were compared, it was established that acid polyacrylamide gel electrophoresis (APAGE) is easier to handle and may provide better resolution than starch gel for differentiation between cultivars, although strict attention to experimental detail is necessary [Autran et al 1979a]. In another study, the use of APAGE provided much better resolution than starch gel, especially among α- and β-gliadins [Metakovsky & Kopus 1991].

Using routine APAGE technique, one operator can obtain electrophoregrams of up to 100 or more samples (wheat genotypes) within one working day (plus overnight for staining gels) [Bushuk & Zillman 1978; Metakovsky et al 1984a; Pogna et al 1986]. It was reported that the relative EM of gliadin polypeptides and therefore the details of the band distribution in the electrophoregram can be influenced by the quality of the main reagents [Lookhart et al 1982], by the type and temperature of the buffer [Autran et al 1979a; Lookhart et al 1985], the concentration of gel polymerization catalysts [Khan et al 1985], the addition of urea to the gel [Mecham et al 1978; Autran et al 1979a], and to some extent by the size and design of the EP apparatus [Autran et al 1979a].

On the other hand, one-dimensional SDS-electrophoresis (SDS-EP) is less suitable for the analysis of gliadin polymorphism (see **1.4.**).

1.3. The standard APAGE procedure

Experience shows that gliadin electrophoregrams of the same wheat genotype may vary in the number, and distribution of bands when using different modifications of the EP procedure. Therefore, there was a need to have a convenient and high-resolution standard APAGE technique for the analysis of gliadin polymorphism which would improve inter-laboratory agreement of results.

At the University of Manitoba (Canada), a convenient one-dimensional APAGE procedure was developed to obtain reproducible results with high resolution for the analysis of gliadin (horizontal EP, Al-lactate buffer, pH 3.1; 6% gel; about 13 V/cm throughout the run; of about 6.5 hours) [Bushuk & Zillman 1978]. This method was recognized to be the most suitable as an international reference procedure for the EP analysis of gliadin [Autran et al 1979a]. It has been adopted by the International Association for Cereal Science and Technology as a standard technique for the analysis of gliadin protein polymorphism [ICC 1995].

Some valuable modifications of this standard method were proposed [Tkachuk & Metlish 1980]. Independently, a similar modification of the standard procedure has been used since 1981 in the Institute of General Genetics, Moscow [Metakovsky et al 1984a]. The modifications included vertical EP and 550V as it was proposed [Tkachuk & Metlish 1980], but two 1 mm thick slabs of 17 wells each (gels 8.3%) for each run, and not so strict control of the temperature of EP [Metakovsky & Novoselskaya 1991]. The coincidence of some details in these two independent modifications indicates their adequacy and feasibility. Direct comparison of two APAGE techniques ([Bushuk & Zillman 1978] and [Metakovsky et al 1984a]) confirmed the existence of some noticeable differences in the distribution of bands along the gliadin electrophoregram: a slightly better resolution of ω-gliadins, and of γ- and β-gliadins was achieved by using these techniques, respectively [Metakovsky et al 1993b].

Modifications of the APAGE procedure for the analysis of gliadin polymorphism [Metakovsky & Novoselskaya 1991] were used throughout the studies carried out in different countries and laboratories in decades from 1981 to 2021 (in Australia, Italy, France, Spain). The method was confirmed to be highly efficient and reliable. Therefore, it is possible to confirm the authors' [Zillman & Bushuk 1979a] words about their technique, that "... reproducibility of the gliadin electrophoregram of gliadin of the same cultivar obtained in many different occasions by different technicians is extremely good".

Various authors have formulated criteria by which one can evaluate the quality of the APAGE procedure [Bushuk & Zillman 1978; Lookhart et al 1982; Khan et al 1985]. In particular, the following test has been proposed: an acceptable level of gliadin resolution is considered to be achieved if, as a result of EP, two intensely bands in the slow γ-zone of the electrophoregram of the well-known Canadian cv Marquis (EM 44.5 and 45.5 [Zillman & Bushuk 1979b]) (Fig. 1) are well separated (do not fuse into one band). Our standard APAGE electrophoregrams meets not only this, but also many more stringent requirements. Indeed, the procedure used of APAGE not only separates these two bands, but also documents slight differences in the EM of each of them. In addition, the technique used permits the resolution, in some wheat genotypes, of a medium-intensity gliadin, differing in its EM among wheat genotypes; it is located between two intensely stained bands in the γ-zone of APAGE pattern [Metakovsky et al 1984a; 2018c] (see **2.5.**).

At least three non-alternatives can be proposed to explain the differences in the intensity of staining of different bands in the APAGE electrophoregram. First, overlapping of different gliadin polypeptides having nearly identical EM in APAGE may lead to an increase of staining intensity of such a complex band. Some storage protein polypeptides with similar properties (MW, EM) could not be separated even with the use of 2D EP. Thus, 18, and 7 spots of the 2D EP pattern were shown to contain two, and three different polypeptides, respectively [Altenbach et al 2020]. Secondly, an intensive band may be encoded by a more intensively expressed gene. The level of expression of different prolamin-encoding genes may differ up to 60-fold [Altenbach et al 2020; Paris et al 2021]. Third, duplicated and amplified genes can strongly increase the intensity of one EP band or spot [Vaccino et al 2002; Metakovsky et al 2018a]. In addition, it cannot be completely excluded that polypeptides of gliadin may differ in their ability to be stained.

Long experience has also shown that variation in the concentration of the gel and/or the pH of the buffer, as well as the addition of urea to the gel might, in some cases, contribute to the separation of overlapping bands in the electrophoregram.

1.4. Use of SDS gel electrophoresis in gliadin research

In contrast to one-dimensional APAGE, one-dimensional SDS-EP, which is successfully used for analysis of various proteins (including wheat storage proteins, glutenins, see 3.2., 4.2.) is hardly applicable to gliadin. Since this method classifies polypeptides by their MW only [Laemmli 1970], it cannot distinguish between several tens of different gliadin polypeptides (having similar molecular weight values), which are produced by one wheat genotype. So, when SDS-EP is used for the analysis of gliadin, a considerable part of the information about its polymorphism is lost [Metakovsky & Graybosch 2006; Uthayakumaran et al 2006].

Of special interest, however, is the APAGE/SDS two-dimensional (2D) separation of gliadin. Since the genetic control and inheritance of many gliadin polypeptides (EP bands) in the first dimension (one-dimensional APAGE) can already be known [Metakovsky et al 1984a; 2018c], the determination, in the second dimension, of the molecular weight of these polypeptides can provide valuable new information [Payne et al 1982; Howes 1986; Metakovsky et al 2018a; b]. For example, the separation of gliadin APAGE bands in the second dimension may be used if the aim is to establish the inheritance of individual gliadin polypeptides and their groups [Metakovsky & Sozinov 1990; Redaelli et al 1994].

The MW of gliadins studied by SDS-EP may be overestimated. Due to the peculiarity of its primary structure, gliadin polypeptides may not undergo sufficient denaturation and, therefore, they do not bind sufficiently to SDS molecules [Hamauzu et al 1974]. For example, the apparent MW (AMW) of ω-gliadins, and of one γ-gliadin studied by SDS-EP were shown to be overestimated by 25-35% [DuPont et al 2000; 2004], and by 25% [Ferrante et al 2006], respectively, in comparison with values obtained by mass spectrometry. In other work, the AMW of one α-gliadin, and of one γ-gliadin were overestimated by 20%, and 30%, respectively [Gobaa et al 2007]. The AMW of the known polypeptide "γ50" present in the cv Marquis [Bushuk & Zillman 1978] is reported to be 32 kDa [Metakovsky et al 1984b], or 54 kDa [Howes 1986].

In genetic research, the accurate determination of the MW of individual gliadins is not the goal. Therefore, it may be accepted that similarity in EM of two polypeptides in the second dimension of the APAGE/SDS 2D EP indicates the similarity of their MW. This assumption may lead to a non-trivial conclusion [Metakovsky et al 2018a; b; 2021b] (see **7.5.**).

It turned out, however, that estimation of the AMW of individual gliadins made through the analysis of APAGE/SDS 2D EP patterns [Metakovsky et al 1984b; 2018a; b] was in acceptable accordance with values for the MW of the corresponding gliadins obtained by equilibrium ultracentrifugation or on the basis of amino acid composition (AAC) [Charbonnier 1974; Ewart 1977; etc.]. For example, the MW (AAC) of two α-gliadins were 22 kDa and 32 kDa [Patey and Waldron 1976], whereas the AMW of a group of α-gliadins calculated from APAGE/SDS 2D EP was in the range 27-33 kDa. The MW of one intensely stained γ-gliadin (ultracentrifuge sedimentation) was found to be 34.6 kDa [Sexson et al 1978] and 34 kDa by 2D EP. However, the AMW of the ω-gliadins calculated from APAGE/SDS-EP was lower (not exceeded 60 kDa) in comparison with results obtained by means of AAC analysis (73-74 kDa) [Booth & Ewart 1969].

The origins of the "unique gliadin doublet", which is present in the electrophoregrams of many Canadian common-wheat cultivars, was studied by two different one-dimensional EP procedures. Gliadins extracted from each seed were studied, firstly, by the APAGE and, secondly, by SDS-EP. It was established that the doublet studied came into the Canadian wheat germplasm from a specific accession of the cv Kanred [Ng & Bushuk 1990]. Later, it was established that the doublet is a part of the block of gliadin EP bands (see **2.2.**) encoded by the allele *Gli-B1d* (see **2.5.**). Importantly, the presence of this allele in the genotype of common wheat correlates with improved flour/dough quality (see **4.7.**).

1.5. Formalisation of gliadin electrophoregrams and their catalogues

The use of APAGE reveals the huge polymorphism of the gliadin family of polypeptides: each wheat genotype produces its own characteristic set of gliadin polypeptides. There are several approaches about to how to compare and document electrophoregrams representing different genotypes. There is a way to describe EP patterns in terms of the relative EM and staining intensity of each band. The distance from the top of the gel to the fastest band is accepted to be 100 units, and the prominent band in the middle of the γ-zone of the common wheat cv Marquis may serve as a convenient intermediate marker. Band staining intensity is designated by a subscript number from 1 to 5 (the strongest) for each band along the electrophoregram [Bushuk & Zillman 1978]. So, the electrophoregram of the cv Marquis consisting of 24 gliadin APAGE bands is described, on the basis of the proposed nomenclature, as follows: ω16.03; ω18.53; ω20.53;

ω23.53; ω26.01; ω30.54; ω37.01; ω38.01; ω39.51; γ44.55; γ45.55; γ47.52; γ50.05; γ53.03; β55.54; β58.04; β60.04; β63.04; β65.54; α71.03; α75.03; α77.03; α82.03; α87.51 [Zillman & Bushuk 1979b]. This nomenclature was broadly used to denote gliadin EP bands in common [Peruffo et al 1985; Pogna et al 1985; Pogna & Mellini 1986; etc.] and in durum [Damidaux et al 1980; Leisle et al 1981] wheats.

Such formalization of gliadin patterns has repeatedly been used [Konarev et al 1970; Gubareva et al 1975; Konarev 1980; MirAli 2000; 2002]. The approach could give a good result when there was a need to compare genotypes in a given group of cultivars [Branlard 1983b; Peruffo et al 1985; Haider et al 2010]. The disadvantage of this approach is that different laboratories use their own modifications of the APAGE procedure, thus providing different levels of resolution and staining intensity. Therefore, electrophoregrams of the same genotype obtained in different labs may reveal different numbers and distribution of detected gliadin bands over the gel zones. Therefore, it becomes impossible to compare the results of the analysis of gliadin polymorphism obtained in different labs [Bushuk & Zillman 1978].

For the same reasons, it is impossible to practise any so-called "universal (standard) electrophoregram" which can be created by combining in one scheme (EP spectrum) all the bands found in the EP patterns of many wheat genotypes studied. Two proposed “universal spectra" suggested for "any wheat" contained only 29 (!) positions [Konarev et al 1970; 1979; Gubareva et al 1975], or 62 positions [Haider et al 2010]. In the spectra based on the analysis of 73 French cultivars, there were 43 positions [Autran & Bourdet 1975] and in the study of 37 cultivars (mainly French), 41 positions [Branlard 1983b]. There were 40 gliadin positions for Italian cultivars [Peruffo et al 1984]. In the study of 70 cultivars of different origin, 55 band positions were observed [Branlard & Dardevet 1985a]. Summarizing all results obtained using this approach, one can state that the polymorphism of gliadin in any group of wheat genotypes studied is very high, but comparison of results obtained in different systems is difficult.

Information about gliadin polymorphism may be stored as a catalogue of gliadin electrophoregrams of cultivars studied [Wrigley et al 1982a]. Several catalogues have been published, either in the form of a description of EP patterns, or in the form of their photographs [Autran & Bourdet 1975; Peruffo et al 1981; 1984; Jones et al 1982; Javornik 1989]. The value of the catalogue of electrophoregrams is determined by the quality of the EP which was used to create it and by the presence of appropriate marker cultivars. The catalogues of 76 Canadian [Zillman & Bushuk

1979b] and 29 Italian cultivars [Peruffo et al 1981] of gliadin electrophoregrams were supplied by the formalized nomenclature (in terms of relative EM) of their bands.

To be able to read (interpret) any APAGE pattern and to compare photos of gel patterns obtained in different runs, it is indispensably to have a universal standard (some well-known genotype) on each gel slab, for example, cultivars (cvs) Bezostaya-1 (abbreviated Bez1 in this book) (Fig. 1-4), Marquis (Fig. 1), or Gabo (Fig. 2). The common-wheat Chinese landrace Chinese-Spring (abbreviated ChS) (Fig. 3) has been and is widely used in many genetic studies. However, this cultivar is not the best choice for exploit it as a universal standard in the EP analysis of gliadin because it produces a quite rare (for the common wheat germplasm studied) set of gliadin proteins. To analyse a new genotype, the standard must be run nearby, preferably on an adjacent lane of the same gel slab. Otherwise, it is nearly impossible to correctly interpret the distribution of bands along the electrophoregram of the genotype under study if it is represented by a lonely strip of gel cut from the slab.

There is a catalogue of several dozens of Canadian and other world-wide cultivars in the form of photographs of gel slabs obtained using the standard conditions of APAGE and the well-known cultivar (cv) Neepawa as a universal standard [Ng et al 1988]. Photos of many Italian cultivars, each accompanied by two markers (local cvs Pembina and E. Morandi), provide insight into Italian common wheat gliadin polymorphism [Peruffo et al 1984].

A more elaborate approach was used for the discrimination of 73 French common-wheat cultivars. EP patterns (one-dimensional acid EP in starch gel) of the same cultivar, obtained in repeated runs, were scanned, errors in the EM for each band caused by day-to-day variation (see **1.6.**) were calculated, and a general table (a type of guide) of step-by-step varietal identification was developed so that each next step depended on the presence (or even intensity) of bands having a certain EM [Autran & Bourdet 1975]. This approach is good for distinguishing wheat genotypes within only a certain (already studied) group of cultivars.

An attempt was made to create a guide for a set of cultivars bred in Russia, using the knowledge of the genetic control of some gliadin bands in an elecrophoregram [Upelniek et al 2013]. However, many cultivars in this guide (catalogue) were shown as single APAGE gel strips in the absence of any standard cultivar nearby.

1.6. The reproducibility of electrophoregrams; the need of using standard genotypes

The human eye is a very accurate tool for comparing multicomponent patterns. However, using only visual analysis compromises the entire EP approach, which, as a result, appears to be aged and outdated. Visual analysis of electrophoregrams takes time and, to some extent, is not very reliable, being subjective [Bell et al 1998]. Therefore, it would be desirable to be able to automatically read (interpret) gliadin EP patterns [Zillman & Bushuk 1979b], and some attempts have been made to apply this approach [Lookhart et al 1982; Sapirstein & Bushuk 1985b].

Two factors seriously limit the application of computer programs for the analysis of gliadin EP patterns. The first factor is the enormous diversity of gliadin polypeptides produced by any wheat genotype, while different gliadins may have similar (apparently identical) EM [Branlard et al 1983b; Metakovsky et al 2018a; b].

The second factor is day-to-day experimental variation of any EP procedure. There are some small but perceptible differences between gel slabs obtained in repeated runs of the same gliadin sample under the same EP conditions, for example, in the distance from the top of gel to the fastest band [Sapirstein & Bushuk 1986]. Moreover, the distribution of some EP bands along APAGE pattern (especially, in ω- and α-zones) may also vary in different runs [Metakovsky & Novoselskaya 1991].

How is it possible to compare gel slabs obtained in different runs? To do that, it would be necessary to calculate the standardized EM for each gliadin band and to minimize the experimental error caused by day-to-day variation. However, in 98 genotypes of common wheat studied, up to 3,500 bands of different staining intensity were revealed by accurate comparison of electrophoregrams. In addition, some important differences in the EM of some gliadins are so small that they can only be reliably confirmed by repeated (in different runs) comparisons on adjoined lanes of the slab [Sapirstein & Bushuk 1985a]. It is impossible to imagine that such a fine difference in EM could be recognized by comparing the results obtained with different runs (on different gel slabs).

Based on the analysis of the gliadin patterns of 43 durum-wheat cultivars, a program was created that was able to identify eight control durum cultivars with a probability of 50-100% [Ruanet et al 2001]. Note that durum wheat has four *Gli* loci (see **7.6.**) and, accordingly, a simpler EP pattern in comparison with common wheat.

Because of day-to-day experimental variation, it is difficult to electrophoretically analyse even incomparably less complex patterns, than that of gliadin [Bell et al 1998]. For example, grain storage proteins known as high molecular weight (HMW) glutenins (see **4.1.**) produced by one common wheat genotype compose a group of only three to five polypeptides (EP bands) differing in their EM in the SDS-EP [Payne & Corfield 1979]. However, some of these polypeptides have similar EM (AMW in SDS-EP) making it difficult, in some cases, to "decipher" even such a simple EP pattern and to identify HMW glutenin alleles in a given wheat genotype (see **4.2.**).

Experience shows that, firstly, only adjoined gliadin electrophoregrams on the same gel slab (originated from neighbouring slots) can be accurately compared when visual analysis is applied. In this case, all differences between two patterns are determined only by the sets of gliadin polypeptides produced by the compared genotypes [Sapirstein & Bushuk 1986; etc.]. Secondly, in order to obtain reliable results, it is advisable to compare a pair of gliadin samples in two or more electrophoretic runs [Autran & Bourdet 1975; Bushuk & Zillman 1978; Payne et al 1987a; Metakovsky 1991; etc.]. With repeated runs, day-to-day experimental variation plays a positive role: it is obvious that two identical polypeptides will be identical under any EP conditions. Day-to-day experimental variation may be observed and the reproducibility of the APAGE pattern of the same cultivar may be checked by comparing 128 different runs of gliadin of the cv Bez1 and of two runs of gliadin of 22 other cultivars [Metakovsky et al 2018c, Fig. S1]. For example, Fig. 1 shows the APAGE electrophoregrams of the cv Marquis obtained in two different runs. The reproducibility, in different runs, of a very fine difference in EM between two APAGE bands produced by two genotypes has also been shown [Metakovsky et al 2021a, Fig.S3].

Scanning the APAGE slabs with subsequent visual analysis of scans does not lead to an acceleration or reliability of the work as compared with visual analysis of slabs or their photographs.

Generally, gliadin polymorphism is discontinuous [Sapirstein & Bushuk 1986] and therefore cannot be correctly described or usefully arranged without reference to genetic data (see **2.4.**). It seems that the creation of a computer program for reading gliadin EP patterns is not advisable, since this would require an unjustifiably large contribution of efforts and funds.

1.7. Summary. APAGE, an adequate technique for detecting gliadin polymorphism

For the identification and differentiation of wheat genotypes, many different approaches have been proposed that exploit the enormous polymorphism of gliadin, the unusual protein of the wheat seed. Each of these approaches may be used to describe a set of gliadin polypeptides produced by a given genotype. However, one-dimensional EP is more suitable for the inspection of a large number of wheat genotypes when there is a need, for example, to routinely identify and distinguish many cultivars.

To compare results of EP analysis obtained in different labs, a standard procedure of one-dimensional polyacrylamide electrophoresis in acid buffer (APAGE) was proposed. The technique is fast and highly productive. One operator can obtain electrophoregrams of up to 100 or more wheat genotypes within one working day (+subsequent night for gel staining). Therefore, the method can be used not only to describe the set of gliadin polypeptides present in a given sample, but also to compare many different genotypes (to examine, for example, progeny seeds from crossing). The results obtained can be documented in the form of a description of the gliadin EP pattern in terms of the electrophoretic mobility of all its bands or in the form of a catalogue of photos. A genetic approach for description of gliadin EP patterns has been developed and will be discussed below.

Visual analysis is adequate for scoring gliadin electrophoregrams. However, for detailed comparison, only neighbouring lanes of a gel slab may be compared. It is advisable to compare a pair of gliadin samples in two or more EP runs because, in repeated runs, the day-to-day experimental variation plays a positive role: two identical polypeptides must have identical EM in any APAGE run, and the difference in EM between two EP bands will be repeated in any run. It is indispensable that each gel slab carries, in addition to the samples under study, one or more standard genotypes.

SDS-EP is much less suitable for describing gliadin polymorphism because too many gliadin polypeptides encoded by one genotype have similar AMW. Of special interest, however, is the 2D APAGE/SDS separation of gliadins. Since the genetic control of gliadin polypeptides (electrophoretic bands) in the first dimension (APAGE) can already be known, the determination, in the second dimension, of the AMW of the same polypeptides can provide valuable new information.

2. MAJOR *GLI* LOCI, *GLI-1* AND *GLI-2*, AND THEIR ALLELIC VARIANTS

2.1. Elementary genomes and the localization of six major *Gli* loci in common wheat

The complex genomes of polyploid wheat cultivated by humans, hexaploid common (bread) wheat *T.aestivum* (genome AABBDD) and tetraploid wheat *T.durum* (genome AABB) consist of elementary genomes A, B and D. These elementary genomes originated from a common genome-progenitor about 2.5 million years ago [Huang et al 2002]. These genomes present in diploid wheat species existing today. For example, *T.monococcum*, and *T.tauschii* have A, and D genomes, respectively. The pathways of origin of the hexaploid wheat, are well known (see **8.1.**).

Each elementary genome (A, B, D) consists of seven chromosomes in a haploid set (2n = 14). Thus, the haploid set of common wheat consists of 21 chromosomes (2n=42). In each elemental genome, chromosomes are numbered from first to seventh with the addition of a genome letter, for example, 1A, 3B, 6D. Chromosomes with the same number but belonging to different genomes, for example, 6A and 6D are called homoeologous. Identical chromosomes present in different wheat species, for example, chromosome 1B in common and durum wheat, or chromosome 1D in common wheat and in diploid *T.tauschii*, are named orthologous. It should be noted that some authors [Dong et al 2010; Zhang et al 2013; etc.], instead of using the term "homoeologous", apply the term "orthologous" to designate loci located on the same chromosomes belonging to different genomes in common wheat, for example, for the *Glu-A3*, *Glu-B3* and *Glu-D3* loci (see **3.3.**) mapped on the chromosomes 1A, 1B and 1D, respectively.

Genetic studies of gliadin have begun with the creation of a series of nulli-tetrasomic lines of the variety ChS in the middle of the 20th century. In each line, a pair of homologous chromosomes, for example 6A, is replaced by homoeologous pair, for example 6B, so that the genotype of this line include four identical homologous chromosomes 6B whereas both chromosomes 6A are absent (line N6A.T6B). Ditelosomic lines lacking a certain chromosome arm were also created [Sears 1954; 1974]. Each wheat geneticist inevitably uses the genotypes created by Dr E.R. Sears (or the results of working with these lines) in his own work.

By comparing electrophoregrams of gliadin of the euploid ChS and its nulli-tetrasomic lines it was established that nearly all gliadin EP bands were encoded at the six *Gli* loci located on the chromosomes of the first (1A, 1B, 1D) and sixth (6A, 6B, 6D) homoeologous groups (loci *Gli-1* and *Gli-2*, respectively). Using ditelosomic lines and other genetic approaches showed that each *Gli* locus is located in the distal part of the short arm of the corresponding chromosome, so that there is practically no genetic linkage between this locus and the centromere [Rybalka & Sozinov 1979; Singh & Shepherd 1988a]. Thus, six *Gli* loci were mapped distally on the chromosome arms 1AS (locus *Gli-A1*), 1BS (*Gli-B1*), 1DS (*Gli-D1*), 6AS (*Gli-A2*), 6BS (*Gli-B2*), and 6DS (*Gli-D2*) [Kasarda et al 1976; Mitrofanova 1976; Payne et al 1982; Payne 1987; Singh & Shepherd 1988a; etc.]. It is believed that, in evolution, the *Gli-2* locus arose from *Gli-1* by translocation of genetic material before the emergence of polyploid wheats [Shewry & Halford 2002; etc.]. However, the *Gli-1* and *Gli-2* loci produce gliadin polypeptides which differ strongly in their amino acid sequences (see **3.1.**). Homoeologous *Gli-2* loci of common wheat encode groups of gliadins differing in their polymorphisms (AMW and EM) (see **2.5.**). There are several so-called minor *Gli* loci in common wheat (see **6.**).

The use of nulli-tetrasomic lines for detecting gliadin polymorphism made it possible to firmly establish that each *Gli* locus produces several different gliadin polypeptides [Wrigley 1970; Kasarda 1984a; Bietz & Burnouf 1985; Akhmedov & Metakovsky 1987; etc.]. Thus, ChS produces much more different gliadin polypeptides than there are *Gli* loci in its genotype. The standard APAGE permitted to detect 3, 5, 4, 5, 6, and 5 EP bands encoded by the *Gli-A1*, *-B1*, *-D1*, *-A2*, *-B2*, and *-D2* loci, respectively, of the variety ChS, [Metakovsky et al 1986a]. The use of 2D EP showed that six major *Gli* loci of a common wheat genotype produce 30-50 different gliadin polypeptides [Payne et al 1982; Redaelli et al 1994; Gobaa et al 2008; etc.].

In the detailed study of seed storage proteins of the cv ChS using a special 2D EP technique, 42 (in the text, 40) gluten protein gene sequences were associated with individual protein spots, including 16 α-, 10 γ-, 1 δ- and 6 ω-gliadin spots (see **3.1.**) and 9 spots LMW glutenins (see **3.2.**) [Altenbach et al 2020].

2.2. Inheritance of gliadin polypeptides and methods of its analysis

Inheritance of gliadins that are produced by a given genotype (variety, cultivar) can be established by studying the progeny from the cross of this genotype. Due to the similarity of EM of some gliadins encoded by parental genotypes, the complete composition of the blocks encoded by each of the six loci can hardly be identified by examining only one given cross [Metakovsky et al 1984a]. The more the sets of produced gliadins of the crossed genotypes differ, the more information on the composition of each block can be obtained.

Inheritance of some gliadin electrophoretic bands by groups was first mentioned in the early 1970s [Doekes 1973], but regular studies of inheritance of gliadins by means of one-dimensional EP in starch gel has begun in Odessa (Ukraine) [reviewed in Sozinov et al 1999]. The analysis of segregating progeny from crossing is the most natural way to determine the inheritance of gliadins of a given genotype.

By analysing the progeny (F2, F3, or higher generations) from crosses of genotypes (cultivars) differing in the set of produced gliadins, it was firmly established that all gliadins (EP bands) controlled by one *Gli* locus are (almost always) inherited together, as one Mendelian co-dominant trait. A group of jointly inherited gliadins encoded by one locus has been called "block" [Sozinov et al 1975a, b; Sozinov & Poperelya 1980]. Later, the inheritance of gliadins in the form of blocks was confirmed in many other laboratories [Bebiakin & Balabolina 1977; Baker & Bushuk 1978; Mecham et al 1978; Mitrofanova 1979; Branlard 1983a; b; Dvořák & Appels 1986; etc.]. Recombination frequency between gliadin-coding genes of a *Gli* locus does not exceed 1% because there were only single recombinant genotypes or their absence among a large number of progenies from many different crosses studied [Metakovsky et al 1984a; Payne et al 1984b; Metakovsky 1990; etc.]. The average frequency of recombination detected between gliadin APAGE bands encoded at the *Gli-D1* locus was 0.35% [Kozub et al 2008b]. The problem of using this approach is the difficulty in identifying heterozygote genotypes, especially at the *Gli-2* loci, encoding faint bands in the α- and β-zones of the electrophoregram. Of great value is the use of near isogenic lines (NIL), which represent higher generations from a given cross and therefore are largely homozygous genotypes.

Another non-alternative approach for determining the inheritance of gliadins is the analysis of non-uniform cultivars. Surprisingly, many common wheat cultivars are non-uniform and represent mixtures of two or more related genotypes (biotypes), differing in the set of gliadins encoded by the *Gli* loci, so that the two biotypes differ at one locus [Tkachuk & Metlish 1980; Metakovsky et al 1991b; 1996a; Poperelya 1996; etc.]. In fact, non-uniform cultivars are later generations originated from the cross performed by the breeder [Metakovsky et al 2019c; 2020]. The comparison of the APAGE patterns of the same non-uniform cultivar gives a comprehensive information about the component composition of some blocks.

Analysis of sets of gliadins produced by closely related genotypes (cultivars) makes it possible to trace, using pedigrees, the inheritance of certain EP bands and their groups. The combination of this approach and analysis of non-uniform cultivars may provide more information about some blocks than analysis of F2 segregating progeny. The problems of this approach are the possible lack of authenticity of the grain sample under study and inaccuracies in the pedigrees (see **10.2.**).

Comparison of gliadins' patterns of substituted lines (SL) and its recurrent parent can be also useful for confirming APAGE band composition of the block identified earlier [Koval et al 1988]. However, the use of SL, in contrast to NIL, to identify blocks of jointly inherited gliadins is less reliable, since, in addition to the difficulty of their development. the genetic background of a SL may differ from the recurrent genotype due to its hidden intra-varietal non-uniformity (authentic biotypes and admixtures) [Metakovsky et al 2019c] and different kinds of errors [Yen & Baenziger 1992]. Intra-varietal heterogeneity (the presence of multiple biotypes (genotypes) originated from initial cross) was discovered in many registered cultivars of world collection [Metakovsky et al 2020], despite the fact that one of the important official requirements for the registration of a new cultivar is its homogeneity [Cooke 1995; Law et al 1998; Cooke & Wrigley 2004; Uthayakumaran et al 2006].

Blocks of jointly inherited gliadins were discovered, besides common wheat, in *T.macha* [Metakovsky & Iakobashvili 1990], *T.spelta* [Iakobashvili 1989; Brezhneva et al 2010], *T.durum* [Kudryavtsev et al 1988; Ruiz & Carrillo 1993; Melnikova et al 2012] (see **8.5.**), *T.boeoticum* [Metakovsky & Baboev 1992b], *T.monococcum* [Metakovsky & Baboev 1992a], and *T.tauschii* [Lagudah & Halloran 1988].

Using F2 progeny from a cross between two genotypes of *T.tauschii*, the tight linkage between γ- and ω-gliadin-encoding genes was established: they were mapped at the distance of 0.69 cM [Hassani et al 2006]. The fact that genes united by one function are inherited together, almost without recombination, is not unusual. For example, all ribosome genes located on the chromosome 6B of common wheat are inherited together as one Mendelian trait [Snape et al 1985; Dvořák & Appels 1986].

2.3. Multiple allelic variants of six major *Gli* loci in common wheat

Sets of gliadins (APAGE bands) produced by the genotypes of cvs ChS and Bez1 differ strongly, so that in some zones of the electrophoregram there are few bands identical in EM in these two cultivars. Analysis of the APAGE patterns of the progeny (F2 and F3) from crosses of these two genotypes confirmed that all EP bands controlled by one locus in the variety ChS are inherited together, as one block. The only exception was one ω-gliadin which was shown to be located on the chromosome 1A, but not inherited within the *Gli-A1*-encoded block [Metakovsky et al 1986; Akhmedov & Metakovsky 1987] (see **6.2.**).

Gliadin bands in the electrophoregram of Bez1 are also inherited in groups (blocks), and there are six independent blocks in this cultivar. On the basis of allelism with blocks of ChS, the chromosomal control of each block in Bez1 has been established. APAGE band composition of blocks in this cultivar was confirmed and specified in several other crosses [Metakovsky et al 1984a] and in studies of non-uniform cultivars. Finally, six *Gli*-encoded blocks of Bez1 was found to be composed of 2, 4, 5, 4, 6, and 5 EP bands encoded by the *Gli-A1*, *Gli-B1*, *Gli-D1*, *Gli-A2*, *Gli-B2*, and *Gli-D2* loci, respectively.

Multiple allelism at the *Gli* loci was firstly discovered by Ukrainian scientists as a result of analysis of progenies from crosses using gliadin starch gel EP. It was established that allelic variants of the same *Gli* locus encode blocks differing in the EM and number of gliadins (EP bands) comprising the block [Sozinov & Poperelya 1980]. For example, there are 17 allelic variants of blocks encoded at the *Gli-A1* locus [Sobko & Poperelya 1986].

In our work, several dozens of crosses of different common wheat genotypes, as well as more than 300 non-uniform cultivars with known pedigrees, in total, were studied. In most of crosses, the cv Bez1 was used as one of parents because alleles for each *Gli* locus in this cultivar were known due to the analysis of the cross ChS x Bez1 (Fig. 2). As a result, multiple allelism for the six major *Gli* loci of common wheat, detected using APAGE, was documented. Any allele encodes at least two gliadin APAGE bands ("a block" of bands) inherited together as a Mendelian co-dominant trait.

It was confirmed that allelic variants of blocks encoded at the same locus differ in the number, EM and staining intensity of gliadins (EP bands) included in the block. As a rule, alternative (allelic) variants of blocks reliably differ from each other, and only a few pairs of blocks differ in the presence or EM of only one EP band. In addition, gliadins composing a block differ in their MW. For example, in cvs Bez1 and Marquis, the *Gli-A1* locus encodes two gliadins with AMW of about 38 kDa and 34 kDa, and three gliadins with AMW of 46 kDa, 47 kDa and 32 kDa, respectively [Metakovsky et al 1984b].

According to the International biochemical nomenclature, allelic variants at one locus are denoted by small letters, starting with "*a*", and all alleles "*a*" are assigned to the ChS variety. Thus, this variety has the *Gli* allelic composition *Gli-A1a*, *Gli-B1a*, *Gli-D1a*, *Gli-A2a*, *Gli-B2a*, *Gli-D2a* (or *a*, *a*, *a*, *a*, *a*, *a*). Other alleles are indicated by the following letters of the alphabet, "*b*", "*c*", etc., and next to "*z*", lettering would go to "*aa*", "*ab*", "*ac*", etc. The cv Bez1 does not have any allele identical to that of ChS, and therefore we assigned the letter "*b*" to all its *Gli* alleles. Thus, *Gli* allele composition of the cv Bez1 is *Gli-A1b*, *Gli-B1b*, *Gli-D1b*, *Gli-A2b*, *Gli-B2b*, *Gli-D2b* (or *b*, *b*, *b*, *b*, *b*, *b*) [Metakovsky 1991a].

The frequency of occurrence of different alleles at one major *Gli* locus in the world-wide common wheat germplasm studied (about 1000 registered cultivars) is highly uneven. For example, the allele *Gli-D1b* was found in more than 40% of all cultivars studied, while *Gli-D1m* occurred in only three cultivars. The most frequent alleles were found to be *Gli-A1f* (about 26% of all cultivars studied), *Gli-B1b* (28%), *Gli-D1b* (41%), *Gli-A2g* (21%), *Gli-B2o* (20%) and *Gli-D2a* (27%), whereas three most frequent alleles at each of the six major *Gli* loci occurred, in sum, more frequently than in 35-70% of all cultivars studied [Metakovsky et al 2006]. Since breeders never selected for alleles at the *Gli* loci, the observed uneven allelic distribution in the world-wide germplasm indicates the existence of factors, other than the breeder's selection, which caused that unevenness (see **9.1.**).

In publications in which starch gel EP was used to identify alleles at the *Gli* loci [Sobko & Poperelya 1986; Sozinov et al 1999; etc.] as well as in our articles published in the 1980s [Metakovsky et al 1984a; b; etc.], the original nomenclature [Sozinov & Poperelya 1980] was used to designate alleles at the *Gli* loci. For example, using this nomenclature, the alleles *Gli-A1b*, and *Gli-B1b* are designated as Gld 1A4 (or Gld-1A-4), and Gld 1B1 (or Gld-1B-1), respectively, and so on [Sozinov et al 1999]. In this book, when quoting early article, these original designations of *Gli* alleles were transformed into adequate designations that meet the requirements of the International biochemical nomenclature. The correspondence of designations in the two nomenclatures has been published [Metakovsky 1991a; Metakovsky et al 1991]. Multiple allelism at the *Gli* loci in common wheat can be studied at the DNA level [Vaccino & Metakovsky 1995; von Büren et al 2000; Zhang et al 2003; Polishchuk et al 2010; Popovych et al 2020; 2021; Metakovsky et al 2021b] (see **7.3.**, **7.6.**).

It should be noted that some important results regarding gliadin polymorphism and genetics (especially the pioneering analysis of the relationship between gliadin alleles and wheat bread-making quality [Sozinov et al 1974; Sozinov & Poperelya 1977; Poperelya et al 1978; etc.] (see **4.6.**) have been published in small Ukrainian local journals and bulletins using Ukrainian or Russian language. This circumstance makes it difficult for the world scientific community to access that still valuable information. In this book, the most interesting results obtained by Ukrainian scientists are quoted. In the list of references, the names of authors, title and the name of scientific journal have been translated in English.

Also, it has been possible, in the text of book, to tentatively identify some *Gli* alleles in some publications, where alleles at *Gli* loci were not identified by the authors of these articles. To do that, we either analysed the APAGE electrophoregrams presented by the authors, or our independent analysis of *Gli* alleles in cultivars studied in these works.

2.4. Catalogues of alleles at the *Gli* loci in common wheat

The known data on the genetic control and inheritance of gliadins (EP bands) make it possible to represent gliadin electrophoregram of any common wheat genotype as a combination of six blocks of APAGE bands. Known information on multiple allelism may be stored as six series of schemes of known allelic variants of blocks, one series for one *Gli* locus. Catalogues of this type were published for EP of gliadin in starch gel [Sozinov & Poperelya 1980] and in APAGE [Metakovsky et al 1984a]. In the version of catalogue made using starch gel, there are 17, 15, 10, 3, 3, and 8 alleles were described at the *Gli-A1*, *-B1*, *-D1*, *-A2*, *-B2*, and *-D2* loci, respectively, 56 alleles in total [Sobko & Poperelya 1986]. When an original method of polyacrylamide gel EP was used, 10, 10, 7, 4, 6, and 7 alleles, were registered in the collection of Ukrainian cultivars, respectively [Poperelya 1996]. However, the catalogues of known alleles presented in the form of series of schemes of encoded blocks, although they gave the general impression of multiple allelism at each of the *Gli* loci (see **2.5.**), were found to be difficult for practice due to the experimental day-to-day variation of EP runs inherent in any EP method (see **1.6.**).

To make it possible the practical application of the catalogue of allelic variants of gliadin blocks, the use of marker (standard) cultivars was proposed. The genotype of each marker cultivar contains one or more alleles at the *Gli* loci, which the given marker cultivar is destined to show. As a result, the catalogue of known gliadin blocks is represented by a set of marker (standard) cultivars. The blocks marked with a given standard cultivar are schematically indicated in its APAGE electrophoregram. Moreover, to facilitate comparison of the EM of different gliadins and to minimize the impact of the day-to-day variation, cv Bez1 was proposed to be used as a "universal standard" on the adjacent lane next to each marker cultivar [Metakovsky 1991]. This cultivar is world-known, and its gliadin EP pattern is easy to interpret (to read), in comparison, for example, with those of cvs ChS or Neepawa. Each of the six blocks of Bez1 was reliably identified through analysis of the progeny from the cross ChS x Bez1. The catalogue described, in total, 111 alleles identified at six *Gli* loci, from 12 (at the *Gli-D1*) to 24 (at the *Gli-A2*) alleles per locus. In total, 49 marker cultivars were proposed, each of them showing from one to six blocks of jointly inherited gliadin EP bands [Metakovsky 1991].

This catalogue was supplemented by new alleles discovered in groups of cultivars previously unexplored for this analysis. Therefore, at the beginning of the 21th century, 148 alleles (from 15 at the *Gli-D1* to 33 at the *Gli-B2*) were known, and schemes of some blocks of gliadin APAGE bands of additional alleles were published [Metakovsky et al 1994; 2000; Metakovsky & Branlard 1998; Ruiz et al 2002a].

Later, weak sides of this catalogue were also realized. In particular, it was found to be difficult to recognize a given allele, represented by one marker cultivar of the catalogue, in other genetic backgrounds (in different wheat genotypes). Moreover, in 2006, 174 alleles at the *Gli* loci were mentioned [Metakovsky et al 2006]. However, for about 25 of them, band composition of encoded blocks remained unpublished and marker cultivars for these alleles were not proposed.

The last published catalogue currently describes, in sum, more than 180 alleles revealed by the APAGE at the six major *Gli* loci in about 1000 common wheat cultivars and lines studied, from 16 alleles at the *Gli-D1* to 45 alleles at the *Gli-B2* loci [Metakovsky et al 2018c]. A simple calculation shows that this number of alleles permits, theoretically, distinction among more than one billion (10^9) wheat genotypes, differing among them at least at one *Gli* locus. Indeed, identification of alleles at six mayor *Gli* loci in world-wide common wheat cultivars showed that there was no one case of the complete coincidence of gliadin genotypes in two or more unrelated cultivars among more than 1000 studied. On the other hand, there were more than 50 cases of identity of gliadin allelic composition between two or more closely related cultivars, such as several Australian cultivars related to Insignia [Metakovsky et al 1990b]; several Canadian cultivars related to Neepawa and Thatcher [Metakovsky et al 1993b]; several Hungarian cultivars related to Bez1 [Metakovsky et al 2018c]; some English [Chernakov & Metakovsky 1994], some French [Metakovsky & Branlard 1998]; some Italian [Metakovsky et al 1994]; some Kazakh [Metakovsky et al 1988], several Russian [Metakovsky et al 1987a] and several Serbian [Metakovsky et al 1991] cultivars having close pedigrees; etc. Gliadin genotypes of cvs Maris-Ranger (UK) and Kincso (Hugary) are identical to their parents Peko (Germany) and Sava (Serbia), respectively [Metakovsky et al 2006].

To show alleles of the catalogue in different genetic backgrounds, 128 marker cultivars have been proposed in the last version of the catalogue, each of them is accompanied by the general standard, cv Bez1, on the adjacent lane of the same slab. In the electrophoregram of each marker cultivar, all six blocks encoded by the alleles present in the genotype of this cultivar were schematically indicated. As a result, some alleles, frequently occurring in common wheat germplasm of the 20th century, were mentioned many times in the catalogue, and on different genetic backgrounds. For example, the alleles *Gli-D1a*, *Gli-D2a*, *Gli-D1b*, and *Gli-B1b* were shown by 44, 30, 28, and 27 marker cultivars, respectively, while 12 other frequent alleles occurred in 11-19 marker cultivars each. However, 32, and 64 alleles, rare in common wheat germplasm, are shown in two, and one marker cultivars, respectively. The number of alleles at the *Gli-2* loci (106 alleles) was found to be almost 75% higher than at the *Gli-1* loci (61). The chromosomes of the sixth homoeologous group also surpass the chromosomes of the first groups in terms of the number of alleles of microsatellites (see **7.5.**, **9.3.**) and their information content [Roussel et al 2004].

It should be noted that our catalogue of gliadin alleles is based almost exceptionally on the analysis of registered common wheat cultivars and includes a few alleles occurring only in some breeding line or was identified in ancient local varieties and landraces.

The article which include the last version of the catalogue of gliadin alleles [Metakovsky et al 2018c] has two Supplements. One of them is the list of alleles at each of the six major *Gli* loci identified for more than 1000 common wheat cultivars and lines. Another Supplement shows APAGE electrophoregrams for 129 cultivars (128 different cultivars + Bez1) with indication, in each of them, schemes of six blocks of jointly inherited gliadin bands. In this book, the references for the first and second Supplements will be designated as [Alleles 2018c] and [Blocks 2018c], respectively.

2.5. APAGE-bands composition of blocks encoded at major *Gli* loci in common wheat

Usually, any block of jointly inherited gliadins consists of more than two EP bands, although there are few exceptions. Some general characteristics of blocks encoded by each of major *Gli* loci are given below. To designate EM of a band, the widely-used nomenclature [Bushuk & Zillman 1978] was used (see **1.5.**). The AMW was calculated from the 2D APAGE/SDS-EP patterns (see **1.4.**).

Gli-A1. The blocks encoded by *Gli-A1*, as a rule, consist of bands of moderate staining intensity located in the γ- and β-zones of the APAGE. Several different blocks include a slow-moving γ-gliadin located between two the slowest strong γ-bands. Among γ-gliadins encoded by alleles at *Gli-A1*, the γ-gliadin with lower AMW has higher EM [Metakovsky et al 2018a]. The allele *Gli-A1o* encoding block composed of one fast-moving γ-gliadin and two ω-gliadins was identified in tetraploid [Metakovsky et al 1989] and diploid [Metakovsky & Baboev 1992a; b] wheat species. Alleles *Gli-A1e, Gli-A1l* and *Gli-A1q* include a slow-moving ω-gliadin (AMW about 41 kDa), the most slow of all *Gli-A1*-encoding ω-gliadins.

In the last version of the catalogue of *Gli* alleles [Blocks 2018c], there are some *Gli-A1*-encoded blocks which include ω-gliadins encoded by so-called minor *Gli* loci (*Gli-A5* or *Gli-A6*) tightly linked to the *Gli-A1* locus [Metakovsky et al 1996] (see **6.7.**; **6.8.**). Therefore, there are some alternative alleles at the *Gli-A1* differing only in the presence or absence of ω-gliadins encoded by a min*or Gli* locus. For example, blocks encoded by alleles *Gli-A1r* and *Gli-A1p*, or *Gli-A1f* and *Gli-A1af* differ only in the presence or absence, respectively, of ω-gliadins encoded at the *Gli-A5*, and *Gli-A6* loci, respectively. Gliadin EP pattern of the cv Marquis (*Gli-A1m*) includes the famous prominent band γ50.05 and a pair of the *Gli-A5b*-encoded ω-gliadins, ω20.53 and ω23.53 [Zillman & Bushuk 1979b].

Employing the *Gli-A5b*- and *Gli-A6b*-encoded ω-gliadins in the *Gli-A1*-controlled blocks enlarges gliadin polymorphism and, to our opinion, increases the value of alleles at the *Gli-A1* locus as genetic markers of the distal part of the short arm of the common wheat chromosome 1A.

Gli-B1. The *Gli-B1*-encoded blocks include one strongly-stained γ-gliadin (the slowest band in the γ-zone of APAGE electrophoregram), some ω-gliadins and some gliadins in the β-zone. Among these blocks, there are at least six variants of the EM of the main γ-gliadin, but, surprisingly, all variants, including the most contrasting ones (for example, *Gli-B1k* and *Gli-B1d*), have very similar, if not identical, AMW of about 41 kDa [Metakovsky et al 2021b].

Blocks produced by some alleles include a pair of faint ω-gliadins encoded by the minor gliadin locus *Gli-B5* (see **6.9.**). As a rule, these alleles are linked to the dominant allele of the *Rg-1* locus controlling the red colour of common wheat spikes and therefore can be called "red-spike" alleles (see **6.10.**).

The allele *Gli-B1b*, which is present in genotypes of both Bez1 and Marquis, encodes a block consisting of one strong γ-gliadin, two bands of moderate intensity (in ω-and β-zones of APAGE) and a faint band in ω-zone. The moderately-stained ω-gliadin named ω-5 [Mecham et al 1978] (the band ω30.54) is one of the longest gliadins (AMW at least 60 kDa) and is one of the most powerful allergens in human food [Tanaka et al 2011].

Gli-D1. Alleles at the *Gli-D1* may encode 2-3 (so-called ChS-type) or 4-5 (Cheyenne-type) slow-moving ω-gliadins and one or two γ-gliadins. The *Gli-D1b*- encoded block, which is present in the genotypes of both Marquis and Bez1, consists of five EP bands, the intensely stained (major) slow-moving γ-gliadin (AMW about 34 kDa), minor fast-moving γ-gliadin (30 kDa), a faint ω-gliadin (the slowest in this block) and two slow ω-gliadins of moderate intensity.

The use of 2D EP permitted to unexpectedly detect two variants of the allele *Gli-D1a* indistinguishable in APAGE. It was shown that there were two slow-moving ω-gliadins (as in cvs ChS or Marquis) or four ω-gliadins (as in cultivars bred in Saratov, Russia). Due to pairwise overlapping, the four ω-gliadins look like two bands in the APAGE, but they can be easily seen in the second dimension [Metakovsky et al 2018a].

The *Gli-D1*-encoded minor (fast-moving) γ-gliadin has, in distinct blocks, one of the four variants, two variants slightly differing in their EM (higher EM, lower AMW), or this γ-gliadin may be amplified (see **5.3.**), or absent [Blocks 2018c].

Gli-A2. This locus encodes nearly all APAGE bands in the α-zone of APAGE. Alleles *Gli-A2a* (ChS), *Gli-A2m* (Marquis), and several more produce a pair of the fastest α-gliadins, α82 and α87.5 (each about 32 kDa), and one or two the fastest gliadins in the β-zone. Some other alleles each encodes a rather compact group of α-gliadins with AMW of 27-34 kDa. These two contrast groups (families, see **7.2.**) of alleles were called ChS-type and Bez1-type, respectively [Kasarda 1980, Kasarda et al 1984a].

The block encoded by the allele *Gli-A2b* consists of four APAGE bands of different staining intensities whereas about 10 gliadin polypeptides of slightly different EM and AMW may be resolved by the 2D EP [Metakovsky et al 2018b].

Monoclonal antibodies (line 230/9) differentiate several *Gli-A2*-encoded blocks [Skerritt et al 1991].

Gli-D2. The *Gli-D2* locus in each its alternative allelic state encodes a block of 3-5 gliadins located at similar distances from one another, like a "staircase", in the β- and α-zones along APAGE pattern: the slowest β, middle β, the fastest β, the slowest α, and middle α. All these bands have similar (moderate) staining intensity and AMW of 32-35 kDa. The blocks encoded at the *Gli-D2* differ in EM and MW (always higher EM, lower AMW) of some of their bands. There are at least three variants of the EM of the fastest *Gli-D2*-encoded α-gliadin (middle α-zone). The allele *Gli-D2w* encodes unique block including a band in the γ-zone of APAGE [Blocks 2018c].

Gli-B2. Almost all blocks encoded at the *Gli-B2* consist of several (up to ten) APAGE faint bands whose EM is difficult to compare. Besides, some of these bands are located in the β-zone of the APAGE where multiple overlapping of bands of moderate staining intensity controlled by other *Gli* loci (especially, *Gli-A1* and *Gli-D2*) is observed. Fortunately, alleles at the *Gli-B2* differ from each other in the EM of several (or even of all) polypeptides that compose the block. Therefore, to identify the allele at the *Gli-B2* in the genotype under study, it may be enough to recognize the combination of gliadins encoded by this locus in the γ-zone of the APAGE, which is available, in many genotypes, for this analysis.

Allele *Gli-B2q* produces, among other gliadins, a unique polypeptide located in the zone of the slowest α-gliadins [Redaelli et al 1994; Blocks 2018c]. Alleles *Gli-B2o*, *Gli-B2c*, *Gli-B2l* encode a moderately-stained gliadin located in the fast β-zone of APAGE. The EM of this polypeptide slightly differs between these three alleles, but each of these particular gliadins has an unusual blue-light colour inherent only in these three gliadins.

It should be noted that the EP band composition of blocks encoded by the same *Gli-2* locus is similar, while blocks encoded by different *Gli-2* loci (*Gli-A2*, *Gli-B2* or *Gli-D2*) differ strongly [Blocks 2018c]. Analysis of DNA sequences also showed that the α-gliadin genes from the same chromosome are more closely related than genes located on different chromosomes indicating their independent evolution in the last 2.4 million years [Huo et al 2018b] (see **8.1.**).

It is important to know that an adequate analysis of alleles at the *Gli-2* loci is only possible when the standard or slightly modified standard procedure of APAGE (see **5.3.**) is used. Some other APAGE procedures can give a good resolution of the ω- and slow-moving γ-gliadins, but not of gliadins located in the β- and α-zones of the electrophoregram. The use of such procedure limits the analysis to the *Gli-1* loci and some minor *Gli* loci encoding ω-gliadins (see **6.1.**) [for example, Kozub et al 2009; 2017; 2020; Sobko et al 2019].

2.6. Secalins of rye can be easily identified by APAGE

The *Gli-B1l* present in the catalogue of *Gli* alleles [Blocks 2018c] is of particular interest. In fact, this designation is only a symbol because it denotes the presence, in common wheat genotype, of the translocation 1BL.1RS carrying the rye genetic material, valuable for wheat breeding. This translocation was also designated *Gli-R1* or *Sec-1* [Miller 1984; McIntosh et al 2003]. Therefore, EP bands composing the *Gli-B1l*-encoded block are not gliadins, but secalins, storage proteins of the rye (*Secale cereale* L.) [Graybosch 2001; Kozub et al 2001] (Fig. 4).

The block encoded by this translocation was originally identified using EP in starch gel and designated Gld 1B3 [Poperelya & Babayants 1978]. This block consists of about 10 EP bands (of secalins) of moderate intensity, most of which are located in the ω-zone of the electrophoregram (their AMW 42-49 kDa), but there are three bands in the γ-zone of APAGE (AMW of each is about 33 kDa) [Metakovsky et al 1984b; Payne et al 1987a]. The MW of polypeptides in the ω-zone produced by 1BL.1RS translocation may be higher, 48-55 kDa [Gupta & Shepherd 1992]. The 1BL.1RS is inherited as a codominant Mendelian trait allelic to other variants of the *Gli-B1* locus.

A series of common wheat genotypes carrying a translocation 1BL.1RS was created in the 1930s in Germany [Lein 1975; Schlegel & Korzun 1997; Kozub et al 2012]. After the WWII, these genotypes were actively used in breeding programs due to the fact that rye material provided wheat resistance to powdery mildew (*Pm8*), as well as to leaf (*Lr26*), stem (*Sr31*) and yellow (*Yr9*) rust [Poperelya & Babayants 1978; Singh et al 1990; Rabinovich 1998; Kozub et al 2001; etc.]. Due to cv Hedgehog and, especially, to cvs Kavkaz and Avrora, the allele *Gli-B1l* is now present in the genotypes of hundreds of world-wide cultivars bred in the second half of the 20th century.

It is known that the 1BL.1RS can negatively affect the quality of wheat grain, as it is believed, due to the lack, in genotypes with this translocation, of locus *Glu-B3* influencing bread-making quality [Dhaliwal et al 1987; Fenn et al 1994; Burnett et al 1995; Lee JH et al 1995; McKendry et al 1996; etc.]. The translocation is currently no longer able to provide a good wheat resistance to all these diseases [Landjeva et al 2006b; Gobaa et al 2007]. Nevertheless, the 1BL.1RS is still of scientific and even breeding interest. Moreover, advanced techniques of chromosomal manipulation associated with assessment of end-use quality may lead to the development of new recombinant lines of wheat genotypes combining the absence of secalins deteriorating end-use wheat quality and the advantages of rye genetic material transferred to wheat [Sharma et al 2020].

The cv Amigo, the first carrying the translocation 1AL.1RS (instead of 1BL.1RS), was bred (1976) in the USA [Koebner & Shepherd 1986]. In this translocation, as well as in the 1BL.1RS one, there are genes which supply wheat plant with resistance to several diseases [Kozub et al 2005; 2012]. The block of the APAGE bands encoded by the 1AL.1RS was called Gld 1A17 [Sobko & Poperelya 1986], or *Gli-A1w* [Metakovsky et al 2018c]. It was shown that the band composition in blocks controlled by 1AL.1RS and 1BL.1RS translocations differs considerably [Berzonsky et al 1991; Kozub et al 2005]. The difference may be explained by the intra-species polymorphism of rye storage proteins [Singh et al 1990]. Indeed, two different rye cvs, Petkus and Insave, were used as a source of the genetic material for transfer to the wheat chromosomes 1B and 1A, respectively [Lukaszewski 1990; Berzonsky et al 1991]. A difference was observed in the composition of the secalins produced by the two wheat genotypes (cv Amigo and the line GRS1201) carrying 1AL.1RS translocation [Graybosch et al 1999].

A spontaneous translocation of the genetic material of rye (the cv Imperial) to the chromosome 1D of common wheat (variety ChS) was reported [Shepherd 1977]. Later, multiple wheat-rye 1RS translocation stocks including 1AL.1RS + 1BL.1RS + 1DL.1RS translocations were created and studied [Gupta & Shepherd 1993].

To identify the presence of the 1BL.1RS translocation in the wheat genotype under study, many different approaches have been proposed, which in itself indicates the importance of this task. There are different techniques of the EP in acid buffer [Sozinov & Poperelya 1977; Berzonsky et al 1991; etc.], SDS-EP [Hussain & Lukow 1994; Graybosch et al 1999; etc.], enzyme analysis [Javornik et al 1991b; William et al 1992], and 2D APAGE/SDS EP [Metakovsky et al 1984b; Payne et al 1987a]. Beautiful pictures are obtained when using hybridization *in situ* of labelled rye DNA with wheat chromosomes [Appels 1982; Lapitan et al

1986; Ribeiro-Carvalho 1997; 2001]. The presence of the 1BL.1RS can be detected using ordinary cytological approaches, C-banding [Gill & Kimber 1977; Graybosch et al 1993a; etc.] Giemsa N-banding [Rayburn & Carver 1988]. Also, HPLC chromatography [Graybosch et al 1993a; etc.], monoclonal antibodies [Howes et al 1989; Graybosch et al 1993b; Andrews et al 1996] and DNA molecular markers [Francis et al 1995; Koebner 1995] were applied.

Characteristic DNA sequences indicating the presence of the 1BL.1RS translocation can be reliably detected in the wheat genome using the RFLP technique and the probe for γ-gliadin sequences. In addition, it was unexpectedly found that rye DNA in the 1BL.1RS translocation contains sequences that are recognized by the pTU1 probe specific for the α-gliadin wheat sequences [Vaccino & Metakovsky 1995].

To detect the presence of the 1BL.1RS translocation, combinations of various methods were also used, for example, of N-banding, APAGE and DNA molecular markers (microsatellites, see **9.3.**) [Landjeva et al 2006b], or APAGE accompanied by the cytological analysis [Sozinov et al 1987b]. Genotypes heterozygous for the 1BL.1RS translocation were identified using a combination of APAGE, C-banding, N-banding and isoelectric focusing of the enzyme GPI [William et al 1992].

Techniques proposed for identifying the rye chromosomal material in the wheat genome are effective, and some of them, without exaggeration, can be called modern and even elegant. Nevertheless, for the purpose of precisely identifying, for example, the 1BL.1RS wheat-rye translocations in a large number of genotypes, the EP technique is the fastest, cheapest and most adequate [Javornik et al 1991b; Gupta & Shepherd 1992]. Indeed, the set of secalins encoded by the chromosome 1R of rye have nothing in common with any block encoded by any *Gli* locus of wheat [Metakovsky 1991; Blocks 2018c]. Therefore, secalin bands can be easily recognized at a glance at the APAGE pattern, even in the absence of adjacent marker genotypes. At the same time, it is easy to determine which of translocations, 1BL.1RS or 1AL.1RS, is present in the genotype under study. For example, if the presence of bands in the electrophoregram encoded by the *Gli-B1l* allele is accompanied by the absence of *Gli-B1*-encoded block, it is certain that the genotype under study carries the 1BL.1RS. There is no problem to distinguish genotypes heterozygous and homozygous for the wheat-rye translocation [Metakovsky & Graybosch 2006], as well as modified 1BL.1RS translocations [Motsnyi et al 2017] and 1R(1B) substitution lines [Motsnyi et al 2009].

Rye (or other foreign) genetic material can be detected in wheat by means of APAGE only if this material contains genes encoding storage proteins. Other methods (*in situ* hybridization) have been applied to discover rye DNA sequences in two genotypes of common wheat landrace Barbela. The genetic material of rye replaced 3.5-5% of the length of the wheat chromosome 2D, did not contain genes encoding secalins [Ribeiro-Carvalho et al 2001] and could not be revealed by APAGE technique.

2.7. Summary. Advantages and natural limitations of the catalogue of alleles

There are six major *Gli* loci in common wheat having 21 chromosomes in its haploid set. These loci are mapped proximately on the short arm of the chromosomes 1A (locus *Gli-A1*), B1 (*Gli-B1*), D1 (*Gli-D1*), 6A (*Gli-A2*), 6B (*Gli-B2*), and 6D (*Gli-D2*). These six loci control nearly all gliadin bands in the gliadin electrophoregram of any common wheat genotype. Each *Gli* locus controls two or more EP bands which can be identified using APAGE. All bands encoded by one *Gli* locus are inherited jointly, as one Mendelian codominant trait (as one block). There is multiple allelism for each of six major *Gli* loci. Alleles of one locus encode blocks that differ in the number, EM and AMW of individual gliadins composing the block. Therefore, there are several levels of polymorphism of gliadins: among polypeptides encoded by a given allele (composing a block), among sets of polypeptides encoded by different alleles at the same *Gli* locus, and among gliadins produced by different wheat genotypes.

There are three main approaches for identification of new variants of blocks: first, the analysis of segregating progenies from crosses of different wheat genotypes; second, analysis of the non-uniform cultivars (representing mixture of lines of high generations from the cross performed by the breeder); third, comparison of closely related cultivars. There are several catalogues of alleles for each *Gli* locus (both for starch gel and for APAGE) differing not only in the number of alleles (blocks) which are shown, but also in the form of representation of these alleles. In the last version of the catalogue for APAGE (2018), there are, in total, 180 alleles, from 16 alleles at the *Gli-D1* to 45 alleles at the *Gli-B2* loci. The *Gli* alleles in any cultivar may be identified by comparison of the electrophoregrams of gliadin of this cultivar and of corresponding marker genotypes (shown in the catalogue) on adjacent lanes of the slab. However, any catalogue of gliadin alleles has natural limitations: first, new alleles will be found in groups of cultivars not studied before, and, second, new alleles may appear as a result of spontaneous mutations at gliadin-encoding genes.

The unusual alleles *Gli-B1l* and *Gli-A1w* encode the blocks of secalins in the APAGE and indicate the presence of the agriculturally important wheat-rye translocations 1BL.1RS and 1AL.1RS, respectively, in wheat genome. Electrophoretic analysis of gliadin is the most common, easy, rapid and reliable way to reveal these translocations in wheat.

3. Gliadins and LMW glutenins: their encoding genes and loci

3.1. Major types of gliadin polypeptides and their encoding DNA sequences

It is accepted to name gliadin polypeptides, their amino acid sequences and their encoding DNA sequences in accordance with zonas of the APAGE electrophoregram. Thus, three strongly different types of gliadin polypeptides, the α-, γ- and ω-gliadins, are in common use [Barak et al 2015]. There are no "β-gliadins" because most of polypeptides in the β-zone of the APAGE have α-type of their amino acid sequence. Also, the presence of a gliadin polypeptide in a given zone of APAGE pattern does not guarantee its type. For example, it was proved that some gliadins in the γ-zone of APAGE have α-type of amino acid sequence [Kasarda et al 1987] and are encoded by the *Gli-B2* locus [Metakovsky 1991] (see **2.5.**). The size and shape of different types of gliadin molecules were determined by assessing a number of their biophysical parameters [Urade et al 2018].

Hundreds of gliadin-coding DNA sequences of various wheat species are already known. It has been firmly established that all gliadin genes lacked introns and all of them have a specific promotor and flanking sequences. A characteristic feature of all genes encoding gliadin is the presence of repetitive sequences of different length and nucleotide composition. There are strong differences between the three main types of gliadin polypeptides in their primary amino acid sequence and domain structure [Kasarda et al 1984b; Okita et al 1985; Sumner-Smith et al 1985; Bartels et al 1986; Scheets & Hedgcoth 1988; Anderson et al 1997; 2001b; Hsia & Anderson 2001; etc.].

ω-GLIADINS. ω-Gliadins compose about 10% of all grain proteins and are the slowest in the APAGE. They differ from other two groups of gliadins in that their amino acid sequences include neither cysteine nor methionine. Besides, ω-gliadins contain the least amount of charged amino acids [Tatham & Shewry 1995] and the highest ratio of glutamine to proline which makes them less hydrophobic than α- or γ-gliadins [Popineau & Pineau 1988; DuPont et al 2000]. Moreover, ω-gliadins, unlike other gliadins, do not have a helical structure [DuPont et al 2000; 2004].

Analysis of the cloned genes showed that a ω-gliadin is composed of the N-terminal (10-11 amino acids) and C-terminal (10-12 amino acids) sequences and a central domain consisting entirely of short repeats (7-8 amino acids) rich in glutamine. The central domain occupies 90-96% of the length of the entire polypeptide [Tatham & Shewry 1995; Hsia & Anderson 2001; Hassani et al 2008]. There are at least four variants of the N-terminal amino acid sequences (four families of ω-gliadins) and many differences in the central domain among different ω-gliadins.

The ω-gliadin polypeptides are generally the longest among gliadins [Shewry & Halford 2002] with MW about 40-65 kDa [Metakovsky et al 1984b; Anderson et al 2009]. The MW of the *Gli-B1*-, and *Gli-D1*-encoded ω-gliadins of the cv Butte-86 ranged from 48.9 to 51.5 kDa, and from 41.9 to 42.8 kDa, respectively [DuPont et al 2000], while the *Gli-A1*-encoded ω-gliadins were smaller [DuPont et al 2004]. In another study, MW of 12 ω-gliadins of this cultivar was in the range of 55.4-66.3 kDa, and pIs of 5.0-6.6 [DuPont et al 2011]. Six of the ω-gliadin genomic sequences of the cv ChS were associated with individual 2D EP spots whose MW was about 41.0-51.5 kDa, and pIs 4.68-6.16 [Altenbach et al 2020].

N-terminal sequencing suggested that some ω-gliadins undergo, in addition to signal peptide cleavage, processing performed enzyme asparaginil-endopeptidase that removed eight amino acids from the N-terminal end of the polypeptide. As a result, one ω-gliadin gene produces two polypeptides differing in their N-terminal sequences [Kasarda et al 1983; DuPont et al 2004; Hassani et al 2008]. The two cloned ω-gliadin-encoding genes were found to differ by a large deletion (nearly 10 kDa of the polypeptide length) in the central domain of one of them [Hsia & Anderson 2001].

Large differences were discovered between the ω-gliadins encoded by different common wheat genomes. The difference between the ω-gliadins encoded by the genome B from the ω-gliadins produced by the genomes A or D is as great as if they were different classes of storage proteins [Anderson et al 2009].

α-GLIADINS. Among the three types of gliadins, α-gliadins (also called α/β-gliadins) are synthesized in the wheat grain in the greatest amount being most commonly consumed by humans [Anderson & Greene 1997].

There are five domains in a typical polypeptide of α-gliadin: N-terminal sequence (7-14 amino acids), followed by two poly-glutamine sequences alternating with two unique sequences [Kasarda 1980; Kasarda et al 1984b; Anderson et al 1984; Sumner-Smith et al 1985; Reeves & Okita 1987; Anderson 1991; Anderson & Greene 1997; van Herpen et al 2006]. Nucleotide sequences encoding poly-glutamine domains are, in fact, microsatellites located within gliadin-coding genes [Devos et al 1995; Lee SJ et al 1995; Li et al 2004] (see **7.5.**, **9.3.**). There are usually six cysteine residues in α-gliadin polypeptide. They are located at the C-end of the molecule and form three intramolecular disulphide bonds [Müller & Wieser 1995]. Ninety individual α-gliadin genes of the cv ChS harboring a promoter region were classified into eleven groups. It was shown that the expression of each α-gliadin gene in developing seeds is differentially regulated by multiple regulatory factors including sunshine duration [Noma et al 2016].

The MW of the vast majority of α-gliadins is in the range of 25-35 kDa [Metakovsky et al 2018b] although some of them reach 40-50 kDa [Harberd et al 1985]. In any common wheat genotype, α-gliadins compose the α- and β-zones of the APAGE, but some of them are present in the γ-zone as well (see **2.5.**). MW of α-gliadins of the cv Butte-86 was in the range of 36.5-50.0 kDa, and pIs of 5.7-7.1 [DuPont et al 2011]. The flour of the cv ChS is expected to contain 26 α-gliadin-encoding full-length gene sequences. Sixteen of them were associated with individual 2D EP spots with the MW 30.0-36.2 kDa and pIs 6.18-8.28 [Altenbach et al 2020].

A peptide was described having the amino acid sequence similar to that of α-gliadins. Its encoding gene is located on the chromosome 6D and has flanking sequences identical to those of heat-shock proteins (Blumenthal et al 1990). This fact may explain why the synthesis of gliadins does not stop at high temperatures (40° C), while the synthesis of glutenins almost completely ceases [Blumenthal et al 1993].

γ-GLIADINS. There are 3-4 domains in the typical γ-gliadin: N-terminal sequence (12 amino acids), domain of repeats (12-15 copies of a heptapeptide rich in glutamine and proline) and the unique sequence (about 125 amino acids) divided into two parts by a small (6-16 amino acids) poly-glutamine domain. In that part of the unique sequence that precedes the poly-glutamine domain, there are six cysteine residues and most of charged amino acids of the entire molecule [Okita et al 1985; Bartels et al 1986; Rafalsky 1986; Scheets & Hedgcoth 1988]. There are usually eight cysteine residues in the molecule forming four intramolecular disulphide bonds [Kasarda 1989; Shewry & Tatham 1997]. Different genomic clones of the γ-gliadin-encoding sequence have up to 95% homology [Scheets & Hedgcoth 1988].

The molecular weight of γ-gliadins is about 30-45 kDa [Metakovsky et al 2018a]. The length of γ-gliadin-encoding sequences of different wheat species ranged from 678 to 1089 base pairs, depending on the number of repeats in the first domain (their number varied from 7 to 22). Most γ-gliadins overlap with α-gliadins in size, but generally γ-gliadins are more basic. In the cv ChS, 10 of 11 γ-gliadin-encoding genomic sequences were associated with individual 2D EP spots with pIs 7.08-8.51 [Altenbach et al 2020].

There were nine different variants (families) for the arrangement of cysteine residues along the γ-gliadin amino acid sequence [Qi et al 2009]. In the cv ChS, 13 distinct sequences were studied, and the genome assignment was successful for only six of them [Anderson et al 2013].

δ-GLIADINS. δ-Gliadins are monomeric polypeptides having, like typical γ-gliadins, eight cysteine residues, but differing from γ-gliadins in details of their amino acid sequence. Their encoding genes were located within *Gli-1* loci, and there is a single δ-gliadin-encoding gene in each of the *Gli-1* loci. In cvs ChS, and Recital, only genes located at the *Gli-A1* and *Gli-B1* loci, and at the *Gli-A1* and *Gli-D1* loci were shown to produce polypeptides, respectively [Anderson et al .2012]. In ChS, two full-length δ-gliadins had predicted size 34.6 kDa and 35.5 kDa and pIs 5.37 and 7.13. It was estimated that δ-gliadins accounted for only 1.2% of wheat flour [Altenbach et al 2020].

Four new classes of storage protein polypeptides were described. These molecules related to gliadins in their encoding nucleotide sequences, but not similar enough to be classified into any of traditional gliadin families in wheat because of lacking a prominent repeat domain which is important characteristic of the gliadins [Anderson et al 2001a].

Generally, α- γ- and ω-gliadins were found to be composed of a number of families, meaning that DNA sequences of members of a given family are more similar than members of different families. Gliadin-encoding genes belonging to the same family differ in nucleotide substitutions, small deletions/insertions, and the number of repeats in the corresponding domains [Kasarda et al 1984b; Okita et al 1985; Sumner-Smith et al 1985; Bartels et al 1986; Scheets & Hedgcoth 1988; Zhang et al 2003]. Some large deletions were described also [Anderson et al 2001b; Clarke et al 2003].

A set of monoclonal antibodies was produced. Some of antibodies were able to bind specifically to gliadins located in α, or α+β, or β+γ, or ω zones of APAGE and reveal some gliadin polymorphism in those zones [Skerritt et al 1991].

3.2. Types of the LMW glutenins: similarity to gliadins and difference from them

In wheat grain, there are two types of glutenin polypeptides (subunits), HMW glutenins (see **4.1.**) and LMW glutenins. In the cv ChS, polypeptides of LMW glutenin comprise 17.3% of gluten proteins, more than HMW glutenins (12.4%). Ten LMW glutenin subunits studied in this cultivar had MW 31.8-41.3 kDa and pIs 7.56-8.88. Three of them undergo N-terminal processing by an asparaginyl endoprotease in addition to signal peptide removal [Altenbach et al 2020].

Polypeptides of glutenin are linked by covalent intermolecular bonds through cysteine composing a network, gluten matrix. To analyse the set of polypeptides synthesized by a given wheat genotype and present in the gluten matrix, it should be separated into individual molecules using a reducing agent (see **1.1.**). The common and simple method of the analysis of glutenins is SDS-EP. The obtained EP pattern is composed of groups of bands differing in MW and intensity of staining. Besides HMW glutenins (see **4.1.**), there are three groups of LMW glutenins differing in their MW, so-called D- (faint and the longest), B- and C-types.

The MW of most of LMW glutenin polypeptides, like gliadins, is accepted to be 30-55 kDa [Payne & Corfield 1979; Kasarda 1989]. However, the AMW of glutenins, determined by the ordinary SDS-EP method, like that of gliadins (see **1.4.**), may be higher when calculated on the basis of analysis of the DNA sequence of the corresponding gene [Gianibelli & Solomon 2003; Liu et al 2010]. For example, the overestimation of the MW of seven different B-type subunits (of *i*-type, see below) was 28-31% [Zhang et al 2004]. It is assumed that the inaccuracy is caused by an insufficient denaturation of the central domain of glutenin polypeptides consisting of repeats, as well as by the peculiarities of the secondary structure of these molecules [Bunce et al 1985]. In addition, some these polypeptides undergo N-terminal glycosylation by xylose [Laurière et al 1996], and that can also influence their MW as determined by SDS-EP.

A profound analysis of LMW glutenins started when two-dimensional separation was firstly applied to these proteins [Jackson et al 1983]. Later, many LMW-glutenin-encoding DNA sequences, including full-length genes from different wheat genotypes, became known [Cassidy et al 1998; Masci et al 1998; Ikeda et al 2002; Zhang et al 2004; etc.]. It is firmly established that LMW glutenin polypeptides of the D-, B- and C-types are completely different.

C-TYPE glutenins. The C-type LMW glutenins have MW about 30-40 kDa and various isoelectric point values [Jackson et al 1983]. Their amino acid sequences are very similar (in fact, nearly identical) to α- or γ-gliadins. However, unlike gliadins, C-type glutenins have an odd number of cysteine residues and, as a result, the ability to form intermolecular covalent bonds [Shewry & Tatham 1997] and integrate in the gluten matrix [Ferante et al 2006]. Most of C-type LMW glutenins are encoded at the *Gli-1* loci (or *Glu-3/Gli-1* complex loci, see **3.3.**) [Tao & Kasarda 1989; Lew et al 1992; Masci et al 2002]. Some C-type subunits are not controlled, however, by this complex locus [Gupta & Shepherd 1993].

D-TYPE glutenins. The D-type LMW subunits are acidic polypeptides with AMW 65-70 kDa (Jackson et al 1983; Jackson 1984), whose amino acid sequence is similar (in fact, identical) to ω-gliadins, but includes a cysteine residue capable of intermolecular interactions [Masci et al 1999; etc.]. The D-type LMW glutenins are encoded at different storage-protein-coding loci, *Glu-3* [Payne et al 1986b; Pogna et al 1995] (see **3.3.**), *Gli-3* [Jackson 1984; Jackson et al 1985; Payne et al 1988b] (see **6.4.**), *Gli-1* and *Gli-2* [Tao & Kasarda 1989; Khelifi et al 1992; Lew et al 1992; Branlard et al 1994; Gianibelli et al 2001] (see **4.10.**).

A good example of a D-type LMW glutenin subunit is one found in different accessions of common wheat. Being insoluble in ethanol, the molecule appeared in the polymeric protein fraction. Its N-terminal amino acid sequence and amino acid composition were typical for the chromosome 1B-encoded ω-gliadins, but it had, unlike ω-gliadins, one cysteine residue in its amino acid sequence which might interact with other glutenin molecules. Its AMW (71 kDa) was slightly higher than that of some long ω-gliadins and significantly longer than any B- and C-type LMW glutenins. The exact location of the gene encoding the synthesis of this polypeptide was not studied [Gianibelli et al 2002a].

B-TYPE glutenins. The B-type subunits are the most frequent among LMW glutenins [Juhász & Gianibelli 2006]. For example, in ChS, 76% of all LMW glutenin polypeptides studied were of the B-type (Masci et al 2002). The B-type glutenins are alkaline polypeptides with MW about 40-50 kDa [Jackson 1984; Jackson et al 1983; 1985]. The MW of the LMW glutenin polypeptides produced by the cv Xiaoyan-54 varied from 33.566 to 45.054 kDa [Dong et al 2010]. Although amino acid sequence of B-type glutenins had domain structure, it differed strongly from C- or D-types LMW glutenins as well as from gliadins. Thus, while C-, and D-polypeptides are very similar to gliadins (to α- or γ-gliadins, and to ω-gliadins, respectively) in their amino acid sequences, the B-type LMW glutenins have almost nothing in common with any gliadin [Bartels et al

1986; Lew et al 1992; Ciaffi et al 1999; D'Ovidio & Masci 2004; etc.].

There are at least 18 families of the N- and C-terminal sequences of the LMW glutenins [Juhász & Gianibelli 2006]. Based on the analysis of nucleotide sequences of about 60 genes and pseudogenes from different common wheat genotypes, all B-type polypeptides were classified into 12 groups (families) belonging to three types (so-called *i*-, *m*- and *s*-types), which differ in C- and N-terminal amino acid sequences [Ikeda et al 2002; Huang & Cloutier 2008]. The genes encoding LMW glutenins of the same type differ from each other in nucleotide substitutions and small deletions and insertions. It was shown that the *i*-type LMW glutenins originated from the *m*-type as a result of deletion of 33 nucleotides at the 5'-end of the encoding sequence [Gao et al 2007], while *m*- and *s*-types differ only in the presence or absence of the first three N-terminal amino acids, respectively [D'Ovidio & Masci 2004].

A special molecular (PCR) marker system was developed for analysis of genes at the *Glu-3* loci [Zhang et al 2011]. The use of these markers for the analysis of recombinant inbred lines made it possible to establish that the *i*-type genes at the *Glu-A3* locus, as well as the *s*-type genes at the *Glu-B1* locus, were inherited together, forming, like gliadins (see **2.2.**), a recombination unit (haplotype) [Dong et al 2010; Zhang et al 2013; Ibba et al 2017a].

A recombination was found between the *m*- and *i*-type genes at the *Glu-A3* locus and between the *m*- and *s*-type genes at the *Glu-B3* locus with estimated genetic distance of 1.3 cM, and 2.7 cM, respectively [Ibba et al 2017a] (in the Discussion of this article, it was erroneously indicated that "recombination was identified between the i-type and s-type genes, with the estimated genetic distance of 1.7 cM"). It was proved that the *m*-type genes at the *Glu-A3* locus were split in two blocks with the genetic distance of 0.6 cM between them. The recombination within the *Glu-D3* locus was also observed with the distance between markers of 3.1 cM and 4.8 cM (in different groups of lines studied) [Ibba et al 2017a]. In other work, the distance between the *m*- and *i*-type genes at the *Glu-A3* locus was calculated to be 1.3 cM [Dong et al 2010]. The same system of PCR markers was applied to segregating populations, and recombinant genotypes at the *Glu-B3* locus were observed [Espí et al 2014].

Surprisingly, on the chromosome arm 7AL of diploid wheat (*T.monococcum*), a silent (non-transcribed) DNA sequence was found recognized by the probe pTdUCD1 specific to B-type LMW glutenins (Dubcovsky et al 1997).

A LMW glutenin polypeptide of the AMW 32 kDa with unusual amino acid composition (less proline, more glycine, cysteine, asparagine and lysine in comparison with ordinary LMW glutenins) was revealed. This molecule is often found among Indian common wheat genotypes and is encoded by a new locus (named *Glu-D5*) situated on the chromosome 7D [Sreeramulu & Singh 1997].

3.3. The *Glu-3* and *Gli-1* loci are tightly linked

The loci that control the synthesis of the B-types LMW glutenin subunits (*Glu-3*) are located distally on the short arms of the chromosomes of the first homoeologous group [Singh & Shepherd 1988a].

The *Glu-3* and *Gli-1* loci are tightly linked and therefore may be considered as one *Glu-3*/*Gli-1* complex locus [Payne et al 1984b; Gupta & Shepherd 1993; Liu & Shepherd 1995].

Frequency of recombination between the *Glu-A3* and *Gli-A1* loci of common wheat, determined by analysis of the progeny of different crosses, was 1.5±0.3% [Rodriguez-Quijano & Carrillo 1996b] and 1.50±1.07% [Nieto-Taladriz & Carrillo 1996]. However, in the test-cross study, no recombination was detected between the *Glu-3* and *Gli-1* loci [Gupta & Shepherd 1988]. Only one recombinant between the *Glu-A3* and *Gli-A1* loci was found among 153 seeds F2 from the cross Bez1 x Anda [Metakovsky et al 1997b]. In durum wheat, the distance between the *Glu-A3* and *Gli-A1* was, in two crosses, 1.2±0.6 cM and 1.7±0.7 cM [Ruiz & Carrillo 1993].

The genetic distance between the *Glu-B3* and *Gli-B1* loci varied, in different crosses, from 0.62±0.44 cM [Payne et al 1984b] and 1.7cM [Singh & Shepherd 1988a] up to 2.6 cM [Khelifi & Branlard 1992] and 2.8±1.3% [Nieto-Taladriz et al 1998]. The distance between these loci, determined using RFLP technique (different probes specific to corresponding DNA sequences) was 4.5 cM [Dubcovsky et al 1997]. In durum wheat, the distance between the *Glu-B3* and *Gli-B1* in three crosses varied from 0.8±0.4 to 4.9±1.1 cM [Ruiz & Carrillo 1993]. The *Glu-B3* locus was mapped between the *Glu-B1* (recombination frequency 45-50%) and *Gli-B1* (2.0±0.8%) loci [Pogna & Mellini 1988]. Moreover, in durum, there are two linked loci (the distance between them is 3.07 cM), *Glu-B3* and *Glu-B4*, both encoding B-type LMW glutenins. The *Glu-B3* locus is shown to locate more distal than *Glu-B4*, and, therefore more close to the *Gli-B1* locus [Liu 1995]. There is evidence pointing to the existence of two LMW-glutenin-encoding loci on the chromosome 1B of common wheat with the estimated genetic distance of 17.5 cM between them [Ibba et al 2017a] (see **6.4.**).

It is generally assumed that there is no recombination between *Glu-D3* and *Gli-D1* loci [Payne et al 1986b; D'Ovidio & Masci 2004]. The C-type LMW glutenin (MW 30 kDa) occurring in Indian common wheat cultivars is encoded by a gene that does not recombine with the *Gli-D1* locus [Sreeramulu & Singh 1997].

There is an unusual *Gli-D1*-encoded gliadin band located in the middle of the ω-zone of the APAGE pattern. This band is somewhat diffuse and often curved (horseshoe-shaped) while neighbouring bands, including those controlled by the *Gli-D1*, of the same electrophoregram have the usual fairly regular (even) shape. This ω-gliadin (AMW of about 48.5 kDa) is present only in blocks encoded by the *Gli-D1i* and *Gli-D1j* alleles. Analysis of 380 F2 plants from crossing cvs Yunnat and B-16 showed that the gene encoding the unusual ω-gliadin recombined with other ω-gliadin-encoding genes of the same locus (allele *Gli-D1j*) with the frequency of 0.78±0.45%. Also, 4094 single seeds were studied and 13 cases of the recombination were observed (0.65±0.18%) [Kozub et al 2003]. Also, it was found that some genes of the *Glu-D3*/*Gli-D1* complex locus of the cv Kazakhstanskaya-6 (has *Gli-D1i*) formed two groups recombining with a low frequency (six recombinant seeds among 520 F2 progeny). The first group was represented by two polypeptides, one of them was the unusual ω-gliadin, and another a C-type LMW glutenin with AMW of 33.5 kDa. The second group also consists of two polypeptides, a strongly-stained γ-gliadin (35 kDa) and a C-type LMW glutenin (29 kDa) [Metakovsky et al 1997b]. So, it was supposed that the first, and the second groups of polypeptides might be produced by the *Glu-D3*, and *Gli-D1* loci, respectively. It can be assumed that the unusual ω-gliadin encoded by the *Gli-D1i* and *Gli-D1j* alleles is probably a D-type LMW glutenin. At the DNA sequence level, there is no clear border between the *Glu-D3* and *Gli-D1* loci [Anderson et al 2012].

In diploid wheat *T.tauschii* (donor of the D genome to common wheat), the distance between the *Glu-D3* and *Gli-D1* loci was determined by means of the RFLP technique. To map these loci, the probes specifically recognizing coding regions of the B-type LMW glutenin gene [Colot et al 1989], and of the γ-gliadin gene, respectively, were used. Thus, the distance between *Glu-D3* and *Gli-D1* loci was calculated to be 0.9-1.7 cM [Spielmeyer et al 2000].

3.4. Polymorphism of LMW glutenins and allelic variants at the *Glu-3* loci

Each common wheat genotype contains 30-40 DNA sequences (transcribed genes, pseudogenes and gene fragments) for LMW glutenin polypeptides [Cassidy et al 1998] (see **5.1.**). Each allelic variant of any *Glu-3* locus encodes, like an allele at the major *Gli* locus, several different polypeptides. All LMW glutenin polypeptides encoded by one *Glu-3* locus are inherited, like gliadins (see **2.2.**), together, as a Mendelian trait [Gupta & Shepherd 1988; 1993; Gupta et al 1994b; Pogna et al 1995; Redaelli et al 1995; etc.].

The nomenclature used for allelic variants of LMW glutenin patterns is the lettering alleles (as for gliadins) and differs from that widely used for HMW glutenins where each individual catalogued HMW glutenin SDS-EP band has its own number [Wrigley et al 2006b] (see **4.1**). Allelic variants of *Glu-3* loci (as well as its homoeologous variants) encode different, sometimes contrast, sets of LMW glutenin polypeptides. For example, the alleles *Glu-B3h* and *Glu-B3i* encode each eight different polypeptides (five of the B-type and three of the C-type). while the allele *Glu-A3e*, which is present in 40% of all cultivars studied, is null [Gupta & Shepherd 1990].

There is a considerable inter-varietal polymorphism. Thus, in the cv Alpe, the *Glu-D3* locus encodes at least 9, 6, and 2 different polypeptides of the B-, C-, and D-type, respectively, and the *Glu-B3* locus of this cultivar produces 6-8 polypeptides of the B-type and only two of the C-type [Pogna et al 1995]. In the cv Neepawa, however, the *Glu-3 loci* encode two molecules of the C-type (*Glu-A3*), seven of the B-type (*Glu-B3)*, and one of the B-type and six of the C-type (*Glu-D3*). In the cv Costantino, the *Glu-A3* locus encodes two polypeptides of the B-type and one of the C-type [Redaelli et al 1995]. In the cv Norin-61, the *Glu-A3*, -*B3* and -*D3* loci encode four, three, and five B-type LMW subunits, respectively [Ikeda et al 2006].

Using an original technique of two-step SDS-EP for analysis of the sets of B- and C-type glutenins in 222 cultivars, 20 alternative variants of pattern of these polypeptides were identified. Each pattern was encoded by one *Glu-3* locus, so that *Glu-A3*, *Glu-B3*, and *Glu-D3* loci controlled 6, 9 and 5 patterns, respectively. In total, 40 different LMW glutenin polypeptides of the B- and C-types were identified, so that one genotype produces 7-16 of them [Gupta & Shepherd 1990]. In the cv Glenlea, however, 30 genes encoding B-type LMW glutenin subunits were registered [Huang & Cloutier 2008]. [Gupta & Shepherd 1990]. The use of more sophisticated methods for the analysis of polymorphism of LMW glutenins makes it possible to reveal significantly more alleles at the *Glu-3* loci. Thus, using PCR markers, it was possible to identify at once five new alleles at the *Glu-D3* and bring the total number of alleles known at this locus to 11 [Appelbee et al 2009]. Using a new modification of the SDS-EP procedure permitted to distinguish two alleles at the *Glu-D3* locus in USA cultivars that seemed to be identical before [Redaelli et al 1997].

In the cv ChS, one, three, and six different polypeptides of the LMW glutenin were produced by the A, B, and D genomes, respectively [Huo et al 2018a]. In 137 Portugal cultivars, 42 different LMW glutenin polypeptides were observed, in total, in their SDS-EP patterns [Ribeiro et al 2011]. In Chinese germplasm, the genes at the *Glu-A3* showed the highest genetic diversity, while genes at the *Glu-D3* were extremely conserved among different wheat genotypes [Zhang et al 2013; Ibba et al 2017b].

Similar number of alleles at the *Glu-3* loci were identified in different groups of cultivars studied. Thus, in the study of 200 wheat genotypes cultivated in France, 5, 11 (including two new alleles not mentioned in the first catalogue), and 4 allelic variants at the three *Glu-3* loci, respectively, were revealed [Branlard et al 2003]. In Argentinian cultivars, 8 (a record number of alleles at the *Glu-A3* locus among all groups of cultivars studied; at least two of them were new as compared with the first catalogue), 7, and 4 variants were mentioned, respectively [Lerner et al 2009], in 182 Indian cultivars, 4, 7, and 3 alleles [Ram et al 2011], and in the USA cultivars, 5, 8, and 5 variants were registered [Shan et al 2007].

In 251 Chinese cultivars, 5 and 8 alleles were found at the *Glu-A3* and *Glu-B3* loci, respectively [Liu et al 2005], while in 64 Chinese cultivars, 3 and 5 alleles at these two loci were registered [Li et al 2010]. Using an original EP procedure permitted to identify 5 and 4 alleles at the *Glu-A3* and *Gli-B3* loci in Ukrainian wheat germplasm studied [Poperelya 1996]. In these three works, alleles at the *Glu-D3* locus were not identified due to the complexity of this task (see **3.5.**). Indeed, the analysis of 1926 cultivars and breeding lines of South Australia, only three alleles at the *Glu-D3* were revealed [Eagles et al 2002]. A set of 16 SNP markers able to detect SDS-PAGE alleles at the *Glu-A3* and *Glu-B3* loci and therefore to capture the diversity of alleles in 88 international lines and 27 Mexican cultivars was designed [Dreisigacker et al 2020].

Thus, a huge polymorphism of LMW glutenin is observed both in the number of alleles found in different groups of genotypes studied and in the number and diversity of polypeptides encoded by a given *Glu-3* locus.

The compete DNA nucleotide sequence of one of the genes located at the *Glu-A3* locus was defined in each of the seven known variants of this locus [Zhang et al 2004]. It was shown that in each allelic state this gene encodes a B-type glutenin having eight cysteine residues and the N-terminal isoleucine (*i*-type). The gene might encode a polypeptide of 388 amino acids, but, in its different allelic states, it has from one to five small deletions in different positions of the polypeptide; therefore, its length is always shorter. In addition, all seven variants of this LMW glutenin differed in single nucleotide substitutions, so that some of the compared pairs differed by a unique substitution, while some other pairs of allelic variants could differ in 37 positions of the amino acid chain [Zhang et al 2004].

The molecular marker system designed for the analysis of the LMW-glutenin-encoding genes and the full-length gene cloning method were used for analysis of the collection of Chinese common wheat accessions. As a result, more than 15 genes were identified, of which from four to six, from three to five, and eight genes were located at the *Glu-A3*, *Glu-B3*, and *Glu-D3* loci, respectively. It was found that the *m*-type LMW glutenin polypeptides were the most abundant in all genotypes studied, and their calculated MW varied from 31.8 to 39.6 kDa. The *s*-type LMW glutenins had the MW ranged from 37.0 kDa to 42.5 kDa, and the *i*-type glutenins had the highest MW (39.2-43.0 kDa) despite lacking the N-terminal sequences. Moreover, it was shown that, in each genotype, there are two-four *i*-type genes and two *m*-type genes at the *Glu-A3* locus, one-three *s*-type genes and two-three *m*-type at the *Glu-B3* locus, and one *i*-type gene and seven *m*-type genes at the *Glu-D3* locus in each single genotype studied. From nine to 13 active genes were present in each accession [Zhang et al 2013].

3.5. The problem of reliable identification of alleles at the *Glu-3* loci

The use of special two-dimensional EP made it possible for the first time to detect and, to some extent, describe multiple allelism at the *Glu-3* loci [Jackson et al 1983].

The first fundamental catalogue of alleles at the *Glu-3* loci was based on the comparison of sets of LMW glutenins of B- and C-types present in 222 cultivars from 32 countries. A marker (standard) cultivar was proposed for each allele in the catalogue composed, in total, of 20 alleles for the three *Glu-3* loci [Gupta & Shepherd 1990] (see **3.4.**). Later, this catalogue was applied to identify alleles at the *Glu-3* loci in several groups of common wheat cultivars.

However, a reliable identification of alleles at *Gli-3* loci in different laboratories has proven to be quite a challenge. For example, in the first work, 14 LMW glutenin subunits (probably, of the B-type) were registered in the cv ChS [Jackson et al 1983; Jackson 1984]. In other works, however, there were either 5 and 4 subunits of LMW glutenin of the B- and C-type [Gupta & Shepherd 1990], or 7 and 10 subunits [Felix et al 1996], respectively, in the same variety. In the cv Insignia, a *Glu-A3*-encoded polypeptide was revealed [Ikeda et al 2006] although this cultivar was proposed [Gupta & Shepherd 1990] as a standard for the null allele (*Glu-A3e*) at this locus. The difficulty in distinguishing alleles *Glu-A3f* and *Glu-A3e* by SDS-EP was reported and criteria for distinguishing newly-proposed alleles *Glu-B3c'* and *Glu-B3b'* from usual *Glu-B3c* and *Glu-B3b* were not given [Branlard et al 2003].

The alleles *Gli-B1i* (occurs only in Australian wheat germplasm), *Gli-B1k, Gli-B1m*, and a few others encode blocks which have similar band composition, and therefore belong to the same family (see **7.2.**, **7.5.**). It was shown that alleles of this family occur in wheat genotypes exclusively together with allele *Glu-B3c* [Gupta et al 1994a], or *Glu-B3i* [Jackson et al 1996]. Surprisingly, however, no one cultivar studied in these two works had the allele *Glu-B3i* or *Gli-B3c,* respectively. Therefore, it seems that the same allele at the *Glu-B3* locus was identified being called, in the two studies, either *Gli-B3c*, or *Gli-B3i* respectively. Therefore, it is not clear, which allele (*Glu-B3c* or *Glu-B3i*) occurs in genotype of different wheat cultivars always together with members of the particular family of *Gli-B1* alleles (see **3.6.**).

Another example of confusion is the unreliability of identification of alternative variants (alleles) at the *Glu-D3* locus. In the study of 182 Indian common wheat genotypes, the identification of a given allele performed by analysis of SDS-EP patterns, was checked by examining the DNA encoding this polypeptide by the PCR technique. As a result, 55 cultivars having, according to the SDS-EP analysis, the allele *Glu-D3a*, were divided into two groups differing in the allelic state of the *Glu-D3* locus, while 122 cultivars with the allele *Glu-D3b* were divided into three groups. It was difficult also to identify, with the use of the SDS-EP, alleles *Glu-A3b*, *Glu-A3e/f* and *Glu-B3b* [Ram et al 2011].

Each *Glu-3* locus is represented by a collection of SNP alleles (specific haplotypes). Using molecular markers, it was established that two known (different) alleles at the same *Glu-3* locus may have identical haplotypes, for example, *Glu-A3e* and *Glu-A3g*, *Glu-B3c* and *Glu-B3d*, *Glu-B3f* and *Glu-B3ab*. In contrast, two haplotypes were distinguished designated, in a set of standard cultivars, as having the same allele *Glu-D3c* [Ibba et al 2017b].

In other work, 15 scientists from six countries compared ability to identify alleles at the *Glu-3* loci in 30 standard cultivars using four different methods: ordinary one-dimensional SDS-EP, 2D EP (isoelectrophocusing/SDS-EP), PCR technique, and MALDI (matrix-associated laser desorption/ionization) mass spectrometry. In total, 25 *Glu-3* alleles were identified in the set of cultivars studied, and only 10 of them were detected by each of the four techniques used. It was concluded that the best results in the differentiation of alleles at the *Gli-A3* and *Gli-B3* loci gave PCR markers, and at the *Gli-D3* locus, mass spectrometry. It was difficult to distinguish the *Gli-A3e* from *Gli-A3f*, or *Gli-A3d* from *Gli-A3g* using SDS-EP, while the pair of alleles *Gli-A3a* and *Gli-A3c* was indistinguishable for the MALDI mass spectrometry [Liu et al 2010]. In 63 Iranian common wheat cultivars, 23 alleles (5, 9, and 9 alleles, respectively) and 47 alleles (8, 17, and 22) at the three *Glu-3* loci were revealed using SDS-EP, and gene-specific PCR primers, respectively [Khoshro et al 2021].

Obviously, a reliable identification of alleles at any *Glu-3* locus using SDS-EP presents more difficulties than of alleles at *Gli-1* locus using APAGE. Apparently the same allele *Glu-B3i*; being transferred to NIL from two different sources (genotypes), showed a contrast influence on quality [Bonafide et al 2015]. That might be caused by difficulties in reliable identification of this alleles, or by some inaccuracy in maintaining the collection of wheat genotypes used for the experiment (see **10.2.**).

The inconsistencies caused by the differences in the methods of analysis make it extremely difficult a comparison of results obtained in different laboratories [Liu et al 2010]. It is believed [Jackson et al 1996; Juhász & Gianibelli 2006] that the base catalogue [Gupta & Shepherd 1990] needs improvement. There are difficulties in identification of alleles at the *Glu-3* loci caused by the absence of a common criteria in the interpretation of the SDS-EP electrophoregrams. It seems that the study of polymorphism of LMW glutenin is now in a similar situation as it was with gliadin until a fairly simple standard EP technique for its study was developed. It cannot be ruled out that it would be difficult to find one simple and effective method for the detection, description and analysis of LMW glutenin polypeptides of different types [Liu et al 2010; Ram et al 2011].

3.6. Correspondence of allelic variation at the *Glu-3* and *Gli-1* loci

The allelic variation at the *Glu-3* loci affects the wheat flour/dough quality (see **4.5.**). However, identification of alleles at the *Glu-3* loci, including those related to quality, is rather unreliable (see **3.5.**). Therefore, for practical purposes, it would be valuable to find a correspondence between the alleles of two tightly linked (see **3.3.**) loci, *Glu-3* and *Gli-1*. It was suggested that, being easier identified, alleles at the *Gli-1* loci might serve as genetic markers for selecting desirable genotype at the *Glu-3* locus in breeding programs because a risk of recombination between the two loci is very low.

Some correspondences do exist between *Glu-3* and *Gli-1* alleles. For example, the presence/absence of two *Glu-D3*-encoded D-subunits of LMW related well to two distinct variants of *Gli-D1*-encoded ω-gliadins (Cheyenne- and ChS-types, see **2.5.**) [Payne et al 1986b]. In another case, the presence of the 42-kDa LMW glutenin polypeptide is always accompanied by the presence of the allele *Gli-B1d* in the genotype (see **4.8.**).

The analysis of 85 Australian cultivars showed a strong one-to-one full correspondence (each allele of the pair occurs, in different genotypes, exclusively with other allele of this pair) only for four pairs of alleles, *Glu-B3c+Gli-B1i* (18 cultivars), *Glu-B3h+Gli-B1d* (4), *Glu-B3d+Gli-B1h* (3), and *Glu-A3d+Gli-A1o* (2). In other cases, only a partial correspondence between the two rows of alleles was observed. For example, combinations of the allele *Gli-B1b* with *Glu-B3b*, or *Glu-B3g* occurred in 46, and 6 cultivars, respectively, and coincidences of the allele *Glu-A3e* with *Gli-A1f*, *Gli-A1m*, and *Gli-A1a* were found in 23, 20, and 1 cultivar, respectively [Gupta et al 1994a].

Pairs of the γ-gliadin specific DNA markers (for example, GliB1.1 and GliB1.2 for the *Gli-B1* locus) were developed that could distinguish (PCR technique) two alternative ("allelic") variants of each *Gli-1* locus [Zhang et al 2003]. Using these markers, only one of the one-to-one matches described [Gupta et al 1994a] was confirmed: allele *Glu-B3c* always occurred together with the marker GliB1.2. The PCR marker GliA1.1 occurred together with *Glu-A3a*, *Glu-A3b* or *Glu-A3c*, while GliA1.2 was found in the same genotype with the alleles *Glu-A3d* and *Glu-A3e*. The GliB1.1 occurred together with *Glu-B3d*, *Glu-B3b*, *Glu-B3c* and *Glu-B3e*, while GliB1.2 was found with *Glu-B3h*, *Glu-B3a* and *Glu-B3* [Zhang et al 2003]. The discrepancy between the results obtained in these two articles dealing with Australian wheat may be explained by the lower resolution of the PCR markers to distinguish alleles at the *Gli-1* loci (only two variants for each locus) as compared with APAGE. At least seven and five alleles at the *Gli-A1* and *Gli-B1* loci, respectively, were registered in Australian cultivars [Metakovsky et al 2019b].

The analysis of the correspondence between alleles at the *Glu-3* and *Gli-1* loci was also carried out on 49 cultivars from six European countries. As a result, a full correspondence was registered for allelic pairs *Glu-A3d+Gli-A1o* (16 cultivars), *Glu-A3e+Gli-A1m* (4), *Glu-B3b+Gli-B1b* (17) and *Glu-B3i+Gli-B1k* or *Gli-B1m* (7). Also, the alleles *Glu-B3d* and *Glu-B3h* occurred with *Gli-B1h* and *Gli-B1d*, respectively (2 cultivars in each case), confirming the results obtained in the analysis of Australian cultivars. At the same time, no unambiguous correspondence was found between the allelic variants at the *Glu-D3* and *Gli-D1* loci. The reason is that among 49 cultivars studied, only two alleles at the *Glu-D3* were identified (*Glu-D3a* and *Glu-D3c*) [Jackson et al 1996]. Similar results were obtained from the analysis of 200 French and other European cultivars: an undoubted correlation was revealed for the pairs of alleles *Glu-B3h+Gli-B1d*, *Glu-B3a+Gli-B1a*, and *Glu-B3j+Gli-B1l*. In most cases, however, the presence in the genotype of an allele at one locus does not guarantee an exact allele at the other locus [Branlard et al 2003].

Thus, comparing the results of the analysis of seed storage proteins in many common wheat cultivars published in several articles, it can be argued that, among the cultivars studied, there are only three cases of an unambiguous and full compliance between alleles at the *Glu-3* and *Gli-1* loci: *Glu-A3d+Gli-A1o*, *Glu-B3h+Gli-B1d*, and *Glu-B3d+Gli-B1h*. In 66 of 72 genotypes studied [Gupta et al 1994a; Jackson et al 1996], the allele *Glu-B3b* occurred together with the *Gli-B1b*. In other hands [Branlard et al 2003], however, the *Gli-B1b* matched with five different alleles at the *Glu-B3*, including *Glu-B3b*.

Thus, probably, the most rigorously proven full correspondence has been observed between members of a particular group (family) of similar alleles at the *Gli-B1* locus (such as *Gli-B1i*, *Gli-B1k*, *Gli-B1m*, etc., see **6.10.**, **7.2.**, **7.5.**) and the only one allele at the *Glu-B3* locus. However, there is a problem to unambiguously identify the allele at the *Glu-B3*. In two independent studies, it was identified either as *Glu-B3c* [Gupta et al 1994a], or as *Glu-B3i* [Jackson et al 1996].

3.7. Summary. Polymorphism of LMW glutenins is more difficult to study than gliadins

There is a considerable polymorphism among gliadins as well as LMW glutenins produced by one common wheat genotype. Three types of gliadins (the α-, γ- and ω-gliadins) are quite different in their amino acid sequences, domain structure and AMW. There are also three different types of polypeptides (subunits) of the LMW glutenins, C-type (their primary structure is similar, if not identical, to α- and γ-gliadins), D-type (similar, if not identical, to ω-gliadins), and B-type (true LMW glutenins). Most of B-type polypeptides are encoded at the *Glu-3* loci tightly linked to the *Gli-1* loci. There is multiple allelism for each of the *Glu-3* loci, but the number of alleles for one locus is less than for *Gli* loci.

Each *Glu-3* locus encodes polypeptides of different types that inherited together, as one trait (as gliadins). There are three types (*i*-, *m*- and *s*-types) of LMW glutenins of B-type. Several levels of polymorphism of LMW glutenins is observed: among polypeptides encoded by a given allele, among sets of polypeptides controlled by different alleles at the same locus (multiple allelism), and among wheat genotypes (cultivars). To reveal and to document the polymorphism of LMW glutenins, several methods were proposed, including different variants of two-dimensional EP. The best results in the differentiation of alleles at the *Gli-A3* and *Gli-B3* loci gave PCR markers, and for alleles at the *Gli-D3* locus, mass spectrometry. The most commonly used method, however, is a conventional SDS-EP.

There is a problem to reliably identify alleles at the *Glu-3* loci and to establish the number and type of polypeptides encoded by a given allele at the *Glu-3* locus. Because of these difficulties, attempts were made to look for a possible coincidence, in wheat genotypes, between the alleles of the *Glu-3* and *Gli-1* loci. Being easier identified, alleles at the *Gli-1* loci might serve as genetic markers for selecting desirable genotype at the *Glu-3* locus in breeding programs because a risk of recombination between the two loci is low. However, an unambiguous correspondence was proven for only three (or four) allelic pairs (*Glu-3* vs *Gli-1*). Confirmation of more correspondences was prevented by the absence of a common criteria in the interpretation of the glutenin SDS electrophoregrams in analysis of alleles at the *Glu-B3* locus.

4. RELATION OF GLIADINS AND GLUTENINS TO END-USE WHEAT QUALITY

4.1. High molecular weight (HMW) glutenin and alleles at the *Glu-1* loci

Generally, a polypeptide HMW glutenin (and its encoding gene) consists of the central repetitive domain (forms 60-80% of the sequence) and the two non-repetitive terminal domains. Each HMW glutenin molecule has two cysteine residues located at opposite ends of the polypeptide. The central domain forms a loose spiral of secondary structure based on repeated β-turn motifs [Forde et al 1985; etc.]. The spiral formed by the central part of the polypeptide might be intrinsically elastic and relate to gluten elasticity. Therefore, it is considered as the most important characteristics affecting the quality of dough [Thatam et al 1985]. The length of the HMW glutenin polypeptide depends on the size of its central domain [Shewry et al 1992; D'Ovidio et al 1994; Hassani et al 2005], and its conformational particularities could be responsible for differences between these polypeptides in their relationship to end-use wheat properties [Goldsbrough et al 1989].

In a wheat seed, HMW and LMW glutenin polypeptides (subunits) are interconnected by covalent disulphide bonds creating a single enormous network (gluten matrix) (see **1.1.**). Several distinct techniques were developed to reveal and analyse polymorphism of HMW glutenin molecules [Payne & Corfield 1979; Payne et al 1981; Sutton 1991; Margiotta et al 1993; Cloutier et al 2000; Gao et al 2010; etc.]. However, a set of molecules of gluten synthesized by a given genotype are still usually (in most of studies) investigated by routine SDS-EP after separation of the gluten matrix into individual polypeptides.

There are two types of HMW glutenin polypeptides encoded at one *Glu-1* locus, the so-called x- and y-subunits, differing in the length of the polypeptide (x-subunit is longer), number of cysteine residues and their position in the amino acid chain [Shewry et al 2006]. Also, x- and y-subunits differ in immunological properties [Curioni et al 1991]. Homoeoallelic y-type subunits (for example, subunits 8 and 12 encoded by *Glu-B1* and *Glu-D1* loci, respectively, see **4.2.**) have greater similarity in

N-terminal sequences than that found between x- and y-type subunits encoded by the same locus [Margiotta et al 1993; Gianibelli & Solomon 2003].

The disulphide bonds always connect x- and y-subunits (neither two x, nor two y) [Tao et al 1992; Shewry & Tatham 1997; Shewry et al 2006]. The formation of intermolecular bonds begins immediately after the synthesis of glutenin polypeptides on ribosomes and continues during grain filling and drying [Shewry & Halford 2002; Naeem & MacRitchie 2005]. The biosynthesis of glutenin in developing grain almost completely stops at high temperature while gliadins continue to be synthesized [Blumenthal et al 1993].

y-Type polypeptides encoded by the *Glu-B1* locus are prone to proteolytic processing at the C-terminal tail leading to the loss of the unique cysteine residue present at the C-terminal domain [Nunes-Miranda et al 2017].

The synthesis of HMW glutenin polypeptides is controlled by the *Glu-1* loci (*Glu-A1*, *Glu-B1*, *Glu-D1*) located on the long arms of the chromosomes 1A, 1B and 1D, respectively, near the centromere [Payne 1987; Singh & Shepherd 1988b]. Therefore, *Glu-1* and *Gli-1* loci are not genetically linked. However, the use of *Glu-1* loci makes it possible to map genes located proximally in the short arm of the chromosomes of the first homoeologous group. For example, the enzyme locus *Gpi-1* recombines with the *Glu-D1* and *Gli-D1* loci with a frequency of 36.2% and 34.5%, respectively [Chojecki et al 1983].

Polymorphism of HMW glutenins is lower than that of gliadins: in total, only about 25-30 polypeptides (subunits of HMW glutenin) are registered in common wheat differing in their EM in the SDS-EP (hence, in their relative AMW) [Payne & Lawrence 1983; Vawser & Cornish 2004; Shewry et al 2006; etc.]. Each catalogued HMW glutenin polypeptide has its own number, for example, the subunit slowest in the SDS-EP has number 1, etc.

There is a multiple allelism at each *Glu-1* locus in common wheat. In the first catalogue of alleles at the *Glu-1* loci, 3, 11 and 7 allelic variants were described for the *Glu-A1*, *Glu-B1* and *Glu-D1* loci, respectively [Payne & Lawrence 1883]. In the more recent catalogue, 39 alleles (5, 21, and 13, respectively) are shown distinguished by SDS-EP [Shewry et al 2006].

There are many publications in which allelic variation at the *Glu-1* loci in different groups of cultivars was studied. For example, in 110 common wheat cultivars registered in Spain [Carrillo et al 1988], as well as in 174 local and old Spanish wheat genotypes [Rodriguez-Quijano &

Carrillo 1994], there were three, seven, and five alleles at the *Glu-A1*, *Glu-B1*, and *Glu-D1* loci, respectively. In 134 Portuguese cultivars, four, ten, and six alleles, [Ribeiro et al 2011], and in 64 Portuguese landraces, three, nine, and four alleles [Rodriguez-Quijano et al 1998], respectively, were registered. In 200 cultivars grown in France, three, eight, and five alleles [Branlard et al 2003], and in 119 Argentinian cultivars, three, six, and two alleles [Lerner et al 2009], respectively, were revealed. In Canada (58 cultivars were studied), there were four, eight, and two alleles [Ng et al 1989], and in the USA HWW cultivars, three, five, and three alleles [Shan et al 2007] were found. In different studies of Chinese cultivars, there were three, four, and three alleles [Li et al 2010], and three, nine, and four alleles [Liu et al 2005]. Three, six and two alleles for *Glu-A1*, *Glu-B1*, and *Glu-D1* loci, respectively, were detected in 107 cultivars of Argentina [Gianibelli et al 2002c]. Therefore, in these 11 independent studies performed on different groups of wheat cultivars and landraces, there were usually three (in two cases, four) alleles at the *Glu-A1*, from five to ten (in one case, three) alleles at the *Glu-B1*, and from two to five (in one case, six) alleles at the *Glu-D1* loci.

A new interesting allelic variant of the Dy10-encoding HMW-glutenin-subunit gene was found and studied. It was shown that the amino acid substitution in the encoded HMW glutenin polypeptide resulted in a post-transcriptional cleavage producing two new peptides having changed number of cysteine residues available for interactions. As a result, these new peptides disrupted the interactions among gluten proteins decreasing the size of glutenin polymers and thereby improving cookie-making quality [Wang et al 2021].

DNA-based molecular markers (SNP) were developed enabling to observe 8, 22 and 9 different alleles at the three *Glu-1* loci, respectively. A set of 17 SNPs were designated as representative markers of the most frequent SDS-PAGE alleles at each locus. The study demonstrated that the SNP markers might provide important additional information on the genetic diversity of HMW glutenins [Ravel et al 2020].

The polymorphism revealed at *Glu-1* loci of common wheat cultivars is lower than at *Gli-1* loci. Indeed, in 140 cultivars bred in the former USSR, only three, four, and three alleles at the *Glu-A1*, *Glu-B1*, and *Glu-D1*, respectively, were found, and more than 70% of all cultivars studied had identical SDS-EP electrophoregram of HMW glutenin [Morgunov et al 1990; Nettevich et al 1991]. In analysis of HMW glutenins of 1380 cultivars from 21 countries, only 17 alleles were registered, three, nine, and five alleles at the three *Glu-1* loci, respectively [Morgunov et al 1993]. Since the publication of the first catalogue of alleles at the *Glu-1* loci, more than 100 allelic variants have been identified in over 7830 cultivars/lines of various wheat species studied [Li et al 2021].

Each allelic variant of the *Glu-D1* locus and most alleles at the *Glu-B1* locus encode two polypeptides (x-type + y-type) inherited together, as a Mendelian trait. For example, the allele *Glu-D1d* encodes subunits 5 and 10 (or 1Dx5 and 1Dy10) while the allele *Glu-D1a* produces subunits 2 and 12 (or 1Dx2 and 1Dy12). A few alleles at the *Glu-B1* locus and alleles *Glu-A1a* and *Glu-A1b* produce one polypeptide only (while gene for y-subunit is silent). In addition, there is a null allele *Glu-A1c* [Payne 1987; Shewry et al 2003; etc.]. Therefore, a common wheat genotype produces a group of three-five HMW glutenin polypeptides (subunits) which can be identified by means of SDS-EP.

4.2. Problems of identification of alleles at the *Glu-1* loci and their nomenclature

Routinely, alleles at *Glu-1* loci are identified visually by a screening of bands in a certain zone of the SDS-electrophoregram [Wrigley et al 2006b]. There are only three-five EP bans in this zone of gel slab, and that permits to identify successfully the known alleles of the *Glu-1* loci [Morris et al 2009]. However, there is some inconvenience in the nomenclature of alleles at the *Glu-1* loci. In most publications; alleles at the *Glu-1* loci are designated through SDS-electrophoretic bands produced by these alleles [Shewry et al 2006; Jin et al 2013], occasionally through lettering of alleles [Vawser & Cornish 2004], and rarely in both ways [Payne & Lawrence 1883]. Such disorder creates certain doubts when comparing results in different publications. While alleles *Glu-B1c* (7*+9), *Glu-B1d* (6+8*), *Glu-B1e* (20), *Glu-B1f* (13+16), *Glu-B1i* (17+18) were reported to be easily identified using SDS-EP [Vawser & Cornish 2004], a difficulty in distinguishing subunits 7 and 7* causes the use, in different publications, of the same allelic designation *Glu-B1c* for both 7+9 and 7*+9 subunit compositions. Moreover, the combination of subunits 17+18 was called either *Glu-B1i* [Vawser & Cornish 2004], or *Glu-B1c* [Ribeiro et al 2013]. It should be also noted that, historically, the allele at the *Glu-B1* locus of the cv ChS has been designated by the letter "*b*" (*Glu-B1b*, 7+8), although, according to the Rules of the International biochemical nomenclature, all alleles of this variety should be designated by the letter "*a*".

Any EP procedure cannot reliably distinguish distinct polypeptides having similar relative EM (in the case of SDS-EP, AMW). To differentiate glutenins with similar AMW, subunits 7 and 7*; 8 and 8*; 12

and 12*; 2, 2* and 2**, etc. were catalogued. However, the existence of subunit 2* in the SDS-EP can be obscured by the presence of subunit 2 in a given genotype, and differentiation between subunits 8 and 8* can be achieved using RP-HPLC, but not by SDS-EP [Marchylo et al 1992]. In other work, two different variants of the subunit 1Bx8 were detected, but one of them was not identifiable by SDS-EP [Gianibelli et al 2002c]. The PCR-based markers were developed allowing discrimination of alleles producing By8 and By8* glutenin polypeptides. To distinguish subunit 7 from 7*, the use of polymerase chain reaction (PCR) is necessary [Cloutier et al 2000; Butow et al 2003]. Other markers can be used to discriminate By9-coding alleles from non-By9 alleles at the *Glu-B1* locus [Lei et al 2006], Two different groups of subunit 7 were observed. One appeared in cultivars with allele 7+8 at the *Glu-B1* locus, a second group of subunit 7 was found in three different alleles [Gianibelli et al 2002c]. The HPLC was applied to show biochemical heterogeneity among *Glu-B1*-encoded subunit 8 and among *Glu-A1*-encoded subunit 2** in common cultivars bred in New Zeeland. These differences could not be recognized by using SDS-EP [Sutton 1991].

The allele *Glu-B1b* (encodes subunits 7+8), is known to be one of four different alleles having similar EMs in SDS-EP, *Glu-B1b* (7+8), *Glu-B1u* (7*+8), *Glu-B1ak* (7*+8*) and *Glu-B1al* (7+8*) [Vawser & Cornish 2004]. Despite a close similarity in their AMW, subunits 7 and 7*, and also 8 and 8*, differ in amino acid sequences and protein characteristics [Sutton 1991; Marchylo et al 1992], and, importantly, contrasting effects on quality [Gianibelli et al 2001] (see **4.3.**). There is probably a need in a new nomenclature system, based on known amino acid sequences of HMW glutenins [Cunsolo et al 2012].

Visual analysis of HMW glutenins in the presence of a standard genotypes on the same SDS-EP slab makes it possible to identify only up to 43% of all HMW subunits of the genotype under study. The use of a number of standard cultivars that represent known HMW glutenins, in combination with digital image analysis software, improved subunit recognition in a set of 226 genotypes bred in the USA up to 73-91% [Morris et al 2009].

A group of HMW glutenins produced by one common wheat genotype have AMW of about 80-120 kDa in the SDS-EP [Payne & Corfield 1979; Payne & Lawrence 1983]. However, this AMW, like AMW of gliadins determined by SDS-EP technique (see **1.4.**), is higher than that calculated from amino acid composition of the molecule. It was shown that AMW of HMW glutenin subunits in 17.3% polyacrylamide gel (PAAG) were generally about 15% higher than the values obtained with

10% PAAG and 32-54% higher than the values calculated from amino acid composition of polypeptides studied [Ng & Bushuk 1989].

The presence, in a newly-studied wheat genotype, of a new band in the "HMW-glutenin-zone" of SDS-EP does not guarantee, without analysis of inheritance of this new polypeptide, a discovery of a new allele at the *Gli-1* locus. For example, in one of genotypes (k-59173) of *T.boeoticum* (a diploid wheat with the genome A), a polypeptide with AMW more than 100 kDa was noted in the zone of SDS-EP pattern typical for HMW glutenins. Surprisingly, this molecule was found to be probably an unusually long gliadin because it was alcohol-soluble and inherited, in the analysis of progeny from crossing of this genotype, linked to polypeptides encoded by the *Gli-A1* locus of *T.boeoticum* [Metakovsky & Baboev 1992b].

Analogously, a polypeptide named T1 was discovered in the HMW-glutenin-zone of the SDS electrophoregram in collection of accessions of diploid wheat *T.tauschii*. Based on its high MW (AMW of about 95 kDa), this polypeptide was firstly considered to be a *Glu-D1*-controlled HMW subunit with a molecular size similar to that of the y-type subunit 10 of *T.aestivum* [Gianibelli et al 2002d]. However, a re-evaluation of this protein showed that T1 is not a component of the polymeric protein fraction. Rather, T1 is an ω-gliadin of unusually high AMW. This suggestion was confirmed by APAGE/SDS 2D EP and analysis of the N-terminal amino-acids sequence of T1. Gene encoding its synthesis was mapped between *Glu-D1* and *Gli-D1* loci [Gianibelli et al 2002d] (see **6.5.**). The presence of monomeric polypeptides in the HMW region of the SDS-EP is discussed (Gianibelli et al 2001).

In *T.tauschii (*accession CPI 110750), an unusually short HMW glutenin of the y-type with AMW 58 000 (SDS-EP) was discovered. However, MW of the polypeptide, calculated on the basis of gene length would be even smaller, around 47 kDa (or 1.38 Kb) [Gianibelli & Solomon 2003] or 45,228 kDa. The polypeptide showed the similarity to the Dy12 subunit present in cv ChS [Hassani et al 2005].

4.3. Allelic variation at *Glu-1*, especially at the *Glu-D1*, affects the end-use quality

The grain quality of any common wheat cultivar is determined by grain hardness, protein content and dough properties [Pasha et al 2010]. Although an increased understanding of environmental effects on the technological quality of wheat is essential, genotype has a higher effect on grain hardness and dough properties [Johansson et al 2020].

For most uses of wheat, dough properties (its strength, extensibility, stability) are the critical elements that determine the quality of the final product (mainly bread). In turn, the wheat flour/dough quality depend on the polymorphism and quantity of glutenin and gliadin which form dough and which are produced by the storage protein loci of a given genotype [Wrigley et al 2006b; Igrejas G & Branlard 2020; etc.]. Due to the presence of free cysteine residues, glutenin polypeptides form intermolecular bonds, and thereby determine the structure of gluten matrix [Kasarda 1989; Tao et al 1992; etc.] (see **1.1.**). The number of cysteine residues and their position in the amino acid sequence of glutenin polypeptide play a key role in determining quality [reviewed in Shewry et al 1992; Shewry & Tatham 1997; D'Ovidio & Masci 2004].

About 40 years ago, it was discovered that genetic polymorphism at the *Glu-1* loci influences the common wheat dough quality [Payne et al 1981; etc.]. This general finding was confirmed in many other studies [Eagles et al 2002; Branlard et al 2003; Liu et al 2005; Shan et al 2007; Lerner et al 2009; Li et al 2010; etc.]. For example, some wheat cultivars bred in Australia consisted of related biotypes (genotypes). It was discovered that genotypes of the same cultivar, differing in the allelic state of the *Glu-1* locus, might also differ in the dough quality [Lawrence et al 1987]. In similar work, 14 of 100 winter wheat lines of diverse origin studied consisted of biotypes. It was found that biotypes of the same line differed in the alleles at the *Glu-1* loci and in numerous quality attributes [Graybosch et al 1994].

Intensive work has been performed with the aim to identify alleles at the *Glu-1* loci in different groups of cultivars and to establish a relation of some alleles to quality. It was established that as much as 69% of variation in bread-making quality among 99 Chinese cultivars depends on polymorphism at *Glu-1* loci [Wang et al 1993]. Up to 60% of the variation in quality (SDS sedimentation test [AACC 1995], in this book, abbreviated SDS-st) in English cultivars [Payne et al 1987a], 68-79% of variation in SDS-st and Alveograph tests in Spanish cultivars [Payne et al 1988a], and 54% of variation in bread-making quality in cultivars grown in Germany [Rogers et al 1989] were associated with polymorphism at the *Glu-1* loci. A part of phenotypic variation of a quality parameter explained by the composition of HMW glutenin reached 35% [Branlard et al 2001]. As a result of the analysis of 243 F5 lines from a cross, 105 French cultivars and 92 world-wide cultivars, it was found that the dough quality (its strength and elasticity) is determined (18-55%) by allelic variation at the *Glu-1* loci [Branlard & Felix 1994]. In dihaploid genotypes obtained from crossing two Canadian cultivars, polymorphism at the *Glu-D1* locus explained up to 50% of the total variation in several parameters of dough tested by mixograph [Huang et al 2006].

In the world collection, more than 90% of all common wheat cultivars have either *Glu-D1d* or *Glu-D1a* in their genotypes [Payne & Lawrence 1983] and the *Glu-D1d* marked by the presence of the HMW glutenin SDS-EP bands 5+10 has the greatest positive effect on the dough quality among all allelic variants of the *Glu-1* loci [Payne et al 1988a; Rogers et al 1989; etc.].

The contribution of different *Glu-1* alleles to the flour quality (SDS-st) was determined studying 84 English cultivars. It was established that cultivars with the allele *Glu-D1d* in the genotype have the highest quality (score 4 points), while the allele *Glu-D1a* (2+12) has a rather negative effect (2 points). The top-rated alleles at each of the three *Glu-1* loci were found to be (in order of intensity of influence) *Gli-D1d* (subunits 5+10), *Gli-B1i* (17+18) and *Gli-A1a* (1) [Payne et al 1987a]. The scores for individual HMW glutenin subunits established in the analysis of English cultivars were applied for calculation of the *Glu-1* scores in Canadian wheat. It was found that, in the set of 67 Canadian-grown cultivars, *Glu-1* scores accounted for 59-69% of the variation in their bread-making quality [Lukow et al 1989].

An obvious positive effect of the *Glu-D1d* on quality was observed in Italian wheat germplasm [Pogna & Mellini 1986; Pogna et al 1989]. The results of the analysis of F4 progeny from five different crosses of the New Zeeland cultivars [Luo et al 2001], of F2 descendants from two crosses of Spanish cultivars [Rodriguez-Quijano & Carrillo 1996b], and of the progeny from crossing Ukrainian cultivars [Chebotar et al 2012a] confirmed the positive effect of the *Glu-D1d* (in comparison with *Gli-D1a*) on the parameters of flour and dough. Out of total of 15 allales at the *Glu-1* loci identified in 110 Spanish common wheat cultivars [Carrillo et al 1988] and 172 old local Spanish wheats [Rodriguez-Quijano & Carrillo 1994], the allele *Glu-D1d* had the greatest positive effect. Similar results were obtained in studies of other groups of cultivars [Dong et al 1991; Manley et al 1992; Branlard & Dardevet 1994; etc.]. Analysis of the NIL lines showed the positive effect of allele *Glu-D1d* (versus *Glu-D1a*) on dough tenacity [Redaelli et al 1997]. The positive impact of the allele *Glu-D1d*, as compared with *Glu-D1a*, on dough strength depends on the genetic background. It is more pronounced in genotypes having weak dough strength and better extensibility [Wang et al 2018].

Polymorphism at the *Glu-1* loci was studied in 90 Canadian Western Red Spring (CWRS) common wheat cultivars created during the 20th century and possessing an exclusively high dough quality. It was discovered that all these cultivars had allele *Glu-D1d* in their genotypes, and that this valuable allele was already present in cvs Marquis (registered in 1907) and Thatcher (1935). During 100 years, a huge number of different donors of useful traits (for protein content or disease resistance, for example) have been used for crosses in Canadian breeding programs, but the alleles at *Glu-D1* (other than *Glu-D1d*), as well as variants of ω-gliadins characteristic of these donor cultivars were not inherited by the CWRS cultivars. It was concluded that there is a relationship between the presence of certain alleles of storage protein loci in the genotype and a high quality of the CWRS cultivars [Depauw et al 2009]. It was shown that subunit 5 (1Dx5) of the pair 5+10 had the greatest effect on quality [Lafiandra et al 1993; Ribeiro et al 2013].

Some alleles at the *Glu-B1* locus relate to dough properties also, for example, *Glu-B1i* and *Glu-B1al*, as well as *Glu-D1d* [Rathan et al 2020]. The positive effect on quality traits of the presence in wheat genotype of alleles encoding subunits 1 (*Glu-A1*), 13+16, 17+18, 7+8 (*Glu-B1*) (about subunits 7+8, see **4.4.**), and 5+10 (allele *Glu-D1d*) was shown. Negative influence had allele *Glu-A1null*, alleles at the *Glu-B1* encoding subunits 7+9 or 14+15, and *Glu-D1a* [Bayram & Korkut 2020]. However, allele *Glu-D1d* may have a significantly larger contribution to quality than "the best" alleles at the *Glu-B1* [Beasley et al 2002].

4.4. Effects of *Glu-1* alleles on quality: fragmentary and somewhat conflicting data

Initially, the *Glu-1* loci were ranked in their influence on dough strength as *Glu-D1*>*Glu-B1*>*Glu-A1* in the study of the F7 lines from the cross of two wheat genotypes [Gupta et al 1994b]. It was shown that, on the genetic background of Chinese wheats, the *Glu-B1* locus (not *Glu-D1*) made the largest contribution to determining mixograph properties [Jin et al 2013]. Analysis of Chinese cultivars and advanced lines showed that polymorphism at the *Glu-D1* had the largest contribution to dough mixing time and mixing tolerance while for the SDS-st test, the ranking of contribution was *Glu-B1*>*Glu-A1*>*Glu-D1*. Alleles *Glu-A1a*, *Glu-B1* (subunits 7+8, see **4.2.**), *Glu-B1h* and *Glu-D1d* were found to be the best for bread-making quality [Liu et al 2005]. Chinese wheat genotypes with the allele *Glu-B1h* (14+15) were found to have higher quality because subunit 14 (1Bx14) provides the formation of longer gluten polymers thereby improving the dough strength and elasticity [Zhao et al 2002].

Among alleles at the *Glu-B1* locus, *Glu-B1b* (7+8) and *Glu-B1i* (17+18) are reported to be the best for quality in common wheat cultivars grown in UK [Payne et al 1987a]. Germain cultivars with alleles *Glu-B1i* and *Glu-B1f* (13+16) showed higher quality in comparison with other alleles at *Gli-B1* tested [Rogers et al 1989]. In the study of 153 German cultivars, the baking quality (the score of the rapid mix test) was positively influenced by HMW glutenin subunits 1, 2*, 7+9, 14+15, 17+18, 5+10 (alleles *Glu-A1a*, *Glu-A1b*, *Glu-B1c*, *Glu-B1h*, *Glu-B1i*, *Glu-D1d*, respectively) [Brönneke et al 2000].

In the study performed on 447 common wheat cultivars worldwide, a positive correlation was established between allelic variants *Glu-A1a*, *Glu-A1b*, *Glu-B1c*, *Glu-B1i*, and *Glu-D1d* and the dough quality (especially, dough strength), while the presence of alleles *Glu-A1c*, *Glu-B1d*, *Glu-B1a*, and *Glu-D1a* in the genotype negatively affected the quality [Oury et al 2010].

The presence of allele *Glu-B1c* (7+9) in the genotype provided better elasticity and swelling of the dough in comparison with allele *Glu-B1u* (7*+8). At the same time, allelic variation at the *Glu-B1* and *Glu-D1* loci did not influence yield, protein content, grain weight, and a number of other valuable wheat traits [Redaelli et al 1997]. By analysing 104 NIL lines created on the genetic background of the cv Yanzhan-1, alleles *Glu-B1h*, *Glu-B1f*, *Glu-D1d*, and *Glu-D1c* were identified ensuring the best dough quality in Chinese wheat [Li et al 2010].

Quality parameters (W, T, etc.) were studied in NIL lines obtained from two crosses between high- and low-quality cultivars. It was found that lines inherited alleles from high-quality parents (*Gli-A1a* or *Gli-A1b*, *Glu-B1c* and *Glu-D1d*) were superior to genotypes having alleles from poor-quality parents (*Gli-A1c*, *Glu-B1d, Glu-D1b* or *Glu-D1a*) [Khelifi & Branlard 1992]. In other genetic background (progeny from the cross of Canadian cultivars), polymorphism at the *Glu-A1* locus did not affect any one of the eight parameters of quality tested [Huang et al 2006]. Predicting flour quality using any scoring system for alleles can be hampered by interactions between different loci encoding storage proteins [Benedettelli et al 1992; Rousset et al 1992; Gupta et al 1994b; Eagles et al 2002; Li et al 2010].

Moreover, in Australian germplasm, the dough quality does not depend on polymorphism at the *Glu-1* loci so strongly [Lawrence 1986]. Only a low correlation (r=0.41) was observed between the quality level predicted from the score system and the experimental data obtained for 102 Australian cultivars. Polymorphism at the *Glu-D1* locus had practically no effect on the dough strength [Wrigley et al 1990; Metakovsky et al 1990b].

The comparison of a predicted quality (based on quality scores for different *Glu-1* alleles) and experimental quality data for 140 Soviet cultivars showed many cases of discrepancy. For example, a cultivar with the highest predicted dough quality showed only intermediate experimental quality [Nettevich et al 1991]. Moreover, about 42% the high-quality cultivars (class Hard Winter Wheat of the USA) have the "bad" allele *Glu-D1a* in their genotypes [Morris et al 2009]. This allele is also present in genotypes of other high-quality cvs Salmone [Vaccino et al 2002] (see **4.5.**), Xiaoyan-6 [Xu et al 2006] (see **4.5.**), and Darius [Branlard & Dardevet 1994] (see **4.7.**). Genotypes with alleles *Glu-A1a, Glu-B1i*, *Glu-D1a*, and with alleles *Glu-A1a, Glu-B1c* (7+9), *Glu-D1a,* were found to have superior mixograph properties, and better viscoelasticity for making traditional Chinese noodle, respectively [Jin et al 2013].

The use of the original procedure of the SDS-EP made it possible to discover a new (according to the authors) allele at the *Gli-B1* locus. The presence of this allele (designated Glt B4) in the genotype significantly improved the flour quality of Ukrainian common wheat cultivars [Poperelya & Blagodarova 1998].

It should be taken in consideration that quality of wheat gluten depends on environmental factors of plant growing as well as on genetics. In the experiment involving 11 locations and 192 cultivars, it was shown that the higher temperature at which wheat plants were grown increases the MW and influences some other parameters studied of glutenin. For example, a temperature difference of 3.5° almost doubles the size of glutenin macromolecule. The characteristics of glutenin studied were not influenced by grain protein content and grain hardness [Branlard et al 2020]. The expression of most α/β-gliadin genes in developing seeds was measured by quantitative RT-PCR using group-specific primers over 3 years and it was established that the expression of most α/β-gliadin and other prolamine genes was correlated with the sunshine duration [Noma et al 2016].

The short interstitial 1DL segment detected through SNP genotyping was introduced in the chromosome 1A of durum wheat. This DNA segment carried *Glu-D1*locus from common wheat cv Courtot (allele *Glu-D1a*) and it was small enough not to be visualized by in situ hybridization. It is suggested that this chromosome carrying a *Glu-1* locus from common wheat can be exploit in durum wheat breeding [Coriton et al 2019].

Another interesting idea how to improve quality was to selectively suppress the activity of *Gli-1* loci in the *Glu-3*/*Gli-1* complex locus increasing thereby a quota of glutenins LMW in the gluten. Therefore, an attempt was made to introduce into wheat genotype, through crossing with a transformed donor, a special dominant trait, RNAi-mediated down-regulation of γ-gliadins. This trait manifests itself causing the disappearance from the electrophoregram of some slow-moving intensely stained γ-gliadin bands encoded by the *Gli-B1* and *Gli-D1* loci (see **2.5.**). At the same time, the synthesis of glutenins increases significantly in transformed genotype. This approach did not give any definite result, since various transformed lines, being compared with the untransformed parental genotype, behaved differently in terms of quality [Gil-Humanes et al 2012]. Moreover, the content of α- and ω-gliadins in the storage proteins of transformed lines increases compared to the normal genotype, and therefore relative content of gliadins remains almost unchanged [Pistón et al 2011]. The effect of "compensatory synthesis" among gliadins was observed in null mutants for *Gli* loci of various cultivars [Metakovsky et al 2021a] (see **7.1.**) In addition, to our opinion, the comparison of the transformed and initial genotypes was not perfect, since, judging by the photos presented in the article (for example, Fig.2A, lines G613 and G622) [Pistón et al 2011], the transformed genotypes produced gliadins (APAGE gliadin bands) obtained from the transforming donor.

Thus, in many publications performed on different groups of cultivars, alleles *Glu-A1a* or *Glu-A1b*, *Glu-B1b*, *Glu-B1i*, *Glu-B1f*, *Glu-B1h*, and *Glu-D1d* were mentioned as being desirable for bread-making quality [Brünneke et al 2000; He et al 2005; etc.]. Another conclusion that can be drawn from the bulk of all somewhat conflicting data on the effects of distinct alleles at the *Glu-1* loci on the dough quality is that the quality scores for *Glu-1* alleles developed for one group of cultivars might not work well (did not predict quality in a given genotype) in other groups of cultivars [Gupta et al 1991; Rousset et al 1992].

4.5. Dough quality may depend on the amount of HMW glutenin produced

During routine genotype-by-genotype analysis of grain or plant material, spontaneous mutants can be found lacking one of the loci controlling the synthesis of wheat storage proteins. Spontaneous null mutation at the *Glu-B1* or *Glu-D1* loci encoding two different polypeptides (of x- and y-types) can be caused by a deletion of chromosome material or by a simultaneous silencing of two distinct genes of a complex locus. The *Glu-A1* locus produces only one polypeptide of glutenin (see **4.1.**) and has null allele (*Glu-A1c*) among its catalogued allelic variants.

Deletion of HMW glutenins reduces dramatically dough strength and bread-making quality. It was shown that flour of common wheat line LH-

11, in which all the genes encoding HMW glutenins silenced (see **5.1.**), was not suitable to make bread [Zhang et al 2018]. On another hand, the replacement (by backcrossing) of the silent 1Ay allele by the expressed allele 1Ay21* conferred a significant positive effects on a range of quality traits, including increase of grain protein content by up to 9% and bread volume by up to 26% [Roy et al 2020].

HMW glutenins of 70 common wheat cultivars of different origin were studied by SDS-EP. Quantity of protein in each band in the HMW-zone of SDS-EP was determined by densitometry, and correlation between the proportion of each of the HMW glutenin band and various flour quality characteristics (SDS-st and four Alveograph tests, including dough strength W, tenacity P, swelling G and extensibility L) were measured. It was established that bands (HMW subunits) 5+10, and 2+12, correlated positively, and negatively, respectively, with SDS-st and parameters W and P of dough. Some positive correlations were also discovered for bands 1, 2*, 9 and 17+18 [Branlard & Dardevet 1985b].

A set of homozygous genotypes (lines) carrying null alleles at the *Glu-1* loci were created by crossing cvs Olympic (null mutant at the *Glu-B1*) and Gabo (allele *Glu-A1c* and null mutant at the *Glu-D1*). The analysis of these genotypes permitted to firmly establish that null at *Glu-1* loci, especially at the *Glu-B1* and *Glu-D1*, causing the absence of some HMW glutenin polypeptides, had a dramatically negative impact on the dough and baking quality. It was suggested that these loci have a greater effect because they produce greater amounts of HMW subunits in comparison with *Glu-A1c* and not because they produce more effective subunits. This raises the question as to whether the different contributions made to the dough quality by different alleles at a locus might also be caused, at least in some cases, by differences in amount rather than differences in effectiveness [Lawrence et al 1988].

The allele *Glu-A1c* (null) is the worst for quality among known alleles at the *Glu-A1* locus [Payne et al 1987a]. The negative effect of allele *Glu-A1c* in comparison with *Glu-A1b* (encodes subunit 2*) was repeatedly reported [Payne et al 1988a; Pogna et al 1989; Rodriguez-Quijano & Carrillo 1996b; Sontag-Strohm et al 1996; Luo et al 2001; Flaete & Uhlen 2003]. Therefore, most of the data indicates that a decrease in amount of HMW glutenin molecules deteriorates rheological properties of gluten. It was shown that observed negative effects of the removal of some HMW subunits on dough strength and other bread-making quality were caused by decrease in the size of gluten polymers [Popineau et al 1994; Gupta et al 1995]. On the other hand, alleles *Gli-D1d* and *Gli-D1a*, contrasting in their effects on quality (see **4.3.**), produce HMW glutenin subunits in similar quantities but there is a differences between these two alleles in the size distribution of the polymeric proteins [Gupta & MacRitchie 1994].

Common wheat genotypes differ in the intensity of transcription of x- and y-subunits. For example, both HMW subunits of the cv Cheyenne and y-subunit of the cv ChS (the *Glu-D1* locus) are synthesized at the same rate, while transcription of the x-subunit in ChS occurs twice as fast [Anderson et al 2007]. The quantitative analyses showed that the contents of HMW subunits varied within a broad range among wheat genotypes, subunits 2, 5, 7, 10 and 12 being the major contributors of HMW glutenins to gluten. Amount of HMW glutenins highly correlated to dough strength and other quality parameters, with x-subunits being more important than y-subunits. HMW subunit 5 (has an additional cysteine residue in its polypeptide chain) and subunit 7 (supplies gluten with the greatest amount of HMW molecules) are important contributors to high dough quality [Wieser & Zimmerman 2000].

There were attempts to double a HMW-glutenin-coding gene in a given genotype using genetic engineering techniques. Sometimes, the introduced gene could express increasing proportion of HMW subunits in gluten [Álvarez et al 2000]. For example, the increase of the number of copies of the gene coding for 1Dx5 subunit in transgenic wheat genotype resulted in increase of quota of this polypeptide from 2.7% to 10.7% of the total gluten proteins and in extremely strong doughs [Rooke et al 1999]. The transgenic line B73-6-1 had four-times increase in the amount of the 1Dx5 HMW subunit, increased hardness and decreased size of kernels, but no changes in yield. The stability and strength of the dough in transgenic line were higher and extensibility was lower. It is suggested that this flour would improve, after blending, a poor-quality flour [Rakszegi et al 2005].

Spontaneous inter-genomic translocations were also used to increase the proportion of certain storage proteins in gluten. For example, a line was created in which a section of the chromosome arm 1DL was translocated to the arm 1AL [Dumur et al 2009]. It was shown that doubling the dose of the "bad" allele *Glu-D1a* (see **4.3.**) in the genotype significantly improves the dough quality (SDS-st, dough strength and mixing resistance) in comparison with original cultivar (Courtot) [Dumur et al 2010], although of the two genes encoding HMW glutenins and located in the transferred genetic material, only one was expressed [Dumur et al 2009]. This finding clearly indicates that dough quality may depend on the amount of HMW subunits.

The flour of Canadian cv Glenlea produce an extremely high dough strength [Marchylo et al 1992; Rooke et al 1999; Butow et al 2003; Juhasz et al 2003]. An addition of this flour can improve the bread-making quality of other flour [Bushuk 1980; Yamauchi et al 2001]. The allele *Glu-B1al*, present in the genotype of this cultivar produce 57% more HMW glutenin [Vawser & Cornish 2004] due to spontaneous duplication of the gene encoding either the subunit 7 (Bx7) [Lukow et al 1992], or subunit 8 (Bx8) [Ribeiro et al 2013] of HMW glutenin. Using this allele, a new class of wheat was created in Canada, the so-called Western Extra-Strong (CWES) wheats [Vawser & Cornish 2004]. The PCR-based markers were developed allowing discrimination of normal Bx7 from over-expressed Bx7 subunits [Lei et al 2006]. An increase in the synthesis of the same, probably, polypeptide due to a similar duplication was observed in several genotypes of the heterogeneous ancient Hungarian wheat Bànkùti-1201 [Juhàsz et al 2003].

Similar duplication (*Glu-B1ar*) was discovered in Australian cv H-45. In this case, however, the enhancement of the synthesis of the subunit did not improve dough quality. Analysis of the nucleotide sequence showed that the duplicated gene of the allele *Glu-B1ar* differs from the analogous gene in the allele *Glu-B1al* by multiple nucleotide substitutions (SNPs), one of which caused the appearance of an additional cysteine in the synthesized polypeptide. Due to this new cysteine, a "good" subunit 1Bx7 (encoded by *Glu-B1al*) was converted in "neutral" or even "bad" for quality (encoded by *Glu-B1ar*) [Gao et al 2012]. The PCR technique was used for identification of variants of the subunit 7. It was discovered that there were at least eight different variants of gene encoding subunit 7, and only one of them showed increased expression [Butow et al 2004].

Overall, both genetic polymorphism at the *Gli-1* loci and quantity of the HMW glutenin polypeptides synthesized by these loci may be responsible for the effects on wheat dough quality [Wrigey et al 2006a].

4.6. Polymorphism of the *Glu-3*-encoded LMW glutenins and end-use quality

Most B-type glutenin molecules encoded at the *Glu-3* loci have eight cysteine residues in their amino acid sequence, but, unlike gliadins, two of them are capable to form covalent intermolecular bonds. Moreover, these free cysteine residues are located, in different glutenin polypeptides, in distinct positions of the chain thereby supplying the molecule with different characteristics. B-type LMW glutenins are assumed to contribute in gluten chain elongation increasing the MW of the gluten matrix [Ikeda et al 2002; D'Ovidio & Masci 2004]. On the contrary, C- and D-types LMW glutenins (see **3.2.**) may function as terminators, interrupting the growing gluten chains. Therefore, each type of LMW gluten in polypeptides can affect the bread-making quality, which depends, generally, on the MW and composition of the gluten matrix [Kasarda 1989; Wrigley et al 2006].

It has been reliably established that allelic variation at the *Glu-3* loci influence dough properties [Gupta et al 1991; Branlard et al 2001; Maucher et al 2009; etc.]. However, owing to ambiguities in the nomenclature of LMW glutenin subunits (see **3.5.**) and to the complexity of the *Glu-3* loci (see **5.**), a relationship between alleles at the *Glu-3* loci and wheat end-use quality has not been adequately determined [Ibba et al 2017c].

There is contradictory information about which specific LMW glutenin polypeptides affect bread-making quality and how they do that.

It was shown that *Glu-A3,* among *Glu-3* loci, had the strongest influence on quality (SDS-st, wet gluten content, bread volume, etc.) in 76 promising bread winter wheat lines [Tsenov et al 2010]. Three genes encoding the LMW glutenin polypeptides, two genes from the genome A and one from the genome D, were expressed in bacteria. It was shown that the genome-A-produced LMW glutenins caused significant increases in peak resistance, mixing time and dough extensibility when incorporated in the flour, while the genome D-produced polypeptides, being less effective for these parameters, decreased significantly the Rmax [Lee et al 1999a].

Variation at the *Glu-A3* locus had a major impact on the dough rheology parameters in the set of elite cultivars, followed by *Glu-B3* and *Glu-D3* [Ibba et al 2017c]. On the contrary, the *Glu-B3* locus made the largest contribution to several quality parameters (SDS-st, mixograph mixing time and mixing tolerance) among all *Glu* loci in the study of 251 Chinese cultivars and advanced lines [Liu et al 2005]. In one more study, the three *Glu-3* loci were ranked, in respect to dough strength (Rmax), as *Glu-B3*>*Glu-A3*>*Glu-D3* [Gupta et al 1994b]. Analysis of the progeny of the cross between two cultivars having contrasting dough quality showed that alleles at each of the three *Glu-3* loci present in the high-quality parent affect positively on dough strength (W) and viscosity (T) [Khelifi & Branlard 1992]. However, only allelic variation at the *Glu-B3* locus (neither *Glu-A3*, nor *Glu-D3*) was found to relate to the dough quality [Nieto-Taladriz et al 1994].

In the study of 280 F3-derived F7 lines of crossing cv Hulberd and a breeding line, genotypes with alleles *Glu-A3c* and *Glu-B3b* showed significantly greater dough strength (Rmax) than with alleles *Glu-A3e* and *Glu-B3c* [Gupta et al 1994b]. However, as a result of analysis of a set of cultivars and lines (48 Australian and 51 from 22 countries), alleles at the *Glu-3* loci were ranked, according to their effects on quality (dough strength, Rmax), as *Glu-A3b*>*d*=*e*>*c* (there is a difference from [Gupta et al 1994b]), *Glu-B3b=i>>g=h>>c*, *Glu-D3b=a>>c,* and with respect to their extensibility *Glu-A3d>b>c>e*, *Glu-D3e>c>b=a=d* [Gupta et al 1991]. Generally, this order was confirmed when analysing the progeny from the cross of cultivars bred in New Zeeland: alleles *Glu-A3d*, *Glu-B3b* and *Glu-D3b* had a positive influence on flour quality [Luo et al 2001].

The analysis of 22 quality-related parameters in cultivars registered in France showed that the presence of the alleles *Glu-B3g* and *Glu-B3f* in wheat genotype correlates significantly with the Alveograph-tested elasticity of the dough [Branlard et al 2001]. The presence of alleles *Glu-A3d* and *Glu-B3d* in wheat genotype correlated with better gluten quality in the study of Chinese cultivars [Liu et al 2005]. The quality tests (extensograph) of two cultivars each composed of biotypes differing by alleles at the *Glu-A3* locus showed that biotypes with allele *Glu-A3b* had significantly greater dough extensibility than those with *Glu-A3c* [Gupta & Shepherd 1988]. The stronger gluten and better end-use quality showed genotypes of common wheat having alleles *Glu-A3b* and *Glu-B3b* [Rathan et al 2020]. In other work, a positive influence on some quality parameters of the allele *Glu-A3b* was confirmed. However, the presence in wheat genotype of the allele *Glu-A3f* was found to associate with even greater positive effect on several quality parameters. This effect was observed on different genetic backgrounds (in the presence of different alleles at the *Glu-1* and *Glu-3* loci) [Tsenov et al 2010]. Comparison of alleles at the *Glu-3* loci using NIL (background is one of Argentinian cultivars) showed that alleles *Glu-A3f*, *Glu-B3b*, and *Glu-B3g* were associated with the highest values in gliadin strength-related parameters, while *Glu-A3a* and *Glu-B3a* were constantly associated with weak gluten and low quality values [Bonafide et al 2015].

The analysis of 25 NIL genotypes made on the background of Chinese wheat showed that *Glu-A3b*, *Glu-A3d*, *Glu-B3g*, and *Glu-D3f* (irrespective to allelic variation at the *Glu-1* and *Glu-3*) were found to be the best in the analysis of mixograph properties [Jin et al 2013]. The statistically significant positive influence on quality traits of alleles *Glu-A3d*, *Glu-A3e*, *Glu-A3f*, *Glu-A3b*; *Glu-B3a*, *Glu-B3b*, *Glu-B3g*, *Glu-D3a*, and the negative effect of *Glu-A3a*, *Glu-A3c*, *Glu-B3f*, *Glu-B3c*, and *Glu-D3c* was established as a result of analysis of 64 common wheat genotypes [Bayram & Korkut 2020]. Polymorphism of LMW glutenins in 63 Iranian cultivars was studied by means of SDS-EP (23 alleles were identified, in total, at the three *Glu-3* loci) and gene-specific PCR primers (47 alleles were distinguished). It was shown that the presence of alleles *Glu-A3a* and *Glu-A3e* in wheat genotype was associated with higher flour quality (SDS-st), alleles *Glu-B3b*, *Glu-B3f*, *Glu-B3d* and *Glu-B3a* influenced positively on protein content and bread volume, and the allele *Glu-D3r* was found to associate with kernel hardness [Khoshro et al 2021].

Comparison of data obtained in different works shows that some alleles at the *Glu-3* loci were repeatedly (in different publications) demonstrate their association with better wheat quality, for example, *Glu-A3b, Glu-A3d, Glu-A3e, Glu-A3f, Glu-B3b, Glu-B3g*. Results of the evaluation of alleles *Glu-A3a* and *Glu-B3f* were contradictory. The inconsistency may appear because the effects of the *Glu-3* loci on quality are modulated by a significant genotype-by-environment interaction [Bonafide et al 2015]. Another problem may be a reliable identification of alleles at the *Glu-3* loci (see **3.5.**).

Genes at the *Glu-3* loci may contribute to the end-use wheat quality not only qualitatively (due to their allelic variation), but also quantitatively, like genes at the *Glu-1* loci (see **4.5.**). There are some evidences suggesting that a higher number of active genes at the *Glu-A3* and *Glu-D3* loci tended to lead to a more elevated SDS-st, although this tendency was influenced by genetic background. It was suggested that selection for high numbers of active genes at the *Glu-3* loci can be used for improvement of bread-making quality [Dong et al 2010].

The effects of individual alleles at the *Glu-3* as well as *Glu-1* loci on the dough quality of common [Branlard & Dardevet 1985b; Sozinov & Poperelya 1985; Payne et al 1987b; Gupta et al 1989; Gupta & MacRitchie 1994; Gupta et al 1994b; Nieto-Taladriz et al 1994; Sontag-Strohm et al 1996; Redaelli et al 1997; Beasley et al 2002; Maruyama-Funatsuki et al 2004; etc.] and durum [Pogna & Mellini 1988] wheat interact and in many cases are additive.

There are some possible mechanisms explaining the difference between alleles at the *Glu-3* locus in their effects on the dough quality.

Primers specific for LMW-glutenin-encoding genes located on the chromosome 1D of common wheat were developed [Zhao et al 2004]. These primers were applied for analysis of allelic variant of the *Glu-D3* locus present in the genotype of the high-quality cv Xiaoyan-6, and the gene encoding the glutenin with the MW of 35 kDa was sequenced. It was found that resulting polypeptide differed from any one of known LMW glutenins in that it had nine cysteine residues in its amino acid chain, three of which can form intermolecular bonds. Therefore, it was suggested that such molecule do not extend or terminate the growing gluten chains, but ensure their branching and thereby contribute to an increase in the mass of the resulting polymer. It was proved that the addition of the purified polypeptide 35 kDa to flour increases the gluten matrix and thereby significantly improves the dough quality [Xu et al 2006].

The quality of a poor-quality cv Harunoakebono was considerably improved by the introduction in its genotype (by development of near-isogenic line) of the LMW glutenin subunits from the extra-strong wheat cv Glenlea. The better quality was shown to associate with five subunits of s-type (see **3.2.**) co-segregating in the progeny of crosses [Maruyama-Funatsuki et al 2005].

Segregation analysis of the F2 progeny from a cross between good (Haruhikari) and a poor bread-making quality cultivars showed that dough strength significantly correlated with one amplified LMW glutenin gene located at the *Glu-B3* locus in Haruhikari. This gene had seven amino acid deletions and six substitutions in comparison with corresponding gene of the poor-quality parent. Some of these differences in the polypeptide chain may cause changes in hydrophilicity of the molecule and relate to differences in dough strength between the two cultivars crossed. No specific gliadin bands encoded by the *Gli-B1* locus linked to *Glu-B3* were identified as promising genetic markers for variant of the *Glu-B3* locus studied [Tanaka et al 2005].

It is broadly assumed that increasing the quota of the B-type LMW subunits in common wheat gluten would improve dough quality [D'Ovidio & Masci 2004]. Indeed, it was shown that "good" allele *Glu-B3b* provides the synthesis of the B-type glutenins in a relatively higher quantity (judging by the staining intensity of the corresponding SDS-EP bands) compared with the allele *Glu-B3c* [Pogna et al 1995]. In other publication, it was shown that pairs of alleles *Glu-A3c* and *Glu-A3e*, and *Glu-B3b* and *Glu-B3c*, contrasting in their effects on dough strength (Rmax), produced B-type LMW subunits in different amounts and thereby affected size distribution and/or quantity of the polymeric protein [Gupta & MacRitchie 1994].

However, a twofold increase in proportion of LMW glutenin in the storage proteins (achieved by transforming immature embryos with microparticles carrying the LMW gene driven by its own promoter) in a line of the cv Bobwhite did not lead to any improvement in quality (SDS-st). It was concluded that the widely accepted correlation between the amount of the gluten polymers and flour properties might not be valid, depending on the qualitative composition of gluten (in fact, on the allelic variation at the *Glu-3* loci) [Masci et al 2003].

On the contrary, the genotype of the high-quality cv Salmone [Borghi et al 1985] include an amplified DNA sequences of some B-type LMW glutenins (DNA fragments SF720 and SF750 detected using a specific probe pLMWTG2). It was shown that genotypes of the NIL lines (F6 from the cross of Salmone with other cultivar) that inherited the amplified sequences demonstrated a significantly higher quality tested by SDS-st, Farinograph stability and Alveograph parameters (dough strength, elasticity and tenacity). The correlation was observed on different genetic backgrounds. It was shown that amplified region located on the satellite of the chromosome 1B and inherited as a normal codominant Mendelian trait. It was not established, however, if amplified sequences were encoded at the *Gli-B3* locus [Vaccino et al 2002].

The gene encoding a subunit named LMW-N13 was transferred from diploid species *Ae. uniaristata* into common wheat. The polypeptide had unusually high MW and an extra cysteine residue compared with typical LMW glutenins. It was shown that the presence of the LMW-N13 in the genotype of common wheat cultivars improves their dough quality due to higher content of gluten polymer and total protein [Du et al 2020].

In the library of transcribed DNA sequences of the common wheat cv Wyuna, a family of unusual genes was revealed. These genes locating on chromosomes of the groups 4 and 7 was shown to encode the synthesis of small polypeptides similar in amino acid sequence to barley ε-hordein (17 kDa) but also to gliadins and LMW glutenins. It was suggested that this new family of the seed-endosperm genes is important in defining the quality characteristics of wheat flour. Rheological analysis showed the protein effected positively on dough strength and other important quality characteristics [Clarke et al 2003].

4.7. Allelic variation at the *Gli-1* loci can affect end-use quality

It was claimed that neither polymorphism of gliadin, nor the amount of gliadin synthesized by a given genotype affects the end-use wheat quality [Gupta & MacRitchie 1994]. Indeed, the analysis of null-mutants for the *Glu-D3/Gli-D1* complex showed that the loss of *Gli-D1* locus (but not of *Glu-D3*) positively affects the dough and baking quality, while the absence of this entire complex, on the contrary, greatly deteriorate it [Lawrence et al 1988].

However, it was shown in early publications that the gliadin fraction of flour may govern bread loaf volume [Finney et al 1982]. Moreover, some statistical correlations between the presence of certain bands in the gliadin electrophoregram and wheat quality were found [Branlard & Rousset 1980; Wrigley et al 1982b; etc.]. Positive correlations were discovered, for example, for APAGE bands 20+25, 23, 35.5, 40 and 48 [Campbell et al 1987], or 31, 35, 37, 47, 59 [Cressey et al 1987]. In other work, the presence of some APAGE bands (including 4 γ- and 7 ω-gliadins probably encoded by *Gli-1* loci) were significantly associated (some of them positively, others negatively) with at least one of technological criteria of quality assessed (SDS-st and Alveograph for dough strength, tenacity and extensibility) [Branlard & Dardevet 1985a]. However, this approach works only in the conditions of EP used in the study because different APAGE procedures give different gliadin patterns, and because quite different gliadin polypeptides, in distinct genotypes, might have hardly distinguishable EM due to a huge gliadin polymorphism. Day-to-day experimental variation of different EP runs should be taken into account also (see **1.6.**).

The most advanced work of this type considered only one pair (two allelic variants) of APAGE bands. It was proved that there was a strong difference between two groups of Italian cultivars: one group had a slow-moving γ-gliadin (γ40, the slowest γ-gliadin in any APAGE pattern) and poor quality while another group had a γ-gliadin of higher EM (γ43.5) and much better quality. Both γ-gliadins had similar staining intensity in APAGE [Pogna et al 1982a; 1985; Pogna & Mellini 1986]. Judging from the list of cultivars mentioned in these publications, it was the *Gli-B1k*, and the *Gli-B1g* [Alleles 2018c] (see **7.5.**) for bad-quality, and good-quality groups, respectively. It should be noted that the two groups of cultivars might differ, besides alleles at the *Gli-B1* locus, also by alleles at other (including *Gli*) genetic loci.

The influence of polymorphism at the *Gli* loci on value of the SDS-st was observed in the group of 143 German and Austrian cultivars. It was concluded that gliadin alleles are suitable for using as markers for prediction of baking quality [Černy et al 1992]. In Australian cultivars, the allelic variation at the *Gli-A1* locus was found to associate with extensibility of dough [Wrigley et al 1990; Metakovsky et al 1990b].

The *Gli-A1* locus. The allele *Gli-A1b* present in the genotype of the cv Bez1 provided better flour quality (SDS-st) than the *Gli-A1f* in the progeny from the cross between cvs Bez1 and Trebishov-1R 100-568. Moreover, it was calculated that the contribution of gliadin polymorphism to the flour quality in this hybrid combination is, in different years of research, 30-58% [Poperelya & Gasanova 1980]. Advantage of the *Gli-A1b* for quality (SDS-st, dough strength W, bread characteristics) was confirmed in studies of progenies from crosses of Bez1 with Dneprovskaya-521 (has *Gli-A1m*) [Sozinov et al 1975a; Sozinov & Poperelya 1985], with Crvena-Zvezda (this cultivar has, probably, the allele *Gli-A1a*) [Poperelya et al 1980], and with Trebishov-1R-100-568 (*Gli-A1f*) [Poperelya & Gasanova 1980].

In the cross between Florence-Aurore (*Gli-A1c*) and Armada (*Gli-A1m*), the progeny with the allele *Gli-A1c* had better quality [Khelifi & Branlard 1992]. A negative influence of the allele *Gli-A1m* was confirmed in the study of F5 progeny from the cross between ChS and ChS (Hope 1A): lines with allele *Gli-A1a* showed better quality (SDS-st) than lines with *Gli-A1m* [Payne et al 1987b].

Results of analysis of descendants of crossing biotypes of the cv Ulla were interpreted as possible indication on positive correlation between the presence of the allele *Gli-A1o* in wheat genotype and quality (SDS-st) [Sontag-Strohm et al 1996]. In other work, genotypes with the complex locus *Glu-A3d/Gli-A1o* (see **3.3.**) demonstrated better quality (SDS-st, mixograph) in comparison with genotypes with *Glu-A3f/Gli-A1f* or *Glu-A3g/Gli-A1f* in NIL lines (grains F7 and F8) from three crosses of four common wheat cultivars. These data were interpreted as indication on the positive effect of the *Gli-A1o* on quality [Flaete & Uhlen 2003]. It should be noted that, in the discussion of this article, it was erroneously stated that a "significant positive effect of the allele *Gli-A1o* on quality was reported by Metakovsky et al (1997)". In fact, in Italian wheat germplasm, cultivars with allele *Gli-A1o* had the least dough strength (20% less, although the difference was not significant) than cultivars with alleles *Gli-A1a* or *Gli-A1f* studied [Metakovsky et al 1997a].

Based on these fragmentary research data, alleles studied at the *Gli-A1* locus may be tentatively ranked for their positive impact on quality (dough strength): *Gli-A1b(=c)>a=f>m*. The results of analysis of the allele *Gli-A1o* are contradictory. At the same time, variation at the *Gli-A1* locus did not appear to contribute towards dough strength in the study of 56 F2-derived F6 families from a cross of two wheat genotypes [Gupta et al 1989].

The *Gli-B1* locus. Descendants from various crosses with the allele *Gli-B1b* from Bez1 demonstrated a significantly higher quality than those with allele *Gli-B1k* [Poperelya et al 1980] or *Gli-B1d* [Sozinov et al 1975a].

Statistical analysis of the data for dough strength (W) for Italian cultivars grown during 20 years in different locations demonstrated a significant positive correlation between the presence of the allele *Gli-B1b* in the genotype of cultivar and their dough quality whereas other alleles studied (*Gli-B1e*, *Gli-B1g*, and *Gli-B1k*) had about 1.6 times less average dough strength value in comparison with *Gli-B1b* [Metakovsky et al 1997a]. Descendants from crosses that keep the allele *Gli-B1b* in their genotypes also demonstrated a higher dough quality than genotypes with the allele *Gli-B1g* [Khelifi & Branlard 1992]. Genotypes with *Glu-*

B3b/Gli-B1b showed better quality than those with *Glu-B3iGli-B1m* [Flaete & Uhlen 2003].

The allele *Gli-B1c* obtained by some newly-bred Eastern-Europe cultivars from Mexican parents is considered to be one of main contributors to the good quality (high dough strength and bread volume) of these cultivars [Poperelya 1996; Zlatska et al 2009]. Indeed, in the cross between cvs Odom (has allele *Gli-B1c*) and Odesskaya (*Gli-B1b*), the best quality showed the descendants that inherited *Gli-B1c* [Poperelya & Blagodarova 1998; Chebotar et al 2012a]. In the cross Prinqual x Marengo, allele *Gli-B1c* from Prinqual also provided better quality (alveograph tests) for the progeny [Nieto-Taladriz et al 1994].

Alleles studied at the *Gli-B1* locus may be tentatively ranked for their (positive) impact on quality (dough strength): *Gli-B1c>b>e=g=k=m* (see also **7.5.**). The results of comparison of effects of the *Gli-B1b* and *Gli-B1d* are contradictory.

The *Gli-D1* locus. It was shown that the presence in wheat genotype of alleles encoding Cheyenne-type APAGE bands provided a better quality as compared with alleles of ChS-type (see **2.5.**) [Sozinov & Poperelya 1980; Branlard et al 1994; etc.].

The observed correlations between allelic variation at *Gli-1* loci and some parameters of quality can be explained (see **4.11.**).

4.8. Molecules encoded at complex *Glu-3/Gli-1* locus affect the end-use quality

Two loci, *Gli-1* (encoding ω- and γ-gliadins) and *Glu-3* (producing most B-type LMW glutenins) are closely linked and may be considered as a complex *Glu-3/Gli-1* locus. The analysis of long DNA genomic sequences showed that there is no obvious border between the *Glu-3* and *Gli-1* loci [Gao et al 2007; Anderson et al 2009; 2012].

It is well known that replacing the complex locus *Glu-B3/Gli-B1* with rye genetic material (1BL.1RS translocation, see **2.6.**) negatively affects the flour/dough quality of common wheat [Graybosch 2001; Kozub et al 2001; Liu et al 2005; etc.]. The degree of influence of this translocation on various parameters of end-use wheat quality depends on the genotype, and even on the growing conditions of the plants [Fenn et al 1994; Landjeva et al 2006b; etc]. Less detrimental but still negative effect on quality was found for the 1AL.1RS translocation [Koebner & Shepherd 1986; Graybosch et al 1993a]. It remains unclear, the loss (as a result of substitution) of which locus, *Glu-3* or *Gli-1*, affects the quality.

There are publications in which it was shown that molecules produced by *Glu-3/Gli-1* complex loci influence end-use wheat quality. For example, it was shown, using molecular markers (RFLP and SSR), that allelic variation at the *Glu-B3/Gli-B1* locus strongly influenced the flour/dough quality (measured by SDS-st and moxograph) in the progeny of the cross between cvs Klein-32 and ChS [Manifesto et al 1998]. The effects of *Glu-3* and *Gli-1* loci were not separated in this study.

NIL lines were developed from the progeny of the cross between a spontaneous mutant genotype of the cv Alpe, carrying null allele at the *Glu-D3/Gli-D1* locus, and other line of the same cultivar. A significant positive effect of this mutation on dough viscosity (P) and negative influence on its swelling (G) were observed. On the background of the allele *Glu-D1a,* the null at *Glu-D3/Gli-D1* increased dough strength. It was suggested that polypeptides encoded at the *Glu-D3/Gli-D1* complex locus play a key role in imparting elasticity to dough, that is the most important quality parameter distinguishes common wheat dough from durum dough [Redaelli et al 1997]. On the contrary, spontaneous null mutation at the *Glu-B3/Gli-B1* identified in the cv San-Pastore increased the value of G, but reduced indices P and W of Alveograph [Pogna et al 1998].

Other spontaneous null mutation at the *Glu-D3/Gli-D1* (line MG-27116) caused the lack of the *Gli-D1*-encoded gliadins and *Glu-D3*-encoded LMW glutenins and a large detrimental effect on flour quality (SDS-st). The magnitude of the effect was even larger than that of the null allele at the *Glu-A1* locus (*Glu-A1c*) in comparison with allele *Glu-A1b* encoding HMW subunit 2* (see **4.1.**). It was suggested that decrease of quality was mainly due to the lack of LMW glutenins produced by the *Glu-D3* locus [Benedetelli et al 1992].

On the other hand, the genotype of the registered French cv Darius include null at the *Gli-D1* locus [Metakovsky & Branlard 1998] and a high flour/dough quality. Moreover, this null mutation, when transferred to another genotype (cvs Sicco and Nap-Hal), significantly improved some indicators of the recipient's flour quality. By analysing the progeny from crossing of Darius with other cultivars, it was found that the mutation had a positive effect on the dough strength and viscosity. It is suggested that the null mutation present in Darius has caused the loss of only *Gli-D1* locus, but not of *Glu-D3*. Therefore, the higher quality provoked by the presence of the mutation in the genotype is due to increase in the ratio of LMW glutenins to gliadins [Branlard & Dardevet 1994].

Analogously, genotypes carrying null mutation at the *Gli-D1* locus (line M1654 and progeny of its crosses), identified through the absence all *Gli-D1*-encoded bands in APAGE electrophoregram, showed higher quality (SDS-st) without changes in protein content [Sozinov & Kopus 1983]. The cv Uni-1 carrying null at *Gli-D1* has a high quality tested for the amount of disulphide bonds in gluten [Netsvetaev et al 2013]. Spontaneous null mutation at the *Gli-A1* or *Gli-D1* locus causing the loss ω-gliadins positively influenced baking quality (the score of the rapid mix test) in German wheat germplasm [Brönneke et al 2000]. Although the presence of the *Glu-D3* locus in these *Gli-D1* nulls, as well as in the lines M1654 and Uni-1, was not studied, the results obtained may be also interpreted through the assumption that the positive effect on quality was caused by an increase in the ratio of LMW glutenin to gliadin in mutant genotypes.

4.9. Which locus, *Glu-3* or *Gli-1*, has greater effect on quality

The key difference between the polypeptides encoded at the *Gli-1* and *Glu-3* loci is that cysteine residues of LMW glutenins are able to form intermolecular bonds, while cysteine present in the molecule of gliadin can form only intramolecular bonds and therefore, theoretically, cannot be involved in gluten matrix by covalent S-S bonds [reviews Payne 1987; Shewry & Tatham 1997; Gianibelli et al 2001; Shewry et al 2003; etc.].

It was assumed that effects of the *Gli-1* loci on quality (see **4.7.**) are not due to the gliadin polymorphism itself, but may be explained by the actions of the LMW glutenin polypeptides produced by the neighbouring (see **3.3.**) *Glu-3* loci [Payne et al 1984a; 1987a; etc.]. In the late 80s and early 90s, this idea was well-accepted, became dominant, so that any relation between allelic variation at the *Gli-1* and the dough quality were explained through the involvement of polypeptides encoded by the *Glu-3* loci.

LMW glutenins were involved to explain the better quality of genotypes having alleles *Gli-A1a* [Payne et al 1987b] or *Gli-B1c* [Nieto-Taladriz et al 1994] in comparison with other allelic variants studied of these *Gli-1* loci. In the cross of cvs Anza (*Gli-D1b*) and Ablaca (*Gli-D1g*), a higher quality (SDS-st) was observed in the progeny having the allele *Gli-D1b*. Although polymorphism of LMW glutenins was not studied in this work, the results were interpreted to mean that it is the allelic variation at the *Glu-D3* locus that affects the quality [Rodriguez-Quijano & Carrillo 1996b].

The idea of the involvement of LMW glutenins in the effects of the *Gli-1* on quality was supported in studies of Italian cultivars. It was shown that genotypes producing the gliadin band γ43.5 in the APAGE (as a rule, it is allele *Gli-B1g*) have better quality than cultivars with slower moving band γ40.0 (*Gli-B1k*) [Pogna et al 1985; Peruffo et al 1985]. The presence of the APAGE band γ43.5 is usually accompanied by the allele *Glu-B3b* (see **3.6.**) associated with a good flour/dough quality while the presence of the APAGE band γ40.0 is accompanied by alleles *Glu-B3c* or *Glu-B3k* [Pogna et al 1995] that negatively affects the quality [Gupta & Shepherd 1988; Gupta et al 1991; etc.] (see **4.6.**). Thus, the difference of effects of alleles at the *Gli-B1* locus on quality can be explained by the presence of certain alleles at the closely linked *Glu-3* locus.

In common wheat, the *Glu-B3*-ecoded LMW glutenin subunit (MW 42 kDa) is one of major polypeptides present in gluten polymer. Its complete amino acid sequence was determined. It was shown that 42-kDa polypeptide has two cysteine residues in its amino acid sequence and, due to that, can serve as an effective glutenin chain extender improving the dough quality [Masci et al 1998; 2000]. Identification of gliadin alleles in the set of cultivars studied in these publications showed that this polypeptide is always present in cultivars with the allele *Gli-B1d* (see **7.5.**), while cultivars having any other allele at the *Gli-B1* locus lack this polypeptide [Branlard & Metakovsky 2006]. In the cross of cvs Nuri-70 (*Gli-B1d*) and UL-82 (judging by the APAGE photos presented in the article, this genotype has *Gli-B1b*), the genotypes that inherited the *Gli-B1d* showed a better quality measured as dough breakdown value [Lagudah et al 1988]. Therefore, it is a case when the presence of the easy-identified (APAGE) allele (*Gli-B1d*) may be used as genetic marker of the valuable 42-kDa-glutenin subunit produced by *Glu-B3* of the *Glu-B3/Gli-B1* complex.

As a result of analysis of the progeny of the cross between KS831957, a hard red winter wheat line producing an extremely strong dough, and Japanese cultivar having poor bread-making quality, the K-32 glutenin subunit (MW 42 kDa) associated with superior bread-making quality was identified. By means of 2D EP it was shown that K-32 consists of two polypeptides (spots) while its allelic variant present in poor-quality cultivar is composed of three components. All these polypeptides had N-terminal amino acid sequence typical for B-type LMW subunits of *i*-type (see **3.2.**) [Maruyama-Funatsuki et al 2004]. The relation between the K-32 and the *Glu-B3*-encoded 42-kDa polypeptide [Masci et al 1998; 2000] was not discussed in this publication.

Attempts have been made to separate the effects on quality of polymorphism at the *Gli-1* and *Glu-3* loci using genotypes with recombination between these two loci.

Almost all durum wheat cultivars have in their APAGE electrophoregrams either γ42 or γ45 gliadin band produced by allelic variants of the *Gli-B1* locus (see **8.5.**). It was established that genotypes with the γ45 have a significantly better dough quality [Damidaux et al 1978; 1980; Kosmolak et al 1980; Leisle et al 1981; etc.]. For example, about 95% of durum wheat genotype grown in Canada in 2001 had the γ45, while all genotypes having γ42 showed a weak dough [Edwards et al 2007]. Moreover, the addition of 25% durum flour (γ45) improved characteristics (P/L, W, bread volume) of common wheat poor-quality flour [Boggini & Pogna 1990]. It was suggested that the good characteristics of the γ45-flour are not conditioned by the gliadin γ45 itself, but by a set of "good" LMW glutenins produced by closely linked *Glu-B3* locus [Payne et al 1984a]. Indeed, increasing proprtions of HMW glutenins were generally associated with weak dough and lower content of gluten polymers in durum wheat [Edwards et al 2007]. To check the assumption, durum cv Berillo was studied. This high-quality cultivar showed the "bad" APAGE band γ42 and a set of "good" LMW polypeptides [Pogna et al 1988; Boggini & Pogna 1989]. It is concluded that it is *Glu-B3* (of the *Glu-B3/Gli-B1* complex locus) that has been responsible for the observed effects of the polymorphism at the *Gli-B1* locus on the dough quality in durum [Pogna & Mellini 1988; Ruiz & Carrillo 1995].

However, what is the key difference between "good" and "bad" LMW glutenins usually accompanying γ45 and γ42, respectively? A special study showed that there was so small difference in amino acid sequences between these two polypeptides that it is difficult to imagine how these differences might have so big impact on quality. Therefore, it was suggested that it is not an allelic variation at the *Glu-3* loci in durum wheat that relates to quality, but, more likely, the amount of the B-type LMW glutenins synthesized by a given genotype [D'Ovidio et al 1999] (see **4.5.**).

Progeny of crosses between common wheat cultivars having strong (had *Gli-B1b*) and poor (*Gli-B1f* and *Gli-B1g*) bread-making quality was evaluated by Alveograph tests and SDS-st. It was shown that any *Glu* allele provided by the high-quality wheats, as well as the allele *Gli-B1b*, had a significant favourable effect on both dough strength and tenacity. Seven genotypes with recombination between *Glu-B3* and *Gli-B1* were revealed in this study. The analysis of these recombinants were interpreted to mean that gliadins produced by the *Glu-B3/Gli-B1*, in contrast to the results on recombinants between *Glu-B3* and *Gli-B1* in durum wheat [Pogna et al 1988; Boggini & Pogna 1989], do affect quality [Khelifi & Branlard 1992].

Therefore, there is no data unequivocally indicating only one mechanism of impact of the *Glu-3/Gli-1* complex locus on quality. Probably, the polymorphism at each of two linked loci, as well as the increase of the proportion in gluten of the B-type LMW glutenin molecules, have their particular contribution to the end-use quality.

4.10. The allelic variation at the *Gli-2* loci relates to end-use quality

Polymorphism at any *Gli-2* locus, like at the *Gli-1* loci, was shown to affect parameters of end-use wheat quality [Sozinov et al 1974; Sozinov & Poperelya 1980; Metakovsky et al 2019a]. For example, in the study of 70 cultivars of different origin, some *Gli-2*-encoded APAGE bands (five α- and three β-gliadins) were associated with at least one of several parameters of dough property tested (Alveograph tests and SDS-st) [Branlard & Dardevet 1985a]. In other work, 162 registered cultivars grown in three experimental locations in France were evaluated for 22 technological tests characterizing the bread-making quality. One of findings was that polymorphism at the *Gli-2* loci accounts for up to 21% of variation in some of these parameters (mixograph, etc.), especially of the extensibility of the dough [Branlard et al 2001]. Using substituted lines Cappelle-Desprez (Bez1) permitted to show that each of *Gli* loci, including *Gli-2* loci, affected the flour quality (SDS-st) [Krattiger et al 1996].

Analysis of aneuploid lines of ChS permitted to revealed that each chromosome of the first and sixth homoeologous groups (therefore, each *Gli* or *Glu* locus) influences SDS-st, but to a different degree. Among *Gli-2* loci, the biggest effect had the *Gli-A2* locus [Rogers et al 1990]. In the progeny from the cross between two Chinese cultivars, a significant relation was established between polymorphism of the chromosome-6A-controlled α-gliadins (*Gli-A2* locus) and the SDS-st value [Li et al 2009]. Polymorphism at the *Gli-A2* locus associated with quality also in the cross of Klein-32 x ChS [Manifesto et al 1998]. Comparison of NIL lines from the cross between the cv Toronit and the line 211.10214 showed that different fractions of gliadins, including α-gliadins (encoded by the *Gli-A2* and *Gli-D2* loci, see **3.1.**) positively affect the elasticity of dough [Gobaa et al 2008]. The mutant of the cv Xiaoyan-81 lacking the *Gli-D2* sequences (*Gli-D2null* called DLGliD2) was proved to have an improved bread-making quality, without penalties in their agronomic traits [Li et al 2018].

In Japanese common wheat cultivars, each of three homoeologous *Gli-2* loci were found to influence dough extensibility (AlveoConsistograph), with the biggest contribution of the *Gli-A2* [Noma et al 2019]. In Australian cultivars, however, genetic polymorphism at the *Gli-A2* locus associated with the Rmax value, but did not correlate neither with pedigrees of cultivars, nor with allelic variation at the *Glu-1* or *Glu-3* loci. Genotypes with alleles *Gli-A2c* and *Gli-A2m* showed higher Rmax values in comparison with genotypes with *Gli-A2a* and *Gli-A2g* [Wrigley et al 1990]. Genetic polymorphism at the *Gli-B2* and *Gli-D2* loci correlated with the dough extensibility in cultivars bred in Australia [Wrigley et al 1990; Metakovsky et al 1990b].

Official data of annual testing for dough strength (W) of 54 registered cultivars (nearly all Italian) in several locations in Italy during 21 years were used to compare different alleles at the same *Gli* locus in their relation to the dough quality. As a result, some statistically significant differences were revealed for the *Gli-A2* and *Gli-B2* loci (and also for the *Gli-B1*). The highest W value were observed for genotypes with the alleles *Gli-A2b* and *Gli-B2c*, whereas alleles *Gli-A2g* and *Gli-B2j* showed the lowest W values. The standardized W value (Wst) for cultivars with the allele *Gli-B2c* was 155.5 units, while Wst for four other alleles considered at this locus varied from 60.2 (*Gli-B2j*) to 99.0 (*Gli-B2h*). Four cultivars studied with the allele *Gli-B2c* were not relatives, and their higher quality could not be explained by the presence, in their genotypes, of valuable alleles at other storage-protein-coding loci. For example, only one of these four cultivars had the allele *Glu-D1d* related, as a rule, to better dough quality (see **4.7.**) [Metakovsky et al 1997a].

The relationship between the presence in the genotypes of certain alleles at the *Gli* loci and dough strength (W) was also studied using two contrast groups of French cultivars, 20 cultivars in each. The two groups were formed by selecting cultivars with the highest and the lowest W values (270-530, and 70-120, respectively) from 80 cultivars studied in total. It was found that the distribution of allelic frequencies between two contrast groups differed strongly (χ^2-test, $P<0.01$) only at the *Gli-A2* and *Gli-B2* loci, and that alleles *Gli-A2t* and *Gli-B2c* occurred significantly more frequently in the group with high W value [Metakovsky et al 1997c]. Altogether, the results obtained in analysis of Italian and French common wheat germplasms indicate a significant positive influence of the allele *Gli-B2c* on dough strength. Moreover, some indications on the association of this allele with higher dough quality were obtained in the analysis of Australian cultivars [Wrigley et al 1990; Metakovsky et al 1990b].

Polymorphism at the *Gli-D2* locus might relate to dough strength (Alveograph) in the study of proteomic profiles of double haploid lines of similar glutenin composition [Gobaa et al 2008]. In the cross between cultivars having alleles *Gli-D2a* and *Gli-D2j* [Alleles 2018c], genotypes F2-F4 which inherited the allele *Gli-D2j* had shorter kneading time compared to genotypes with the *Gli-D2a* [Martin & Carrillo 1999]. Double mutant line developed by the cross of two spontaneous mutants, at the *Gli-A2* (cv Reader) and *Gli-D2* (Saratovskaya-29), showed low indices of the dough quality (viscosity, swelling, and strength). This finding indicates the contribution to dough quality of α-gliadins produced by *Gli-A2* and *Gli-D2* loci [Pogna et al 1998].

In some publications, any influence of polymorphism at the *Gli-2* loci on bread-making quality was denied because it was not found, for example, in the cross studied [Payne et al 1987b; 1990]. However, alleles at *Gli-2* locus compared in a given study might not differ in their relation to quality. Indeed, it would be even strange to expect that any two haphazardly taken alleles would significantly differ in their influence on quality parameters studied. More probably, that only a very few, single alleles would have noticeable impact on quality. Some alleles at the *Glu-3* loci did not differ in their influence on quality in some groups of cultivars (see **4.6.**). Moreover, even among HMW glutenin subunits playing a key role in the forming gluten matrix, a few alleles at the *Glu-1* loci have well-documented and strong impact on the dough quality in groups of cultivars studied, for example, *Glu-D1d* being compared with the *Glu-D1a* (see **4.3**). The influence of several other alleles at the *Glu-1* loci either rather diminutive, or contradictory [Weegels et al 1996; Li et al 2010] (see **4.4.**).

Obviously, the correlation observed between polymorphism at the *Gli-2* loci and bread-making quality cannot be explained through the linkage of the gliadin- and LMW-glutenin-encoding genes (see **4.7.**) because no *Glu* loci were found on the chromosomes of the sixth homoeologous group.

4.11. Some glutenin-like polypeptides are encoded at the *Gli* loci

The amino acid sequence of ω-gliadins, and of α - and γ-gliadins, are nearly identical to that of D-type, and C-type LMW glutenins, respectively. The difference (almost the only) between gliadins and LMW glutenins is that amino acid chain of gliadin polypeptides do not include a cysteine residue capable of intermolecular interactions [Masci et al 1993; etc.] (see **3.2.**). Therefore, any gliadin polypeptide can acquire the properties of the glutenin subunit to bind to the gluten matrix if an additional cysteine appears in its amino acid chain, for example, as a result of mutation in the corresponding gene. Therefore, glutenin-like molecules influencing the dough quality may be encoded at the *Gli* loci [Tao & Kasarda 1989; Lew et al 1992; Masci et al 2002].

Six out of 50 α-gliadin-encoding genes of the cv ChS had an additional (seventh, see **3.1.**) cysteine residue, suggesting that these α-gliadins acquired the property of binding other proteins through intermolecular disulphide bands (Noma et al 2016). In other study of the same cultivar, five gliadin polypeptides having one additional cysteine residue were discovered. Thus, two different polypeptides of α-gliadin possessed seven cysteine residues instead of six, two γ-gliadins had nine instead of eight, and one polypeptide of ω-gliadin acquired one cysteine. Gliadin polypeptide containing an odd number of cysteine residues acquires the ability to incorporate into the gluten polymer. Such chain-terminating gliadins accounted for 4.5% of all gluten of the flour [Altenbach et al 2020]. In durum wheat cv Svevo one of four expressed γ-gliadin genes encodes polypeptide contained nine (instead of eight) cysteine residues [Paris et al 2021].

The C-type LMW glutenins tend to separate together with gliadins in many fractionation procedures. It was found that glutenin-like gliadins were bound to the polymer fraction of gluten [Masci et al 1993; Sreeramulu & Singh 1997]. The γ-gliadin-like molecules are linked to polymers by covalent disulphide bonds [Keck et al 1995]. Every fourth sequenced γ-gliadin-encoding gene of the cv Cheyenne controls the synthesis of a polypeptide which has one cysteine residue in its amino acid sequence that can form intermolecular covalent bonds. Therefore, these polypeptides are actually C-type LMW glutenins (Anderson et al 2001a). It is assumed that a polypeptide (glutenin or glutenin-like gliadin) with an odd number of cysteine residues can interrupt (terminate) the elongation of growing chains in the gluten polymer [Tao & Kasarda 1989; Lew et al 1992; Masci et al 1993; Gianibelli et al 2002a; etc.]. Thus, glutenin-like gliadins can, to certain extent, affect the formation of the gluten matrix and, thereby, the bread-making quality.

A naturally mutated γ-gliadin having additional (ninth) cysteine in its amino acid chain was discovered in durum wheat cv Lira. This cysteine is located in the N-terminal polypeptide sequence and is able to form intermolecular covalent bond [Ferrante et al 2006]. The cultivar is composed of two biotypes strongly differing in their dough quality, and the quota of this glutenin-like γ-gliadin among storage proteins is much higher in the poor-quality biotype [Masci et al 1995]. Due to additional cysteine, the mutant γ-gliadin can interrupt the elongation of gluten polymer chains and thereby negatively affect the quality [D'Ovidio et al 1995].

Some C-type molecules have α-type-gliadin sequence and are encoded at the *Gli-2* loci. For example, two C-type LMW glutenins are encoded at the *Gli-D2* locus of the variety ChS [Felix et al 1996]. In this cultivar, up to 95% of the C-type glutenins are in fact "mutant" α- and γ-gliadin polypeptides judging by their N-terminal amino acid sequences, and 25 of these molecules were studied. It was shown that 2, 5, 5, 3, 4, and 6 of them were encoded at the *Glu-A3*/*Gli-A1*, *Glu-B3*/*Gli-B1*, *Glu-D3*/*Gli-D1*, *Gli-A2*, *Gli-B2*, and *Gli-D2* loci, respectively [Masci et al 2002]. Two α-gliadins encoded by the allele *Gli-A2b* of the cv Alpe are probably C-type LMW glutenins [Pogna et al 1995]. The allele *Gli-B2ah* of the cv Kharkovskaya-6 encodes several gliadins and at least one C-type LMW subunit with AMW 37 kDa [Metakovsky et al 1997b].

The glutenin-like ω-gliadins (D-type LMW) are also discovered. For example, a D-type LMW subunit with AMW 71 kDa was, in fact, the ω-gliadin having an additional cysteine residue. It occurred in a larger amount in gluten polymers of smaller size. It is suggested that this molecule terminates gluten chains thereby reducing their molecular mass [Gianibelli et al 2002a]. It was proved that "mutant ω-gliadins" of cv ChS have ω-gliadin type N-terminal sequence and one cysteine in their amino acid chain per molecule. Besides, dimers of LMW D-glutenins were observed [Masci et al 1999]. In the cv Recital, two such polypeptides were discovered, one encoded by the genome B, and the other by the genome D [Anderson et al 2009].

The cv Newton is composed of two genotypes (bt1 and bt2) differing in the composition of their *Gli-D1*-encoded ω-gliadins (ChS-type and Cheyenne-type, respectively) and in their quality parameters (the bt2 shows better quality). Besides, the bt1 produces a particular D-type LMW glutenin polypeptide absent in the bt2, while bt2 has ω-gliadin absent in bt1. It was suggested that it was the D-type polypeptide that reduced the bt1's dough quality by interrupting the elongation of gluten polymer [Masci et al 1991]. However, our analysis of the APAGE electrophoregram of Newton presented in this publication (Fig. 7) showed that its two biotypes might differ not only in the allelic state of the *Gli-D1*, but also of the *Gli-B2* and, probably, of the *Gli-D2* loci. The allelic difference at the *Gli-2* loci might contribute to the quality parameters of biotypes studied (see **4.10.**).

New type of storage protein genes was identified in the library of cDNA clones of the old Hungarian variety Bánkúti-1201. Each of eight genes studied of this type produce chimeric (recombinant) molecules which have γ-gliadin, and B-type LMW glutenin sequences at the N-terminal, and C-terminal ends of the encoded polypeptide, respectively. The chimeric polypeptides differed in size (MW), gliadin/glutenin ratio and cysteine content. Such polypeptides might influence bread-making quality [Nagy et al 2005].

4.12. Polypeptides "*d4*" and "*d7*" and their relationship to quality

The EM of classical gliadin in APAGE does not depend on agents that break disulphide intermolecular bonds. On the contrary, glutenin can be studied by any EP method only after treatment of gluten with an agent that dissolves it into individual polypeptides. However, in the alcohol (gliadin) fraction of wheat seed protein, there are some ω-gliadins strongly bound to glutenins. For their analysis, a special procedure of two-step APAGE+SDS-electrophoresis was developed [Khelifi et al 1992; Branlard et al 1992]. Among the molecules revealed by this technique the most interesting were the polypeptides named "*d7*" and, especially, "*d4*".

These polypeptides are found in the so-called D-zone of the SDS EP pattern, that is where D-type LMW glutenins and ω-gliadins are located [Jackson et al 1983]. Taking into account a close similarity of the primary structure of D-type LMW glutenins and ω-gliadins, it may be suggested that polypeptides *d4* and *d7* are subunits of the D-type LMW subunits. Indeed, these polypeptides disappeared from the D-zone of SDS-electrophoregram, if the alcohol-soluble faction of wheat seed protein was not treated, before EP, with an agent that destroys disulphide bonds [Khelifi & Branlard 1992; Branlard et al 1994].

Comparison of data on the presence of the *d7* and alleles at the *Gli-1* loci in the same genotypes [Alleles 2018c] made it possible to establish that this polypeptide is *Gli-D1*-encoded and that it corresponds to the slowest of ω-gliadins that can be observed in the APAGE. Therefore, this polypeptide is an integral part of blocks controlled by alleles of the Cheyenne-type (*Gli-D1g*, *Gli-D1j*, and *Gli-D1o*) [Blocks 2018c].

The *d4* (AMW of about 55.5 kDa) is one of the longest alcohol-soluble molecule of wheat proteins [Metakovsky et al 1984b; Branlard et al 1994; Metakovsky et al 1997b] and is encoded by the *Gli-B1* locus [Khelifi et al 1992]. This polypeptide has N-terminal amino acid sequence coincident with the SRI sequence type of ω-gliadins encoded by gene located on the chromosome 1B but, unlike ω-gliadins, can form disulphide aggregates with glutenins [Nieto-Taladriz et al 1998].

Identification of alleles at the *Gli-B1* locus in wheat genotypes with and without the *d4* [Alleles 2018c] permitted to established that this polypeptide is produced only by cultivars with alleles *Gli-B1c*, *Gli-B1e*, *Gli-B1f* and *Gli-B1g*. These allelic variants of the *Gli-B1* locus not only encode blocks with similar APAGE band composition [Blocks 2018c], but also is shown to have a remarkable similarity in their DNA sequences along the locus (see **7.5.**) [Popovych et al 2020; Metakovsky et al 2021b]. However, the only common APAGE band present in each of these blocks is the fast-moving ω-gliadin. Both this ω-gliadin and *d4* are absent in genotypes with other alleles at the *Gli-B1* locus, for example, *Gli-B1b* or *Gli-B1d* [Blocks 2018c].

The identity of *d4* and the fast-moving ω-gliadin was firmly confirmed by the analysis of two spontaneous mutants found in cvs Leningradka and Albidum-43. The first mutant lacked several *Gli-B1e*-encoded APAGE bands and the *d4* polypeptide, although one of the gliadins encoded by this allele (the slow-moving ω-gliadin) is preserved in the mutant. The second mutant lacked only a *Gli-B1e*-encoded slow-moving ω-gliadin but the fast-moving ω-gliadin and *d4* polypeptide were retained in the APAGE and SDS patterns, respectively [Metakovsky et al 1997b].

The *d4* occurred in genotypes having alleles *Glu-B3g* (Branlard & Metakovsky 2006) and *Glu-B3f* (Jackson et al 1996). Thus, this polypeptide, having properties of both gliadins and the D-type LMW glutenins might be encoded either at the *Gli-B1* (alleles *Gli-B1c*, *Gli-B1e*, *Gli-B1f*, *Gli-B1g*), or *Glu-B3* (*Glu-B3g*, *Glu-B3f*) loci. There is no discrepancy because it is possible to assume that *d4* is an ω-gliadin which acquired an additional cysteine in its amino acid composition together with the properties of the LMW glutenin subunit as a result of mutation at the *Gli-1* locus, as it was suggested (see **4.11.**).

As a result of the analysis of the F2 progeny (210 individuals) from crossing of cultivars Ablaca (does not produce *d4*) and Prinqual (has allele *Gli-B1c* and *d4*), eight recombinants were found between the gene encoding *d4* and the *Gli-B1c* locus [Nieto-Taladriz et al 1998]. This level of recombination frequency (2.8±1.3%) corresponded well to the distance between *Glu-B3* and *Gli-B1* reported in other studies (see **3.3.**). Is, therefore, the *d4* encoded at the *Glu-B3* locus? In this case, however, one must reject an identity of the *d4* and the fast-moving ω-gliadin and expect a noticeable number of recombinants between *d4* and other gliadins encoding at *Gli-B1* locus, for example, intensive γ-gliadin. However, no supposed new combination of gliadins (a result of such recombination) was observed neither in published catalogues of gliadin electrophoregrams

[Peruffo et al 1981; 1984; Ng et al 1988], nor as a recombinant block of gliadin APAGE bands [Blocks 2018c], nor among the genotypes of progeny from crosses studied [Metakovsky et al 1984a].

We assume that genes encoding the d4 and d7 glutenin-like ω-gliadins (in fact, D-type LMW glutnins) are located at the *Gli-B1* and *Gli-D1* loci, respectively. Such molecules may influence, in theory, negatively on the dough quality interrupting (terminating) the elongation of growing chains in the gluten polymer (see **4.11.**). Nevertheless, genotypes producing slow-moving ω-gliadins of Cheyenne type (and therefore the *d7*) have a higher flour quality compared to the genotypes with alleles of the ChS type, for example, *Gli-D1a* or *Gli-D1b*. This correlation is believed to be due to the presence of ω-gliadins bound to gluten [Bietz et al 1975; Branlard et al 1992]. Such correlation was observed in a group of cultivars bread in the former USSR [Sozinov & Poperelya 1980], in French cultivars [Branlard et al 1994] and in cultivars bred in USA and Canada [ref. from Branlard et al 1994].

The polypeptide *d4* also has a positive effect on common wheat end-use quality [Branlard et al 1994]. For example, relation to quality (rapid mix test) of polypeptides located in the D-zone of SDS-EP was studied in 153 German cultivars. It was found that higher quality was shown by genotypes that produced bands *d4* (more precisely, *d2+d4*), *d3* and *d7* [Brönneke et al 2000]. Similar work carried out on 181 common wheat cultivars of various origin and 949 lines from Swedish breeding programs also showed that genotypes producing bands d7 and, especially, *d4* in D-zone have the best quality (SDS-st) [Johansson 1996].

The mechanism of the positive influence of the polypeptides d4 and d7 on the end-use common wheat quality remains unclear [Branlard et al 1994].

4.13. Summary. Mechanisms of the influence of glutenins and gliadins on quality

Due to the presence of free cysteine residues, the *Glu-1*-encoded HMW glutenin polypeptides form intermolecular bonds, and thereby determine the structure of gluten matrix. So, genetic polymorphism at the *Glu-1* loci influences common wheat end-use quality. Among alleles at *Glu-1*, the greatest positive effect showed *Glu-D1d* marked by the presence of HMW glutenin bands 5+10 in SDS EP. There are some alleles at the *Glu-A1* and *Glu-D1* having positive effect on the dough quality in some of groups of cultivars studied. The quality score system for *Glu-1* alleles elaborated for a group of cultivars did not always predict quality in other groups of cultivars. Also, differences in amount of produced HMW subunits rather than differences in effectiveness of different glutenin polypeptides might be responsible for relation of *Glu-1* alleles to quality.

Most B-type glutenin polypeptides (encoded at the *Glu-3* loci) have eight cysteine residues in their amino acid sequence, but, unlike gliadins, two of them are capable to form covalent intermolecular bonds supplying the molecule with ability to be involved in glutenin chain elongation thereby improving the dough quality. Alleles at the *Glu-3* loci were ranked according to their influence on dough strength. On the contrary, C- and D-types LMW subunits (frequently encoded at the *Gli* loci) may function as terminators, interrupting the growing gluten chains. There is only fragmentary information about which specific LMW glutenin polypeptides affect bread-making quality and how they do it. Increasing the quota of the B-type LMW glutenins in gluten may improve the dough quality.

Theoretically, gliadins, due to their primary structure, cannot be involved in gluten matrix by covalent S-S bonds. Nevertheless, there are at least several alleles at *Gli* loci presence of which in wheat genotype associated with good flour (dough) quality (*Gli-B1c*, *Gli-B1d*, *Gli-D1g*, *Gli-D1j*, *Gli-A2t*, *Gli-A2b*, *Gli-B2c*). Several possible mechanisms of their action is suggested.

Firstly, effects of the *Gli-1* loci on quality are not due to the gliadin polymorphism itself, but may be explained by the actions of the B-type LMW glutenin polypeptides produced by the *Glu-3* loci tightly linked to *Gli-1*. However, this mechanism cannot explain the actions of gliadins encoded at the *Gli-2* loci. Secondly, any *Gli* locus may produce glutenin-like gliadins which appear as a result of some mutation in "true" gliadin-encoding gene. So, the influence of *Gli* loci on the gluten properties may be due to the producing glutenin-like molecules (in fact, C- and D-type glutenins). Thirdly, the *Glu-1*-encoded polypeptides such as *d4* and *d7* which have properties of both gliadins and glutenins. There are data that these molecules can positively affect quality. The mechanism of their action remains unclear, however. Fourthly, the direct influence of "true" gliadins and of the ratio of synthesized gliadins and glutenins on quality are not finally excluded.

5. THE STRUCTURE OF *GLI* AND *GLU* LOCI

5.1. *Gli* and *Glu* loci contain transcribed and silent sequences

Genomes of wheat are of the most complex known to science. For example, the common wheat genome is up to 17 billion base pairs (about 17 000 000 Kb), and most of it is not occupied by transcribed genes [Bennett & Smith 1976; Bennett & Leitch 1995; Gill & Friebe 2002]. The genome size of *T.tauschii*, the diploid donor of the genome D into common wheat (see **2.1.**), is about 4.02Gb, of which 90% is repetitive sequences [You et al 2011]. The final assembly that composed 96% of the genome sequence of the cv ChS contained 15 344 693 583 bases [Zimin et al 2017].

The results of analysis of the inheritance of APAGE gliadin bands (see **2.2.**) indicate that, in each *Gli* locus, there are several transcribed genes. These genes may produce distinct types of gliadin polypeptides (for example, any *Gli-1* locus encodes γ- and ω-gliadins). That was confirmed by analysis the genomic DNA sequences of the *Gli* loci [Harberd et al 1985; Gao et al 2007; Wicker et al 2010; Anderson et al 2012]. Alternative (allelic) variants of one *Gli* locus differ considerably in the set of encoded gliadin polypeptides which can be revealed by using APAGE (see **2.3.**). For example, the allele *Gli-D1k* encodes two ω- and one γ-gliadin, while allele *Gli-D1j* produces five ω- and two γ-gliadins.

Inter-varietal polymorphism is described in the number of DNA sequences encoding gliadins [Anderson et al 1984; Harberd et al 1985; Reeves & Okita 1987; etc.]. For example, in cvs ChS, and Butte-86, three, and six ω-gliadin-encoding sequences, respectively were found [DuPont et al 2004]. Genotypes of cvs ChS, and Cheyenne have 60, and 150 copies of α-gliadin sequences per haploid genome, respectively.

There are 39, and 24 copies of γ-gliadin sequences identified by RFLP analysis in ChS, and in durum cv Langdon, respectively. Also, 35 and 17 sequences for LMW glutenins were found in ChS and Langdon, respectively [Sabelli & Shewry 1991]. In the cv Cheyenne, as many as 150 α-gliadin sequences were revealed. The number of copy of any DNA sequence in the genome can change as a result of unequal crossing over and gene conversion [Anderson et al 1997].

The number of gliadin polypeptides synthesized by a given genotype is significantly less than the number of gliadin-coding DNA sequences in this genotype. For example, out of 50 individual α-gliadin genes of the cv ChS harboring a typical promoter region, having basic domains structure and six conserved cysteine residues (see **3.1.**), 16, 16, and 18 were, respectively, located on chromosome 6A, 6B, 6D [Noma et al 2016]. However, only 7, 12, and 7 of them, respectively expressed in developing endosperm of ChS [Noma et al 2018]. In another study performed on this cultivar, 47 α-, 14 γ-, 5 δ-, 19 ω-gliadin genomic DNA sequences, and 17 LMW glutenin sequences were revealed. However, only 26, 11, 2, 5 and 10 of them, respectively, could produce corresponding full-length polypeptides while remaining genes were probably pseudogenes [Altenbach et al 2020].

2D EP procedure, in which the analysis is based mainly on net charge of polypeptides, separates gliadins of a single cultivar into 30-46 components [Autran et al 1979b]. It was possible, using 2D EP, to reveal 16 different gliadin polypeptides produced by the cv Cheyenne [Lafiandra et al 1984] whereas DNA of this cultivar contains, in total, about 100 gliadin-coding sequences [Reeves & Okita 1987; Anderson et al 1997]. One-dimensional APAGE reveals about 20-25 gliadin bands in one genotype [Blocks 2018c], and 2D APAGE/SDS, up to 40 individual spots [Redaelli et al 1994]. Proteome technique permits to reveal more than 600 common wheat grain polypeptides in the progeny of a cross of two cultivars, but only 67 of them were identified as gliadins (29 α-, 17 β-, 14 γ- and 7 ω-gliadins) [Gobaa et al 2008].

The difference between the number of gliadin polypeptides produced by one genotype and the number of gliadin-coding DNA sequences found in this genotype indicates that the majority of these gliadin DNA sequences are silent being represented by pseudogenes and gene fragments. Indeed, the quota of non-transcribed α-gliadin DNA sequences among all α-gliadin sequences present in wheat genomes is from 72% (genome A) to 95% (genome D) [van Herpen et al 2006]. In the cv ChS, 14 of 19 ω-gliadin and 7 of 17 LMW glutenin-coding sequences were pseudogenes with little or no expression, while 11 of 14 γ-gliadin genes were intact [Huo et al 2018a]. Out of 44 α-gliadin-encoding sequences studied in the cv ChS, 18 were found to be pseudogenes [Huo et al 2018b].

In durum wheat, only six of the 19 α-gliadin-coding sequences studied were transcribed, five were pseudogenes, and eight were fragments of gliadin-coding genes [Gu et al 2004b]. Of the eight ω-gliadin-coding sequences studied of the chromosome 1B of durum wheat, two were transcribed, two were pseudogenes, and four were gene fragments (the 3' portion of ω-gene) [Gao et al 2007; Anderson et al 2009]. Four of six genes found at the *Glu-A1* locus were pseudogenes [Gu et al 2004a]. So, there are strong evidences confirming a high proportion of silent sequences (pseudogenes and gene fragments) among all storage-protein sequences.

5.2. Transposable elements between and within *Gli* and *Glu* loci

It was established that 70-90% of the nucleotide sequences of wheat genomic DNA are transposable elements (TEs) [Huang & Cloutier 2008; Wicker et al 2011; etc.]. For example, TEs constituted 68,2% of the diploid wheat *T.tauschii* [Li W et al 2004]. The TEs, being dispersed throughout the genome, is a part of all wheat storage protein loci *Glu* and *Gli* [Wicker et al 2003; Li W et al 2004; Charles et al 2008]. At the *Gli-2* loci, 70% of DNA located between the α-gliadin-coding sequences are shown to be different variants of TEs. The distance between neighbouring sequences encoding α-gliadins depends on the length of blocks of TE [Gu et al 2004b].

Comparison of the intergenic region covering the *Glu-1* loci in common wheat, durum wheat and *T.tauschii* revealed that all homoeologous (between genomes) and orthologous (between species) variants of this region are composed of blocks of nested retroelements. It was found that the region studied is generally conserved, although a few no shared TEs were detected between orthologous variants present in genomes A and D [Gu et al 2006].

The TEs occupy the most of DNA of the *Glu-3* loci. In common wheat Glenlea, the DNA of the *Glu-3* loci contain 30 distinct variants of TEs, each of them repeats in the genome from one to 24 times [Huang & Cloutier 2008]. In durum, the large blocks of repetitive DNA separate LMW-coding sequences owing to the rapid amplification of TEs [Gao et al 2007].

In durum chromosome 1A, the stretch of TEs of 19 different types includes an unusual structure of eight distinct TEs nested in one another. This structure spans a region of over 140 kb [Gu et al 2004a]. In *T. monococcum,* 70% of the contiguous 211 Kb sequence studied of the chromosome 1AS was occupied by TEs of 10 distinct types including a complex structure of 14 TEs nested in one another. It was discovered that the sequence studied was restructured many times in evolution, due to activity of TEs [Wicker et al 2001]. Chromosome arms 1AS and, especially, 1BS are zones of unusually high activity of mobile elements resulted in appearance of chromosome-specific pseudogenes [Wicker et al 2011].

TEs not only represent the bulk of the gnome's DNA, but are also the most dynamic, changeable fraction of it. For example, the DNA sequences covering orthologous *Glu-A3* locus were compared in durum (contig of 142 Kb) and *T.monococcum* (contig of 285 Kb). It was found that conservation of these sequences was restricted of small regions containing the LMW glutenin genes, whereas more than 90% of compared sequences showed a dramatic difference between the two species. The diversity exceeded 90% of the length of DNA sequence studied and was represented by duplications/deletions (of the size from several nucleotides up to 4-5 Kb), as well as nucleotide substitutions. The divergence of non-transcribed sequences was explained through activity of TE [Wicker et al 2003]. The *Ha* locus controlling grain hardness in the genome D of hexaploid wheat was found to be 29 Kb shorter than in the D genome of its diploid progenitor *T.tauschii*, principally because of TE insertions and two large deletions caused by illegitimate DNA recombination [Chantret et al 2005].

Movements of TE can cause mutations of transcribed genes [MacPhee 1991]. Silencing any gliadin gene and its transformation into a pseudogene (see **5.1.**, **7.4.**) may be caused by different mechanisms, including movements of TE (Charles et al 2006; Gao et al 2007). Probably, TE have also played a crucial role in the differentiation of the three genomes (A, B, and D) and their subsequent transformations during wheat evolution from diploid to polyploid level [Jiang et al 2004; Senerchia et al 2013; etc.] (see **8.1**).

5.3. Duplications/amplifications of gliadin- and LMW-glutenin-coding sequences

There is an abundance of duplicate genes in plant genomes, up to 65% of sequences known in plant genome have a duplicated copy. The complete genome sequence of diploid wheat *T.tauschii* (accession AL8/78 havig the D genome closely related to common wheat) was found to possess an unprecedented amounts of very similar repeated sequences and a greater number of dispersed duplicated genes than any other sequenced genome [Luo et al 2017]. The *Gli-2* loci contain a large number of tandemly duplicated genes and highly repetitive DNA. These duplications occurred in the time after divergence of the A, B and D genomes [Huo et al 2018b].

Gene duplication is one of the most important evolutionary processes that generate genetic diversity and functional novelty [Panchy et al 2016].

A single gene in a species could have multiple variants after duplication and subsequent divergence in the genomes of different species. Indeed, the three wheat elemental genomes, A, B and D, in spite of having a common genome-progenitor (see **8.1.**), differ significantly from each other. For example, a full set of the prolamin-encoded genes located on the short arm of the chromosomes 1A, 1B, and 1D, was represented by 18, 29 and 51 sequences, respectively. In addition, genes of the genome A showed lower expression level, while the B and D genomes contributed similarly [Huo et al 2018a].

There were eight different α-gliadin-encoding genes within *Gli-D2* locus of cv Xiaoyan-81, two of them were represented each by a tandem duplication [Li et al 2018]. Eleven active γ-gliadin gene sequences were revealed in the genome of the cv ChS, two of them encode identical polypeptides and were assumed to be the result of a recent gene duplication [Anderson et al 2013]. Some α-gliadin-encoding sequences originated as a result of amplification of an initial sequence [Harberd et al 1985]. DNA sequences of several α-gliadin pseudogenes were compared and their close relationship were revealed. It was assumed that the original pseudogene may arise as a result of spontaneous mutation that caused the appearance of a stop codon. Subsequent mutations, duplications, and gene conversion transform the original pseudogene into a family of similar pseudogenes [Anderson 1991].

In the chromosome 1AS of diploid wheat *T.monococcum*, there are three sequences recognized by the low copy probe SER159, while in tetraploid durum wheat there is only one such sequence. It was concluded that there was an amplification of this sequence in some genotypes of diploid wheat. Besides, in the same DNA region, a big (54 Kb) duplication was found which included, in particular, a LMW-glutenin-encoding gene. The two copies of the duplicated sequence are nearly identical (similarity 98%) indicating a recent origin of this duplication [Wicker et al 2003] probably through so-called illegitimate recombination [Devos et al 2002].

In *T.tauschii*, duplications of genetic material have occurred many times in the region of chromosome 1DS harbouring LMW glutenin- and γ-gliadin-encoding sequences [Spielmeyer et al 2000]. Duplications in the *Gli-2*, and *Glu-3* loci of common wheat led to the appearance of many copies of α-gliadin-encoding genes [Gu et al 2004b], and LMW-encoding genes [Huang & Cloutier 2008], respectively.

Amplification of one individual gene within the *Gli-D1* locus of common wheat turned the allele *Gli-D1b* into the *Gli-D1d*. Indeed, the only difference between blocks encoded by these alleles is a considerable over-production of the fast-moving γ-gliadin in the block encoded by the

Gli-D1d [Blocks 2018c]. The difference between blocks encoded by *Gli-D1b* and *Gli-D1d* were registered using APAGE and 2D APAGE/SDS [Metakovsky et al 1993c]. The over-production of the polypeptide was caused by amplification of corresponded DNA sequences. The size of the amplified sequence was about 1.25 Kb judging by the RFLP pattern obtained using the restrictase *Rsa*I [Vaccino & Metakovsky 1995].

In common wheat, a significant inter-varietal polymorphism in the number of copies of ribosomal DNA was observed. They arose, probably due to spontaneous duplication and amplification of individual genes [Dvořák & Appels 1986].

The duplication of genetic material may be caused by spontaneous translocation of genetic material [Perera et al 1983; Lange et al 1987; Friebe & Gill 1994; Badaeva et al 2007]. Homoeoallelic translocations of the *Gli-D1* locus on the chromosome arm 1AS [Akhmedov et al 1988; Metakovsky et al 1990a] and of the *Glu-D1* locus on the chromosome 1AL [Dumur et al 2010] were registered.

Duplication and amplification of some HMW-glutenin- and LMW-glutenin-coding sequences can strongly influence the dough quality of common wheat (see **4.5.**; **4.6.**).

5.4. Collinearity of homoeologous and orthologous genes within the *Glu* and *Gli* loci

The question of gene collinearity in homoeologous and orthologous chromosomes is of general scientific importance. The collinearity is almost always retained among related species [Keller & Feuillet 2000], although some events at the molecular level (in particular, TE's movements) can violate this collinearity [Akhunov et al 2003]. Collinearity of genes in plant genomes generally decreases with increasing evolutionary distance while the actual number of genes remains more or less constant [Wicker et al 2010].

The collinearity of about 100 prolamin and resistance genes harbouring distally on the short arms of chromosomes 1D of common wheat (cv ChS) and *T.tauschii* was studied. It was documented that gene rearrangement events, including gene duplication and deletion, resulted in considerable erosion of gene collinearity among orthologous sequences studied due to the presence of five large structural variations with sizes of 100-2000 Kb. About 44% of *T.tauschii* sequences and 71% of common wheat sequences were not shared [Huo et al 2018a].

The DNA sequences covering homoeologous loci *Glu-A3*/*Gli-A1* and *Glu-B3*/*Gli-B1* of durum cv Langdon were compared and the absence of the collinearity of some genes in homoeologous DNA segments was discovered. For example, a gene encoding the disease resistance protein (*TdRGL*) is located proximally, and distally from LMW-encoding gene on the chromosomes 1A, and 1B, respectively. Also, in the region studied of the chromosome 1B, eight ω-gliadin-encoding sequences were found while several γ-gliadin-encoding sequences were located within the same region of the chromosome 1A. Therefore, sequence comparisons of the orthologous regions reveal that gene movement could be one of the major factors contributing to the violation of micro collinearity between homoeologous A and B genomes in wheat [Gao et al 2007]. Also in durum (cv Svevo) a considerable difference was found in the order of genes within *Gli-1* loci. At the *Gli-A1* locus, the δ-gliadin-encoding gene was located between two γ-gliadin genes while at the *Gli-B1* locus, two δ-gliadin sequences were located apart from γ-gliadin ones [Paris et al 2021].

The sequences of two long contigs of the chromosome 1AS (including sequences belonging to the *Glu-A3* locus) of *T.monococcum* and *T.durum* were compared. It was discovered that each contig contained five putative genes, only two of them (including LMW-glutenin-encoding one) were conserved between the two species. The observed divergence of the two sequences studied was explained through activity of TE [Wicker et al 2003]. Linkage maps of chromosomes of the first homoeologous group of common wheat, *T.tauschii* and *T.monococcum* were compared using the RFLP technique and seven inconsistencies in the order of markers were registered [van Deynze et al 1995].

On the other hand, comparison of the 277 Kb sequence covering the *Glu-A1* locus in four wheat species (two polyploids and two diploids) showed the complete collinearity of nine different genes and even TEs located in the DNA region studied [Dong et al 2012]. Also, each of the three *Glu-1* regions (*Glu-A1*, *Glu-B1* and *Glu-D1*) was compared between common and durum wheats and *T.tauschii* studied, and a strict collinearity of genes as well as sequences in intergenic regions across different wheat ploidy levels was observed. Some difference between orthologous genes was found, however, in the number and composition of TEs located in the intergenic regions [Gu et al 2004a; 2006]. The linear order of the RFLP genetic markers was maintained between chromosome 1B of common wheat and chromosome 1D of diploid wheat *T.tauschii* [Kota et al 1993].

The violation of the gene collinearity and the presence of different sets of sequences in the homoeologous and orthologous chromosomal regions are always explained as a result of TEs movements (see **5.2.**).

5.5. Alien genes within the *Gli* and *Glu* loci

At the *Glu-3/Gli-1* complex loci, active genes are separated by large intergenic and highly recombinogenic regions (see **5.6.**) comprised of large blocks of repetitive DNA and several non-prolamin genes [Ibba et al 2017a]. For example, some "foreign" genes (genes which do not encode storage proteins) were found to locate between DNA sequences for LMW glutenins, γ- and ω-gliadins in the comparative study of the *Glu-A3/Gli-A1* and *Glu-B3/Gli-B1* complex loci of durum wheat. Sequences for UDP-glycosyltransferase, cyclophilin-like protein, GPI-anchored proteins and some genes with unknown function were registered. Some of these genes can be normally transcribed. Not all of these genes, however, were present in both loci compared [Gao et al 2007].

Non-gliadin genes that can be also expressed (for example, polypeptides F2D10.32, zink finger protein, NBS/LRR and some others) were found between genes encoding α-gliadins located within *Gli-2* loci in durum [Gu et al 2004b].

In the *Glu-1* loci of common, durum and diploid wheat, a sequence of pseudogene of unknown globulin was discovered between genes for HMW glutenins encoded at the *Glu-1* locus [Gu et al 2004a; 2006]. The presence, between sequences for HMW-glutenin-encoded genes, of this pseudogene was confirmed in the study of a large genomic region carrying the *Glu-A1* locus in four wheat species of different ploidy [Dong et al 2012].

Genes that are not related to storage proteins are present in abundance between the storage-protein-encoding genes in the genomic DNA region sequenced (3100 Kb) covering *Glu-D3/Gli-D1* locus of diploid wheat *T.tauschii* [Anderson et al 2012].

5.6. The distance between active genes within one *Glu-3/Gli-1* locus

The analysis the component composition of gliadin in progenies from crosses of different common wheat genotypes showed that the genetic distance between the most frequently recombining genes located in a given *Gli* locus does not exceed 1% (see **2.2.**). It could be suggested that the distances (expressed in number of nucleotide base pairs, bp) between genes located within one *Gli* locus are small, that gliadin genes inherited together is a "cluster", rather compact group.

The average size of one storage protein gene, with all its regulatory sequences, is 0.8-1.5 Kb [Shewry & Halford 2002; Blatter et al 2004].

Among 12 LMW-glutenin-encoded genes studied of the cv Glenlea, the longest consisted of 1167 bp [Huang & Cloutier 2008]. Sequencing of large sections of DNA showed that physical distances (bp) between adjacent transcribed genes producing gliadins [Gu et al 2004b], LMW glutenins [Wicker et al 2003] or HMW glutenins [Gu et al 2006] are very large, much more than the size of one gene. Thus, storage-protein-encoding genes inherited together are not "clustered".

It was noticed that two genes encoding gliadin almost never occurred inside the same cloned sequence of genomic DNA, many times the length of one gene [Rafalski 1986; Reeves & Okita 1987]. Sequence of Cheyenne genomic DNA showed that α-gliadin-encoding genes were located at so big distances (bp) from one another that, among 17-Kb clones (the size at least ten times of the length of one gene), only one of many clones studied contained two α-gliadin sequences, and the 48-Kb clone also had only one α-gliadin sequence. Twelve of 20 genomic clones studied were located at the *Gli-A2* locus. Therefore, it was suggested that the size of the *Gli-A2* locus (was called *Gli-2A* in the publication) is at least 2000 Kb, but possibly, it is much larger [Anderson et al 1997].

The homoeologous storage-protein-coding loci of common wheat differ in size, the *Glu-D3* being significantly larger than *Glu-A3* or *Glu-B3* [Huang & Cloutier 2008]. The genetic size of the *Glu-A3*, *Glu-B3*, and *Glu-D3* loci in the cv Xiaoyan-54 was estimated to be 1.7 cM, 0.6 cM, and 4.9 cM, respectively. The longest genetic distance found for *Glu-D3*, correlated with the finding of more LMW-GS genes in this locus. In this work, the order of six LMW-glutenin genes within the *Glu-D3* locus was established [Dong et al 2010].

A region (3100 Kb) of the genomic DNA of diploid wheat *T.tauschii* was sequenced along with the complex locus *Glu-D3*/*Gli-D1* located in it. In this region, three groups of sequences were discovered, encoding LMW glutenins (proximal group, locus *Glu-D3*), γ-gliadins, and ω-gliadins (distal group). The distance between groups was 320 Kb and 480 Kb, respectively. One more group of sequences encoding LMW glutenins (was named *Glu-D4*) was located at the distance of about 1300 Kb proximally from the *Glu-D3* [Anderson et al 2012].

The storage protein-encoding genes are located unevenly along the *Gli* loci. It was found that 86% of all transcribed sequences are assembled on the chromosomes 1A, 1B and 1D into five "clusters". These clusters together occupy only 10% of the total length of each of these chromosomes [Gill et al 1996]. In the cloned sequence (211 Kb) of the chromosome arm 1AS of *T.monococcum*, three genes (a "cluster") were located within a strip of 31 Kb while the other two genes were at a large distance from each other and from the cluster [Wicker et al 2001]. α-Gliadin-encoding sequences were also unevenly distributed along a big cloned region studied (700 Kb) of durum: clones of 130 Kb might carry from one to four of such sequences [Gu et al 2004b]. One of clones (265 Kb) studied of the *Glu-3*/*Gli-1* region harboured two LMW-glutenin-encoding genes and five γ-gliadin-encoding sequences [Gao et al 2007].

It should be noted that the homoeologous *Gli-2* loci of common wheat produce distinct sets of α-gliadins. Almost all polypeptides produced by the *Gli-A2* have similar EM (APAGE) and AMW, whereas the *Gli-B2*-encoded gliadins are much more different from each other, and the *Gli-D2*-encoded gliadins, as a rule, have nearly identical AMW, but differ strongly in their EM (see **2.5.**). Therefore, probably, the organization of transcribed gliadin genes within homoeologous *Gli-2* loci may differ considerably. However, there are only a few publications (for example, [Anderson et al 1997]) in which an attempt was made to indicate the genome from which the DNA region of the *Gli-2* locus was sequenced.

5.7. Physical and genetic distances between transcribed genes

It would seem logic that the greater the physical distance between two genes expressed in bp, higher recombination frequency between them should be observed. However, this is not true [Snape et al 1985]. Paradoxically, the smaller the physical distance between two genes (bp), the more often these genes recombine. Moreover, recombination events are distributed unevenly along the chromosome.

It was firmly established that both the saturation of the physical length of chromosome with genes and the frequency of recombination between them increase significantly in the distal part of the chromosome arm [Keller & Feuillet 2000; Gill & Friebe 2002; Dvořák et al 2004; etc.]. On the contrary, genes located proximally, especially in the centromere region, are separated by extended DNA sequences and hardly recombine. As a result, the calculated genetic distances between markers (genes) located proximally are small, while in the distal part of the chromosome arm, they are large [Sourdille et al 2004].

Eighteen homozygous stocks carrying different deletions were used to construct a physical (cytogenetic) map of common wheat chromosome 1B. Comparison of this map with the RFLP-technique-based genetic linkage map showed a striking difference of the two maps in the relative distances between genetic markers: those markers that clustered in the middle of genetic map were physically located in the distal regions of both arms of the chromosome [Kota et al 1993]. The distance between two markers located on either side of the centromere of chromosome 1A was 8.4% of the length of genetic map and 56.6% of the physical length of this chromosome [van Deynze et al 1995]. The genetic distance between the centromere and the locus *Nor-B2* is only 0.3-2.2 cM, but two-thirds of the length of the short arm of the chromosome 6B [Dvořák & Appels 1986].

It was shown that 82% and 17% of recombination events occurred in two rather short regions of the distal part of the chromosome 1BS (called 1S0.8 and 1S0.5, respectively). These two regions together occupy only 14% of the physical length of the chromosome while only less than 1% of recombination events occur within 85% of the chromosomal length. The region 1S0.8 is shown to harbour many genes controlling agriculturally important traits [Sandhu et al 2001].

Physical distances (bp) between genes inherited together are large even in the chromosome regions of the high gene density, reaching 55-120 Kb per cM of genetic distance [Gill et al 1996; Gill & Friebe 2002].

The *Glu-3* loci are located distally (see **3.3.**) on chromosomes, in the zone of high recombination frequency. However, the ratio of physical to genetic distance al the *Glu-A3* locus was higher than 100 Kb per cM [Ibba et al 2017a]. The average distance between two neighbour B-type LMW-glutenin-encoding genes, inherited as a Mendelian trait, was 81 Kb, although in some cases this value was much higher (Huang & Cloutier 2008). At the *Glu-D3* locus, the estimated physical distances between two recombining markers ranged from 13 Kb to 515 Kb [Ibba et al 2017a]. No recombination was detected among three *i*-type LMW-glutenin-encoding genes within *Glu-A3* locus despite the fact that two of them were separated by more than 70 Kb. The average ratio of physical to genetic distance for the *Glu-D3* locus is probably higher than 100 Kb per cM. For example, the recombination was not found for two pairs of genes located within this locus at the distance of 16 Kb one from the other. It was suggested, that, in this case, the absence of recombination is likely to be related to the short physical distance between genes studied [Dong et al 2010].

Two transcribed genes inherited together of the locus *Glu-B1* are located close to centromere (see **4.1.**) and, therefore, in zone of rare recombination. The distance between them exceeds, in durum wheat, 150 Kb [Gu et al 2004a]. Physical distances between genes located in the zone of low recombination frequency reach the size of 22 000 Kb per 1 cM of genetic distance [Gill et al 1996].

In durum wheat, the neighbour LMW-glutenin genes were shown to be separated from each other by distances reaching 100 Kb and more, and the distance between two genes of 10 Kb (up to ten times longer than the size the gene itself) was considered to be very short and these genes to be clustered. The separation of the LMW-glutenin genes may be explained, in particular, through the activity of TEs. The gliadin-coding genes were generally more clustered, although the distance between two neighbour gliadin genes was not less than 10 Kb [Gao et al 2007].

In *T.tauschii*, the distances between the closest RFLP-based markers in the distal part of the chromosome 1D (that is, where the *Glu-D3/Gli-D1* complex locus is located) were 20-50 Kb per 0.9-1.7 cM for distal, and 270 Kb per cM for more proximal parts of the region studied. Highly recombinogenic regions were identified where the ratio of physical to genetic distance was estimated to be less than 20 Kb per cM [Spielmeyer et al 2000].

Genetic variation in common wheat correlates with the frequency of recombination, so that the number of alleles at distally-located loci are much higher than at loci situated more proximally on the same chromosome arm [Huang S et al 2002]. In six species of diploid wheat, the relationship was investigated between the level of polymorphism of 52 (in total) genetic markers, the location of each marker on the chromosome, and the frequency of recombination between markers. Each marker was represented by DNA sequence having a single copy in the genome. It was found that the more distally a marker was located, the more polymorphic it was and the more often it recombined with other markers. As a result, interspecies DNA divergence is most pronounced in the distal regions of chromosomes [Dvořák et al 1998a].

5.8. Summary. The unique structure of any allelic variant of any *Glu* or *Gli* locus

Each alternative (allelic) variant of the same *Gli* locus has a very complex structure and sus unrepeatable composition and design. One locus consists of distinct transcribed as well as silent sequences in different proportions, produces distinct types of gliadin polypeptides (for example, γ- and ω-gliadins). A big and variable set of diverse types of transposable elements (TEs) occupy up to 90% of the nucleotide sequences of any storage-protein-coding locus. Homoeologous (between genomes) and orthologous (between species) variants of the chromosomal region covering *Glu-1* loci differ in the composition of blocks of TEs. The DNA of the *Glu-3* loci in common wheat contain 30 distinct variants of TEs, each of them repeats in the genome from one to 24 times. TEs not only represent the bulk of the gnome's DNA, but are also the most dynamic, changeable fraction of it. TEs might have a crucial role in the differentiation of the three genomes (A, B, and D) and their subsequent transformations during wheat evolution from diploid to polyploid level.

Duplications of genetic material have occurred many times in the course of evolution in the chromosome 1DS harbouring LMW glutenin- and γ-gliadin-encoding sequences. There is an intervarietal polymorphism in the number of copies of some gliadin sequences.

The violations of collinearity of homoeologous and orthologous genes within *Glu* and *Gli* loci were reported. In addition, some alien transcribed genes were discovered within the *Glu-3*/*Gli-1* complex loci between genes encoding LMW glutenins, γ- and ω-gliadins.

The distance (in DNA bp) between neighbouring expressed genes inherited together within a *Glu-3*/*Gli-1* locus are so large (up to 100 Kb and more) that the distance of 10 Kb (up to ten times longer than the size the gene itself) is considered to be very short and these genes to be clustered. The gliadin-coding genes are generally more clustered than LMW-glutenin genes. α-Gliadin-encoding sequences are unevenly distributed along a 700-Kb region studied of durum so that clones of the size of 130 Kb might carry from one to four of such sequences. However, the genomic affiliation of the *Gli-2* clones studied almost never was identified. The homoeologous storage-protein-coding loci of common wheat differ in their size, the *Glu-D3* being significantly longer than *Glu-A3* or *Glu-B3*.

The calculated genetic distances among markers (genes) located proximally (close to the centromere) are small, while in the distal part of the chromosome arm, they are very large. The physical distances between storage protein genes reach 55-120 Kb per 1 cM of genetic distance. About 86% of transcribed sequences occupy only 10% of the total length of chromosome. Thus, 82% and 17% of recombination events occur in two rather short regions of the distal part of the chromosome 1BS that occupy only 14% of its physical length, and only less than 1% of recombination events occur within 85% of the chromosomal length.

6. MINOR *GLI* LOCI IN COMMON WHEAT

6.1. Some ω-gliadins (APAGE) are not encoded at the *Gli-1* or *Gli-2* loci

The general picture of the storage-protein-coding loci becomes even more complex when so-called minor *Gli* loci is considered.

The first data suggesting that one chromosome can carry two or more *Gli* loci, which recombine at a noticeable frequency, was obtained in analysis of the inheritance of 20 EP gliadin bands in the F2 progeny from the cross of common wheat cvs Mayo 54 x Joss. It was shown that one gliadin band in the ω-zone of the APAGE (designated N44) was controlled by two genes located on the short arm of the chromosome 1B and recombined with the frequency about 25% [Branlard 1983a; b]. This finding might be explained so that two ω-gliadins, encoded by recombining genes, had similar EM and might overlap composing the band N44 under the conditions of APAGE used. At that time, however, it was believed that there is no recombination between gliadin-encoding genes located on one chromosome. It was all the more surprising that one band of the APAGE electrophoregram is controlled by two genes [Payne et al 1984c].

Unambiguous data about recombination between two *Gli* loci on the same chromosome have come from Ukraine. As a result of analysis of F2 grains from the cross Gernot x Bez1, a new *Gli* locus was found, which recombined readily (31.25±3.17%) with the *Gli-A1*. Each parental cultivar had a gliadin band in the ω-zone of the starch acid gel which was not included in blocks encoded by the *Gli-1* or *Gli-A2* loci, and the ω-gliadin in Gernot had a lower EM than in Bez1 [Sobko 1984]. This short article published in a small local journal opened up a new field of research in the genetics of common wheat storage proteins and it has become classic. The additional (minor) locus on the chromosome 1A discovered by Sobko (1984) was later named *Gli-A3* [Payne et al 1988b].

A locus probably homoeologous to *Gli-A3* and named *Gli-B3* [Payne et al 1988b] was discovered in the study of the inheritance of gliadins in the substituted line ChS (Thatcher 1B). This *Gli* locus (authors' name *Gld-B6*) encoding an ω-gliadin (designated B30) was mapped at a distance of 28.1 cM from the *Gli-B1* locus [Galili & Feldman 1984]. In the catalogue of allelic variants of blocks encoded at the *Gli-B1* locus and identified in starch gel EP [Sozinov & Poperelya 1980; Sobko & Poperelya 1986], there are two blocks, named Gld 1B7 and Gld 1B2, differing only in the presence or absence, respectively, of one ω-gliadin encoded, as it is known now, by the *Gli-B3* locus.

A strong proof of the existence of minor *Gli* loci, recombining with the major *Gli-1* loci, on the chromosomes of the first homoeologous group was obtained in the analysis of common wheat cultivars having intra-varietal non-uniformity (multi-line cultivars consisting of two or more related genotypes, the origin of which can be traced back to the progeny from initial cross performed by the breeder [Metakovsky et al 2019c; 2020]). Using seed-by-seed analysis of a non-uniform cultivar, genotypes may be found that have identical alleles at each of the six major *Gli* loci but differ in presence or absence of a gliadin band in the ω-zone of the APAGE [Metakovsky et al 1986a]. Cultivars with this type of the intra-varietal non-uniformity occur in different countries, for example, in Bulgaria (cv Vega), Croatia (Irena), France (Pactole), Kazakhstan (Bogarnaya-56), UK (Torfrida), etc. [Alleles 2018c]. The Russian cvs Bezenchukskaya-98, and Saratovskaya-210 each consists of two genotypes differing in the presence or absence of one gliadin encoded at the *Gli-A3*, and *Gli-B3* loci, respectively [Metakovsky et al 1985b; 1996]. Four biotypes (genotypes) of the Kazakh cv Tselinogradka were identical in alleles of all six *Gli-1* and *Gli-2* loci, but differed from each other in the presence/absence of two ω-gliadins, one of them controlled by the *Gli-A3* and another by the *Gli-B3* loci [Metakovsky et al 1986a].

6.2. Locus *Gli-A3* and its allelic variants

The fact that the locus named *Gli-A3* is located on the chromosome 1A was confirmed in the analysis of the progenies from the cross Bez1 x ChS. In ChS, there is an ω-gliadin that is located on the chromosome 1A but not under control of the *Gli-A1* locus (see **2.2.**). In the APAGE electrophoregram of the cv Bez1, there is also one ω-gliadin which is not encoded by either *Gli-1* or *Gli-2*. ω-Gliadin in ChS has higher EM (APAGE) than that of Bez1. It was proved that genes encoding these ω-gliadins were allelic. Recombination frequency between the *Gli-A1* and *Gli-A3* loci was 0.36±0.04 [Metakovsky et al 1986a; Akhmedov & Metakovsky 1987].

In the crosses of Bez1 with cvs Gernot and Felix, the *Gli-A3* locus was mapped just between *Glu-A1* and *Gli-A1* loci: the recombination frequency (on average over two crosses) between *Glu-A1* and *Gli-A3* was

27.3±2.7%, and between *Gli-A3* and *Gli-A1*, 31.9±2.2% [Sobko et al 1986]. In other crosses, *Gli-A3* locus recombined with the *Glu-A1* and *Gli-A1* with the frequency of 31.5±3.1% and 25.6±2.7%; 38.6±6.1% and 20.3±4.6%; 33.8±2.5% and 19.6±1.7%; 38.1±6.4% and 34.9±6.1%, respectively [Metakovsky et al 1996] and about 28.7% and 30.9%, respectively (Payne & Metakovsky, 1986, unpublished). Also, to identify *Gli* loci, the RFLP approach and the specific probe pcP387 [van Deynze et al 1995] were used. As a result, the *Gli-A3* locus was mapped in the middle of the short arm of the chromosome 1A at the same distance (17-20 cM) from the centromere and the *Gli-A1* locus [Dubcovsky et al 1995]. It was suggested that polymorphism at the *Gli-A3* would be low and that the most frequent allele would be null [Payne et al 1988b].

The ω-gliadin encoded by the *Gli-A3* is located in the middle of the ω-zone of APAGE and, in some genotypes, may overlap with ω-gliadins encoded by other *Gli-1* loci. Therefore, comparison of the EM of ω-gliadins encoded by different alleles at the *Gli-A3* was performed using APAGE and 2D APAGE/SDS-EP. In total, about 15 cultivars and progenies of their crosses were studied with the aim to characterize EM and AMW of ω-gliadins produced by alleles at the *Gli-A3*. The intriguing result of the study is a close similarity (identity) of the AMW (about 41 kDa) of ω-gliadins producing by different alleles at the *Gli-A3* and having different EM [Metakovsky et al 1996]. On the contrary, among α- and some γ-gliadins, allelic variant of a polypeptide having higher EM usually shows lower AMW [Metakovsky et al 2018a].

At least four alleles were revealed, the amount is not so small for a locus encoding a single polypeptide. The allele *Gli-A3a* encoding the fastest (of the four) ω-gliadin was identified in common wheat cvs Безенчукская-98, ChS, Courtot, Prinqual and in biotypes of Tselinogradka. The allele *Gli-A3b* is present in Bez1 and its descendants. The slowest ω-gliadin is under control of the allele *Gli-A3c* and identified in cvs Anda, Felix, Gernot and, probably, Costantino. The null allele *Gli-A3d* (absence of the ω-gliadin with AMW of 41kDa encoded by the *Gli-A3*) is present in biotypes of Tselinogradka and Bezenchukskaya-98, in cvs Kharkovskaya-6, Richelle, and, probably, in Magnif-27 and Moulin [Metakovsky et al 1986a; 1996].

An unexpected interpretation was given to the results of the analysis of F2 progeny from the cross of common wheat cvs Ariana-8 and Cajeme-71. It was claimed that the *Gli-A3* locus encodes the synthesis of a minor γ-gliadin while the distance between the *Gli-A3* and *Gli-A1* loci was calculated to be 21.03±3.59% [Nieto-Taladriz & Carrillo 1996]. Obviously, the reason that the polypeptide encoded by the *Gli-A3* locus was named "γ-gliadin" is that its inheritance was studied using SDS-EP and its AMW was found to be too small for ω-gliadin. However, it has been firmly established that any polypeptide encoded at the *Gli-A3* locus is located in the middle of the ω-zone of the APAGE although it has unusually low (for ω-gliadins, see **3.1.**) AMW of 41 kD [Metakovsky et al 1996].

Locus *Gli-A3* (allele *Gli-A3a)* was discovered in hexaploid wheat *T.macha*. In crosses between cv Bez1 and genotypes of *T.macha*, the frequency of recombination between the *Gli-A1* and *Gli-A3* loci was 0.20±0.04 when the variety *T.macha* var. *ericzjanae* was involved in the cross [Metakovsky & Iakobashvili 1990], and 0.24±0.05 when var. *subletshchumicum* was used (Iakobashvili & Metakovsky, unpublished).

In durum wheat, the locus orthologous to the *Gli-A3* of hexaploid wheat, encodes the synthesis of one ω-gliadin and was mapped at a distance 14.0±4.6 cM from the *Gli-A1* [Ruiz & Carrillo 1993].

Allelic variation at the minor *Gli-A3* locus associates, in common wheat, with good quality (SDS-st) [Poperelya & Gasanova 1980] (see **4.7.**).

6.3. Locus "*Gli-A4*" and heterohomologous chromosomes

The existence of the *Gli-A4* locus was justified by the fact that this locus, found in the DM-111 genotype, recombined with the *Gli-A1* with the frequency of 10.0±2.4% [Redaelli et al 1992] while the usual distance between the *Gli-A1* and *Gli-A3* loci was, in several different crosses, 20-40% (see **6.2.**). In the cv Bezenchukskaya-98, the locus encoding the ω-gliadin which was not a part of any block encoded by major *Gli* loci and recombined with the *Gli-A1* locus with comparatively low frequency, about 13% [Metakovsky et al 1986a]. Thus, it was firstly assumed that the new locus *Gli-A4* was located on the chromosome 1A of common wheat between *Gli-A3* and *Gli-A1* [Redaelli et al 1992; Pogna et al 1994]. Although this conclusion was later refuted [Metakovsky et al 1996; 2006], it is interesting to discuss the problem of differences in recombination frequency between two genetic markers observed in different crosses. Indeed, the assumption about the existence of the *Gli-A4* was erroneously based only on its lower frequency of recombination with the *Gli-A1* locus as compared with that between *Gli-A3* and *Gli-A1* loci.

It is known that the frequency of recombination between two genetic markers can be influenced by the external conditions under which the recombination takes place. For example, recombination between telomeric end of the 6BL chromosome arm and the gene *b2* located in the centromeric region of the 6BL varied from 45% to 18% when plants were grown at the 21°C, and 32 °C, respectively [Fu & Sears 1973]. Frequency of recombination between genes encoding α-amylase on the chromosome 6B was found to be 20.6±2.2% [Nishikawa et al 1981] or 4.08±0.96% [Ilichevsky et al 1989]. In the same cross Bez1 x ChS, recombination frequency between *Gli-A1* and *Gli-A3* was 35.3±3.7%, and 25.3±4.4% in F2 seeds, and F2 plants, respectively [Metakovsky et al 1986a].

Moreover, in self-pollinated plants, the frequency of recombination can be influenced by an interesting phenomenon, the divergence of homologous chromosomes. It was shown that homologous chromosomes may differ considerably, among common wheat genotypes, in heterochromatin banding which may be revealed using differential chromosome staining. For example, the karyotypes of cvs ChS and Diamant differed drastically in the morphology of chromosomes, as well as in both heterochromatin content and its localization on the chromosomes [Zurabishvili et al 1979]. Only four (3A, 5A, 2D, 7D) of 17 homologous chromosomes studied were found to be identical in all five cultivars studied, ChS, Cheyenne, Hope, Timstein and Wichita. The highest level of polymorphism was observed between the chromosomes of the genome B [Endo & Gill 1984]. Comparison of 35 genotypes revealed polymorphism for almost all 21 common wheat chromosomes. For example, at least four morphological variants of the chromosome 1A were described, differing in the presence of telomeric heterochromatin in each of the arms and in the size of the pricentromeric heterochromatin [Friebe & Gill 1994].

Divergence of homologous chromosomes in wheat may be caused, for example, by homologous gene conversion [Morrell et al 2006], due to activity of TE (see **5.2.**), and by translocations [Perera et al 1983; Lange et al 1987; Friebe & Gill 1994; Badaeva et al 2007].

In F1 hybrids from crosses between common wheat cultivars, the pairing of homologous chromosomes belonging to the genome A and, especially, the genome B, and the formation of chiasmata during meiosis are significantly reduced in comparison with parental forms, where true homologues (two identical chromosomes) are mated [Dvořák & McGuire 1981; Dvořák & Chen 1984; Dubcovsky et al 1995]. Homologous parental chromosomes present in a hybrid, are called "heterohomologous" to distinguish them from a pair of true homologous chromosomes in the parental genomes [Dvořák & McGuire 1981; Crossway & Dvořák 1984]. Divergence of heterohomologous chromosomes, which manifests itself, in particular, in the arrangement of heterochromatin regions, reduces the pairing of diverged homologues in inter-varietal F1 hybrids [Endo & Gill

1984] and decrease the recombination frequency between genetic markers on the heterohomologous chromosomes. The reduction of pairing occurs evenly along the entire chromosome [Dubcovsky et al 1995], but it is especially pronounced in the short arms of the chromosomes 1B and 6B, where it decreases by 40% [Dvořák & McGuire 1981]. The portion of the chromosome 6B, that does not recombine in the hybrid of common wheat cultivars, accounts for 82% of the entire length of this chromosome [Dvořák & Chen 1984] (see **5.7.**). Decreasing pairing of chromosomes 1B and suppression of recombination between markers located on these chromosome was observed in hybrides from crossing two tetraploid wheats, *T.durum* and *T.dicoccoides* [Dubcovsky et al 1997].

The difference between heterohomologous chromosomes is due to dissimilarity in their noncoding portion of nucleotide sequence [Crossway & Dvořák 1984]. This dissimilarity is represented by different sets of repeated sequences, mobile elements, deletions/insertions of DNA sequences [Dvořák et al 2004] and by single nucleotide substitutions (SNS) [Ravel et al 2006] in two homologous but distinct chromosomes. The rate of the DNA changes (DNA divergence) in non-transcribed regions is high [Wicker et al 2001; Isidore et al 2005], three-four times higher than in transcribed genes or their introns [Dvořák et al 2006].

Orthologous chromosomes differ more strongly than heterohomologous chromosomes in the same species [Isidore et al 2005; Dubcovsky et al 1995; 1997]. Even more obvious differentiation was observed between homoeologous chromosomes of common wheat. For example, in the cv Renan, the distance between genes encoding x- and y-subunits of the HMW glutenin in its *Glu-A1*, *Glu-B1*, and *Glu-D1* loci was 220 Kb, 150 Kb, and 50 Kb, respectively, and DNA sequence of the intergenic regions had little in common between homoeologous [Gu et al 2004a; 2006] (see **5.2.**). Thus, heterohomologous, for example, chromosome 1B in different common wheat cultivars, in terms of the degree of differentiation are intermediate between true homologous (two identical chromosomes in a diploid genotype) and orthologous chromosomes.

The degree of differentiation (no-identity) between heterohomologous chromosomes may differ among common wheat genotypes (cultivars). Therefore, the recombination frequency between two genetic markers in a given cross may depend on the degree of differentiation that exists between the heterohomologous chromosomes in the genotypes used for crossing. Indeed, the distance observed between the *Nor-B2* and *Gli-B2* loci was 2.0±1.6 cM, and 10.2±4.3 cM when it was measured in the crosses of ChS with Timstein and Cheyenne, respectively [Dvořák & Chen 1984; Dvořák & Appels 1986]. The distance between *Gli-A1* and *Gli-A3* loci varied from 7 cM to 27 cM [Dubcovsky et al 1997]. At the same time, in four identical, but independent crosses between two Kazakh cultivars, the frequency of recombination between *Gli-A1* and *Gli-A3* were very similar: 19.9%, 19.5%, 18.5%, 21.1% [Metakovsky et al 1996].

The assumption about the existence of the *Gli-A4* locus was made as a result of the analysis of F2 progeny from the cross DM-111 x Perzivan-1. The DM-111 is the F8 line from the cross of two null mutants of Italian cvs San Pastore (*Gli-B1 null*) and Alpe (*Gli-D1 null*). Therefore, the DM-111 lacked all gliadins encoded by the *Gli-B1* and *Gli-D1* loci [Redaelli et al 1992]. On another hand, the genotype of the cv Perzivan-1 includes the translocation of the *Gli-D1g* locus on the chromosome 1A [Akhmedov et al 1988; Metakovsky et al 1990a]. Therefore, its chromosome 1A carrying an additional genetic material differed considerably from ordinary chromosome 1A.

It is known that chromosomal rearrangements strongly affect the frequency of crossing over [Lange et al 1987]. Therefore, it is reasonable to assume that the unusualness of the chromosome 1A of the cv Perzivan-1 could affect the frequency of recombination between two genetic loci studied (*Gli-A1* and *Gli-A3*) located on this chromosome. Moreover, the ω-gliadin encoded by the assumed *Gli-A4* locus was proved to be identical in its EM and AMW with ω-gliadin controlled by the *Gli-A3b* allelic variant of the *Gli-A3* locus [Metakovsky et al 1996]. This is not surprising: the cv Bez1 possessing the *Gli-A3b* allele (see **6.2.**) is one of parents of the cv Alpe and therefore the grandmother of the DM-111 line. It can be firmly concluded that ω-gliadin, presumably controlled by the *Gli-A4* locus in the DM-111, is actually encoded by the allele *Gli-A3b*, obtained by this line from Bez1.

The frequency of recombination between the two markers differed significantly when analysing the progeny from crosses between different inbred rye lines. For example, the recombination between *Glu-R1* (analogue of the *Glu-A1*) and *Gli-R3* (analogue of the *Gli-A3*) was, in two different test-crosses, about 33.8±3.2% and 13.7±2.9%, and between *Gli-R3* and *Gli-R1* (analogue of the *Gli-A1*), about 12.0±2.2% and 7.5±2.2%, respectively [Carrillo et al 1992].

Thus, to identify a new locus, the recombination frequency cannot be used as the only criterion; it is necessary to check the allelism of this locus with other known loci on the same chromosome. So, it seems controversial to say that durum wheat, like common wheat, has minor *Gli* loci on the chromosome 1A, including the *Gli-A4* [Autran et al 1995]. Unfortunately, the authors did not present any evidence to that statement. On the contrary, no evidence was found, using the RFLP approach and specific probes, for the existence of the *Gli-A4* locus (between the *Gli-A3* and *Gli-A1*) neither in common wheat, nor in durum. It was concluded that this locus is probably synonymous with *Gli-A3* [Dubcovsky et al 1997].

6.4. Locus *Gli-B3*/*Glu-B2* and its variants in common and durum wheat

The *Gli-B3* locus, being mapped just between the *Glu-B1* and *Gli-B1* loci is obviously homoeologous to the *Gli-A3* locus of common wheat. There are two alleles, null (*Glu-B3a*), and *Gli-B3b*, which occurs on the background of various alleles at the *Gli-B1* locus and encodes a polypeptide (AMW of about 50 kDa) located in the central part of the ω-zone of the APAGE electrophoregram. In cv Neepawa, this polypeptide has all characteristics of a typical ω-gliadin [Dachkevitch et al 1993]. Being firstly found on the chromosome 1B of the cv Thatcher, the *Gli-B3b* was located on the distance of 28.1 cM from the *Gli-B1* (allele *Gli-B1d*) [Galili & Feldman 1984]. Later, the distance between the *Gli-B3* and *Gli-B1* loci was studied in several crosses and was found to be 21.7±2.2% (Bogarnaya-56, *Gli-B1q*) [Seitova et al 1986], 29±7% (Bezenchukskaya-98, *Gli-B1q*) [Metakovsky et al 1986b], 22.1±3.6 cM (Neepawa, *Gli-B1d*) [Dachkevitch et al 1993], and 23.7±2.8% (Tselinogradka, *Gli-B1e*) [Metakovsky et al 1997b]. It was proved that the polypeptide produced by the allele *Gli-B3b* is typical gliadin [Dachkevitch et al 1993; Metakovsky et al 1997b].

The *Gli-B3* locus encoding a typical ω-gliadin was found to present in *T.macha var. subletshchumicum*. This locus was mapped in the cross between this variety of *T.macha* and cv Bez1. Thus, the recombination frequency between the *Glu-B1* and *Gli-B3* loci was 0.18±0.05 and between the *Gli-B3* and *Gli-B1* loci was 0.27±0.06 (Metakovsky & Iakobashvili 1990).

Some of variants of the *Gli-B3* locus, however, encode the synthesis polypeptide having characteristics of glutenins. The locus producing the LMW glutenins of the D-type (see **3.2.**) was firstly discovered as a result of the analysis of the cross ChS x Sicco. Alleles of this locus in these cultivars encode D-type glutenins differing in their isoelectric point. Therefore, it was possible to map the new glutenin-coding locus between the *Glu-B1* and *Gli-B1* loci at the distance of 16.1±4.7% and 21.0±4.7% from them, respectively. This locus was named *Glu-B2* [Jackson et al 1985]. Later, it has been recognized that the *Gli-B3* and the *Glu-B2* are the same locus [Payne et al 1988b]. The D-type LMW glutenin encoded at the *Gli-B3* in common wheat is probably a glutenin-like ω-gliadin originated as a result of spontaneous mutation (see **4.11.**).

The peculiarity of the *Gli-B3* is that there are indications that some allelic variants of this locus may produce distinct types of glutenin molecules. Thus, in the common wheat cv Kharkovskaya-6, the *Gli-B3*, recombining with the *Gli-B1* with the frequency of 24.8±4.5%, encoded a glutenin polypeptide with AMW of 37,5 kDa which is too short to be a D-type LMW glutenin. It cannot be excluded that this molecule is a glutenin of the C-type, or even B-type [Metakovsky et al 1997b]. Therefore, in hexaploid wheat, the *Gli-B3* may produce, in contrast to *Gli-A3*, molecules of the D-type LMW glutenins or rather short glutenin-like polypeptide.

Similar (rather short) glutenin molecule is encoded by the locus *Glu-B2* located at the distance of 17 cM from *Gli-B1* in durum wheat [Nieto-Taladriz et al 1997]. The use of the RFLP approach (the pTdUCD1 probe specific for the B-type-LMW-glutenin-encoding DNA sequences) for the analysis of the progeny of the cross of durum wheat genotypes made it possible to prove the presence of the B-type LMW glutenin DNA sequences in the *Gli-B3*/*Glu-B2* locus of durum wheat [Dubcovsky et al 1997]. Therefore, allelic variants of the *Gli-B3*/*Glu-B2* locus, which recombines with the *Gli-1* locus with a frequency of about 20%, may encode (in durum, unlike common wheat) either ω-gliadin, or the B-type LMW glutenin [Ruiz & Carrillo 1993; Liu 1995, Liu & Shepherd 1995] (see **8.4.**). It was suggested that the *Gli-B3*/*Glu-B2* complex locus may consist, in durum wheat, of the two parts, so that, in some genotypes, its "gliadin part" is expressed, and in others, the "glutenin part" (Ruiz & Carrillo 1993). Nothing is known if both parts of this complex locus, if they exist, may express simultaneously. No influence of the *Gli-B3* locus on the durum wheat quality was reported [Ruiz & Carrillo 1996].

Probably, distinct variants of the *Gli-B3*/*Glu-B2* locus in polyploid wheat might originate from different genotypes of a diploid donor(s) of the genome B (see **8.4.**).

6.5. Locus *Gli-D3*

The storage-protein-coding locus called *Gli-DT1* was mapped between the *Glu-D1* and *Gli-D1* loci on the short arm of the chromosome 1D in diploid wheat *T.tauschii*. The polypeptide encoded by *Gli-DT1* was named T1 and found to be an ω-gliadin with unusually high AMW of about 95 kDa. To map the *Gli-DT1*, F2 grains were studied of the cross between common wheat cv Egret and an artificial hexaploid obtained from the cross between a known durum cultivar and the accession AUS18913 of *T.tauschii*. Thus, the map distance between the *Gli-DT1* and *Glu-D1*, and between *Gli-DT1* and *Gli-D1* loci was calculated as 40.20 cM and 13.18 cM, respectively. Based on these genetic distances, it was assumed that the locus *Gli-DT1* is homoeologous to the locus *Gli-A4* of common wheat. [Gianibelli et al 2002b]. Considering the fact that the *Gli-A4* is actually the *Gli-A3* locus (see **6.3.**), it can be tentatively accepted that the *Gli-DT1* locus of *T.tauschii* is homoeologous to the *Gli-A3* and *Gli-B3* loci of common wheat.

In other work, F2 individuals from the cross between two genotypes of *T.tauschii* were studied to create, using RFLP technique, a detailed genetic linkage map of the chromosome 1DS. The DNA probes specific for coding regions of γ-gliadin, and LMW glutenin genes were used to detect the *Gli-D1*, and *Glu-D3* loci, respectively. No gliadin-encoding locus possibly homoeologous to the *Gli-A3* and *Gli-B3* of common wheat was found between *Glu-D1* and *Glu-D3*/*Gli-D1* loci of *T.tauschii* [Spielmeyer et al 2000]. However, no probe specifically recognizing DNA sequences for ω-gliadins were used in this work.

Results of analysis of 543 F2 seeds of the cross of cvs Ablaca and Cajeme-71 were interpreted as a discovery of a new (*Gli-D4*) locus on the chromosome 1D in common wheat. It was assumed that this new locus encoded the fast-moving γ-gliadin (γ-53.5) of the cv Cajeme-71 and was mapped proximally from the *Gli-D1*, between the *Glu-D1* and *Gli-D1* loci [Rodriguez-Quijano & Carrillo 1996a]. Taking into account the genetic distance between *Gli-D4* and *Gli-D1* loci registered in this work (10.1±2.4 cM), it can be suggested that *Gli-D4* is probably homoeologous to the *Gli-A3* and *Gli-B3* (see **6.3.**). However, there are some inconsistencies of these results and some other data on analysis of minor *Gli* loci in common wheat.

First, the *Gli-A3* (see **6.2.**) and *Gli-B3* (see **6.4.**) loci control, in all genotypes of common wheat and other wheat species studied, the synthesis of ω-gliadins located in the middle part of the ω-zone of APAGE electrophoregram while the *Gli-D4* locus of Rodriguez-Quijano & Carrillo (1996a) encoded fast-moving γ-gliadin. Second, using the probe pcP387 that recognizes DNA sequences encoding γ-gliadin [van Deynze et al 1995], only one locus (*Gli-D1*), located distally on the chromosome arm 1DS of common wheat was identified. No loci located between the centromere and the *Gli-D1* (like *Gli-D4*) were found [Dubcovsky et al 1997]. In addition, the frequency of recombination between genes encoding fast-moving γ-gliadin and other genes of the *Gli-D1* locus did not exceed 1% [Metakovsky 1990].

Although some explanations for the discrepancies cannot be excluded (for example, inter-varietal polymorphism of the locus *Gli-D1*), the existence, on the chromosome 1D of common wheat, of the *Gli* locus homoeologous to *Gli-A3* and *Gli-B3* cannot be considered, to our opinion, to be finally proved.

6.6. Confusion in the nomenclature of alleles at the *Gli-3* loci

The homoeologous loci *Gli-A3* (see **6.2.**) and *Gli-B3* (see **6.4.**) were mapped between the HMW-glutenin-coding locus *Glu-1* and the complex locus *Glu-3*/*Gli-1* on the short arms of chromosomes 1A and 1B, respectively, in common wheat [Galili & Feldman 1984; Sobko 1984; Sobko et al 1986; Metakovsky et al 1986a; 1996; 1997b; Seitova et al 1986; Dachkevitch et al 1993; Dubcovsky et al 1995] as well as in other wheat species studied [Metakovsky & Iakobashvili 1990; Ruiz & Carrillo 1993; Dubcovsky et al 1997; Nieto-Taladriz et al 1997]. The recombination frequency between *Glu-A1* and *Gli-A3* varied, in different crosses of common wheat, within 27-39% and between *Gli-A3* and *Gli-A1* within 10-35% depending on the genotypes involved in the cross (see **6.3.**). The distance between *Glu-B1* and *Glu-B3* was about 20%, and between *Gli-B3* and *Gli-B1* within 21-30%.

Nevertheless, it was claimed in some publications that there is "...the genetic evidence prior to this study that indicated *Gli-1*, *Gli-3*, and *Glu-3* loci were tightly linked" [Gao et al 2007, p.179] and that "The *Gli-3* locus ... is tightly linked to the *Glu-3* locus" [Anderson et al 2009, p.398]. In these otherwise brilliant studies, the linkage between the *Gli-3* and *Gli-1* loci was not studied. Therefore, to our opinion, it could be only a problem of nomenclature of the *Gli* loci. It can be suggested that, in these articles [Gao et al 2007; Anderson et al 2009], the loci *Gli-1* was erroneously called "*Gli-3*".

There are some results presented in these publications indicating that, in fact, the complex locus *Glu-3*/*Gli-1* (not *Gli-3*) has been studied. For example, the sequence studied harbours a gene encoding a long ω-gliadin called ω5. This ω-gliadin-encoding gene is an indispensable part of the allele *Gli-B1b* (see **2.5.**). In the same sequence studied, there are, besides genes for ω-gliadins, also genes encoding γ-gliadins and LMW-glutenins [Gao et al 2007; Anderson et al 2009]. The presence of tightly linked genes encoding LMW-glutenins, ω- and γ-gliadins is an undeniable characteristic of the complex loci *Glu-3*/*Gli-1* (see **3.3.**).

On the other hand, the designation *Gli-A3* and *Gli-B3* [Payne et al 1988b] has long been used in genetic studies on gliadin [Singh & Shepherd 1988a; Dubcovsky et al 1995; 1997; Metakovsky et al 1996;

etc.]. Moreover, the existence, in common wheat, of the locus *Gli-D3* mentioned by Anderson et al (2009) has not still proved (see **6.5.**).

6.7. The minor *Gli-A5* locus in wheat

The loci *Hg-1* and *Gli-A1* are linked. In the cross Bez1 (*Gli-A1b*, *hg-1*) x Rimpau-WW (*Gli-A1l*, *Hg-1*), the frequency of recombination between *Gli-A1* and *Hg-1* was 3.88±1.00% [Sobko & Poperelya 1982] while in the cross between cv Little-Club and the line Prelude-Lr10 was 3.95±1.38% [Howes 1986].

In winter durum wheat, *Gli-A1* and *Hg-1* recombined with the frequency (in three crosses) of 2.79±0.84% [Panin & Netsvetaev 1986].

The minor *Gli-A5* locus was identified as a result of analysis of progeny from six different crosses (in total, 1355 F2 seeds), each of which involved the cv Salmone (*Gli-A1l*, *Hg-1*). The distance between the *Hg-1* and minor *Gli-A5* loci varied in six different crosses from 0.0% (one cross) up to 5.5±2.0% while the distance between the *Gli-A1* and *Gli-A5* loci was, in this study, less than 1% [Pogna et al 1993; 1994]. These data situate the *Gli-A5* between *Gli-A1* (proximal) and *Hg-1* (distal), closer to the *Gli-A1*.

There are two known alleles at the *Gli-A5* locus, *Gli-A5a* (null) and *Gli-A5b* encoding two faint ω-gliadins of relatively low EM [Pogna et al 1993] and AMW of about 46 kDa (closer to the top of gel) and 47 kDa [Metakovsky et al 1984b]. The two bands encoded by the *Gli-A5b* are included in some *Gli-A1*-encoded blocks, for example, *Gli-A1m* (see **2.5.**).

It was observed that allele *Hg-1* occurs in common wheat cultivars almost exclusively in combination with the allele *Gli-A1l* or *Gli-A1p* (genotypes *Gli-A1l*, *Gli-A5b*, *Hg-1* and *Gli-A1p*, *Gli-A5b*, *Hg-1* [Koval et al 1986]. The isogenic lines (nine backcrosses) created on the background of the cv Novosibirskaya-67 (*Gli-A1f*, *hg-1*) carried in their genotypes both *Gli-A1p* and *Hg-1* obtained from donors Pyrothrix-28 or к-20551 [Koval et al 1988]. On the contrary, the combination *Gli-A5b*, *hg-1* occurs in the presence of some other alleles at the locus *Gli-A1*. For example, the *Gli-A5b*-encoded ω-gliadins are produced by the genotype of cv Marquis (*Gli-A1m*, *Gli-A5b*, *hg-1*).

There is an indication that a locus orthologous to *Gli-A5* of common wheat is present in *T.monococcum* [Metakovsky & Baboev 1992a]. A high level of homology of the distal part of the chromosome arm 1AS was established between *T.monococcum* and *T.aestivum* [Kuspira et al 1989].

6.8. Locus *Gli-A6* and the order of *Gli* loci on the 1AS chromosome arm

As a result of the analysis of the set of gliadins (SDS-EP) produced in different dihaploid lines obtained from the cross of cvs Courtot (allele *Gli-A1k*) and ChS, a recombination was observed (1.94±1.00%) between γ- and ω-gliadin-encoding genes of the *Gli-A1* locus of Courtot. The locus encoding the ω-gliadin was firstly called *Gli-A5* [Felix et al 1996].

Block encoded by the locus *Gli-A1f* is composed of one ω-gliadin, one γ-gliadin, and one band in the β-zone of the APAGE. In the cross of cvs Bez1 (*Gli-A1b*) and Anda (*Gli-A1f*), it was found that gene encoding ω-gliadin recombines with other genes of the locus, and with *Gli-A3* locus, with the frequency of 4.1±1.2%, and 26.1±3.7%, respectively [Metakovsky et al 1996]. It was established that the ω-gliadins produced by alleles *Gli-A1k* and *Gli-A1f* are identical, and the corresponding locus was named *Gli-A6*.

There are three known alleles at the *Gli-A6* locus. The variety ChS and most of cultivars studied have allele *Gli-A6a* (null). Allele *Gli-A6c* encoding ω-gliadin is present in cvs Courtot (*Gli-A1k*), Anda (*Gli-A1f*), Prinqual (*Gli-A1f*) and many others [Metakovsky et al 1996].

The comparison of sets of gliadins produced in common wheat cultivars by means of the APAGE/SDS-EP confirmed that genotype of cv Saratovskaya-29 (*Gli-A1f*) includes the allele *Gli-A6c* and that this allele encodes the ω-gliadin with the AMW of about 40 kDa [Redaelli et al 1994, Fig. 2 (spot 45); Payne & Metakovsky, unpublished]. On the other hand, the genotype of cvs Bezenchukskaya-98 (*Gli-A1m*) and Neepawa (*Gli-A1m*) include the allele *Gli-A6b* encoding the ω-gliadin having higher EM and higher AMW (of about 42.5 kDa) in comparison with that produced by the *Gli-A6c* [Dachkevitch et al 1993, Fig 1A, spot N15].

The rare allele *Gli-A1y* (cv Mironovskaya-Yubileinaya) encodes a block composed of all bands encoded by alleles *Gli-A1b* and *Gli-A6c* [Blocks 2018c]. This cultivar originated from the cross between Bezostaya-4 (*Gli-A1b*, *Gli-A6a*) and Mironovskaya-808 (*Gli-A1f*, *Gli-A6c*), and the allele *Gli-A1y* might appear as a result of recombination between *Gli-A1b* and *Gli-A6c*. Two similar recombinant genotypes were registered in analysis of 102 F2 seeds of the cross Prinqual (*Gli-A1f*, *Gli-A6c*) x Moulin (*Gli-A1ab*, *Gli-A6a*) [Metakovsky et al 1996].

The allelism between *Gli-A6c* and *Gli-A6b* was established in the cross between cvs Bezenchukskaya-98 (*Gli-A1m*; *Gli-A5b; Gli-A6b*) and Saratovskaya-210 (*Gli-A1f*; *Gli-A5a*; *Gli-A6c*) [Metakovsky et al 1986a]. The results indicate that, in this cross, the *Gli-A1* and *Gli-A5*, and the *Gli-*

A1 and *Gli-A6* loci recombined with the frequency of about 1%, and 5±1%, respectively. Moreover, both *Gli-A5* and *Gli-A6* loci were mapped distally from the *Gli-A1* [Metakovsky et al 1986a]. Therefore, the chromosome 1A carries, in addition to other genes, at least six storage-protein-encoding loci situated in the order: *Glu-1*- Centromere - *Gli-A3* - *Glu-A3*/*Gli-A1* - *Gli-A5* - *Gli-A6* - *Hg-1* [Metakovsky et al 1986a; b;1996].

The existence of two closely linked *Gli* loci (at the distance of 1.2cM) distally on the short arm of the chromosome 1A was proved using the RFLP technique. One of these two loci is obviously the *Gli-A1*, but it remains unclear whether the second locus is the *Gli-A5* or *Gli-A6* [Dubcovsky et al 1997].

6.9. Locus *Gli-B5* and two families of alleles at the *Gli-B1*

The group of EP (APAGE) bands encoded by the allele *Gli-B1s* includes a pair of faint ω-gliadins. The locus *Gli-B5* was firstly identified in the analysis of the progeny from six crosses of the red-spike cv Salmone having this allele. It was discovered that this allelic variant of the *Gli-B1* locus consists of two parts that can recombine with each other at a low frequency. The part producing the pair of ω-gliadins was named *Gli-B5*. Another part of the locus encodes the rest of bands of the block. The distance between these two parts of the *Gli-B1s* is 1.4 cM [Pogna et al 1993]. Later, it was proved that the *Gli-B5b* controls a characteristic pair of ω-gliadins present in several *Gli-B1*-encoded blocks. The distance between the *Gli-B1* and *Gli-B5* loci was 1.5±0.6% [Pogna et al 1993], or 3.5±1.4% [Nieto-Taladriz et al 1998], when cvs Alpe (*Gli-B1k*, *Gli-B5a*), or Prinqual (*Gli-B1c*, *Gli-B5b*) were involved in crosses, respectively.

The *Gli-B5* locus is obviously homoeologous to *Gli-A5*. Similar to the *Gli-A5*, the *Gli-B5* has two alleles, one of them is null (*Gli-B5a*), and another (*Gli-B5b*) encodes a pair of faint slow-moving ω-gliadins. Using 2D EP, the AMW of these ω-gliadins was determined to be about 47.5 kDa and 48.5 kDa [Metakovsky et al 1997b]. Thus, similar to *Gli-A5*, the faster (APAGE) polypeptide of the pair encoded by *Gli-B5* has a higher AMW. ω-Gliadins encoded by the *Gli-B5b* have higher AMW and slightly higher EM in comparison with gliadins encoded by the *Gli-A5b* [Pogna et al 1993].

The pair of ω-gliadins encoded by *Gli-B5b* is accepted to be a part of blocks produced by many alleles (10 of the 25 mentioned in the catalogue) at the *Gli-B1* locus [Blocks 2018c] (see **2.5.**). Among these 10 alleles, the *Gli-B1c*, *Gli-B1i*, *Gli-B1k*, *Gli-B1m* are the most frequent in common wheat germplasm [Alleles 2018c]. On the other hand, genotypes having the *Gli-B1* alleles which are the most frequent in the common wheat germplasm studied (for example, *Gli-B1b*, *Gli-B1d*, *Gli-B1e*, *Gli-B1f*, *Gli-B1g*) carried allele *Gli-B5a*. Therefore, there are two contrast families of alleles at the *Gli-B1*, first, linked to *Gli-B5b*, and the second, linked to *Gli-B5a*. Each of these two contrast groups of alleles (red and white) is composed of 3-4 families of alleles differing considerably from each other being analysed by the RFLP and other techniques of analysis of the DNA sequences (see **7.5.**).

It should be remarked that, in early publications, the term "*Gli-B5b*" was erroneously used to designate null allele [Pogna et al 1993, 1995; Metakovsky et al 1994], although null is present in the variety ChS and therefore must be named *Gli-B5a*.

A locus orthologous to *Gli-B5* probably also exists in durum wheat (Kudryavtsev et al 1996).

6.10. Families of "red-" and "white-spike" alleles at the *Gli-B1* locus

The locus *Rg-1* providing a red colour to wheat spike is located distally to the *Gli-B1* on the chromosome arm 1BS [McIntosh & Cusick 1987]. The genetic distance between *Gli-B1* and *Rg-1* was about 1.06±0.45% [Poperelya et al 1980] or 1.8±0.8% [Payne et al 1986a]. The distance between the *Gli-B5* and *Rg-1* varied, in six crosses, from 0.0% (three crosses) up to 4.8±1.4% [Pogna et al 1993]. Isogenic lines were produced on the background of a white-spike cv Novosibirskaya-67 (*Gli-B1e*, rg-*1*) by nine backcrosses with selection for red colour of spike. These isogenic lines acquired (as it was aimed) red spikes (*Rg-1*), but in addition alleles at *Gli-B1* (+*Gli-B5b*) from red-spike donors used. So, nine backcrosses were not enough to separate the *Gli-B1* (+*Gli-B5*) from the *Rg-1* locus [Koval et al 1988].

In winter durum wheat, the frequency of recombination between *Gli-B1* and *Rg-1* loci was 5.44±1.71% (Panin & Netsvetaev 1986).

It was noticed that some alleles at the *Gli-B1* (*Gli-B1c, Gli-B1i, Gli-B1k, Gli-B1m*, etc.) occur almost always together with the *Gli-B5b* and red colour of spikes (see **6.9.**). Therefore, these alleles may be called "red alleles" at the *Gli-B1*. In contrast, alleles *Gli-B1b, Gli-B1d*, *Gli-B1e, Gli-B1f,* etc. may be called "white alleles" at the *Gli-B1*.

Indeed, seventy white-spike common wheat genotypes of different origin were studied and only one of them had allelic combination *Gli-B1k* (red allele) + *Gli-B5b*. Also, only one of 30 red-spike genotypes (represented by four botanical varieties) had alleles *Gli-B1b* (white allele)

+ *Gli-B5a* [Koval et al 1986]. Analogously, seven and one of 38 Australian cultivars studied had *Gli-B5b*, red spike and "red alleles" *Gli-B1i* and *Gli-B1c*, respectively. Each of other 30 cultivars studied had *Gli-B5a*, white spike and "white allele" of the *Gli-B1* (mainly *Gli-B1b*) [Metakovsky et al 1990b].

Rare combinations between *Gli-B1*, *Gli-B5* and *Rg-1* loci, observed in some registered cultivars of common wheat, made it possible to establish the order of these genetic loci on the chromosome 1BS.

In Italian germplasm, white-spike (*rg-1*) cvs Pandas and Libra had a "truncated" version (lacking the *Gli-B5b*-encoded bands) of the block produced by red-allele *Gli-B1k*. On the contrary, in genotype of the red-spike cv Este, there was a combination of the white allele *Gli-B1g, Gli-B5b* and *Rg-1*. So, allelic combinations observed in cvs Pandas, Libra and Este might appear as a result of recombination between the *Gli-B1* and the pair *Gli-B5* + *Rg-1*. The combination of red-allele *Gli-B1k* + *Gli-B5b* + *rg-1* observed in cvs N. Strampelli, Costantino and some their descendants indicate the recombination between the pair of loci *Gli-B1* + *Gli-B5* and the *Rg-1* locus [Metakovsky et al 1994].

In Spanish wheat germplasm, 16, and 39, cultivars had typical contrast combinations red- allele *Gli-B1* + *Gli-B5b* + *Rg-1*, and white-allele *Gli-B1* + *Gli-B5a* + *rg-1*, respectively. However, there were a few recombinant combinations, for example, *Gli-B1*(white) + *Gli-B5a* + *Rg-1*, or *Gli-B1*(red) + *Gli-B5b* + rg-*1*. These combinations testify a recombination between *Gli-B1* + *Gli-B5* and *Rg-1* [Metakovsky et al 2000]. The recombination between *Gli-B1* and *Gli-B5* was observed in local variety Hembrilla-Corta-de-Alta-Montaña [Ruiz et al 2002a].

Therefore, the only possible arrangement of the three loci in the satellite of the arm 1BS is *Gli-B1* (proximal) - *Gli-B5* - *Rg-1* (distal) [Pogna et al 1993, Metakovsky et al 1994].

Thus, the three loci mapped on satellite of the chromosome 1BS (*Gli-B1*, *Gli-B5*, *Rg-1*) are not only closely linked, but occur, in the vast majority of common wheat genotypes, in two definite combinations, the first, *Gli-B1*(red) + *Gli-B5b* + *Rg-1*, and the second, *Gli-B1*(white) + *Gli-B5a* + rg-*1*. The origin of these two contrast complexes can be traced to diploid donors that contributed their elementary genomes A, B and D into hexaploid wheat.

We suggest that such complexes arose as a result of natural selection among genotypes of wild polymorphic donors of genomes. At the diploid level, a composition of allelic variants of tightly linked loci in a given chromosome's region could have an adaptive value for the plant. The selected complexes can provide plants with better adaptation to specific environments [Golenberg & Nevo 1987; Metakovsky et al 1987a; Nevo 1994; Sozinov et al 1999]. In the self-pollinating species, the composition of the favourable poly-allelic complex could be relatively stable due to known low frequency of cross-pollination, although new variants may be produced by gene conversion [Morrell et al 2006]. Alternative variants of a poly-allelic complex can exist in parallel (in different plants) and evolve separately from each other (see **7.5.**).

Similar, albeit less unequivocal, results were obtained in analysis of allelic combinations in durum wheat. It was found that the presence of the *Gli-B1*-encoded gliadin EP band γ-45 is combined with black colour of awns in 90% of cultivars studied [Koval et al 1986]. Incomplete correspondence may be due to the fact that the colour of awns in durum is determined by three genes located in different chromosomes [Panin & Netsvetaev 1986].

6.11. Locus *Gli-D5*

There is some low level of recombination (about 1% and less) between genes encoded γ- and ω-gliadins (APAGE) at the *Glu-D3*/*Gli-D1* complex locus [Metakovsky et al 1997b; Kozub et al 2003]. It was suggested that there was a recombination between *Glu-D3* and *Gli-D1* loci (see **3.3.**), although the possibility could not be excluded that the documented recombination occurred between the *Gli-D1* and supposed *Gli-D5* locus.

The results of analysis of F2 progeny of the cross between cvs Ablaca and Cajeme-71 were interpreted in favour of the existence of the *Gli-D5* locus homoeologous to *Gli-A5* and *Gli-B5* in common wheat [Rodriguez-Quijano & Carrillo 1996a]. In this publication, two new *Gli* loci were described, and one of them (named *Gli-D5*) encoding an ω-gliadin (ω-17.9) of the cv Ablaca. The locus was mapped on the chromosome 1DS at the distance of 3.7±0.8 cM from the *Gli-D1* locus, distally to it. In the same publication, other new locus, named *Gli-D4*, was mapped proximately to the *Gli-D1*. The distance between *Gli-D4* and *Gli-D5* was found to be 29.0±3.5% [Rodriguez-Quijano & Carrillo 1996a]. Taking into account the genetic distance between *Gli-D5* and *Gli-D1* and distal position of *Gli-D5* relatively to *Gli-D1*, it was suggested that *Gli-D5* is homoeologous to the *Gli-A5* and *Gli-B5*. However, at least one detail complicates this interpretation.

The cv Ablaca, in accordance to its APAGE photos shown in the article [Rodriguez-Quijano & Carrillo 1996a], had, in this experiment, allele *Gli-D1g*. This allele encodes block (of Cheyene type) consisting of four ω-gliadins and two γ-gliadins [Blocks 2018c]. All these six EP bands are inherited together, as one Mendelian trait (see **2.5.**). Assuming that one of ω-gliadins of this block is controlled by other locus (*Gli-D5*) which is located in about 3.7±0.8 cM from the *Gli-D1*, one can expect a tangible number of cases of recombination between this locus *Gli-D5* and the *Gli-D1*. The distance of 3.7 cM is enough to produce many different recombinants among progenies from hundreds crosses, in which cultivars with the Cheyenne-type ω-gliadins would be inevitably involved, performing each year in breeding centres around the world. Thus, wheat genotypes encoding truncated and otherwise modified blocks would appear. However, no such modified blocks are present in published catalogues of gliadin electrophoregrams of many wheat cultivars studied from different countries [Autran & Bourdet 1975; Peruffo et al 1981; 1984; Jones et al 1982; Ng et al 1988; Javornik 1989; Blocks 2018c; etc.].

Therefore, the existence of the *Gli-D5* locus homoeologous to *Gli-A5* and *Gli-B5*, as well as of the *Gli-D4* claimed to be homoeologous [Rodriguez-Quijano & Carrillo 1996a] to the *Gli-A3* and *Gli-B3* loci in common wheat (see **6.5.**), is still to be confirmed.

No additional gliadin-encoding locus, besides *Gli-D1*, was found distally to *Gli-D1* on the short arm of the chromosome 1D in the cross between cv ChS and a disomic substitution line of *T.tauschii* chromosome 1D in ChS. The order of genetic loci studied *Glu-D1* - *Gli-D1* - *Rg2* - *Lr21* was established [Jones et al 1990].

6.12. Summary. Minor *Gli* loci are part of wheat-gliadin polymorphism

A strong proof of the existence of minor *Gli* loci, recombining with the major *Gli-1* loci, on the chromosomes of the first homoeologous group was obtained in the analysis of common wheat cultivars having intravarietal non-uniformity (multi-line cultivars). Four biotypes (genotypes) of the Kazakh cv Tselinogradka were identical in alleles of all six *Gli-1* and *Gli-2* loci, but differed from each other in the presence/absence of ω-gliadins controlled by *Gli-A3* and *Gli-B3* loci.

There are at least four alleles at the minor locus *Gli-A3* including null allele (*Gli-A3a*). Alleles at the *Gli-A3* encode ω-gliadins having different EM but identical AMW of about 41 kDa. On the contrary, the *Gli-B3* has only two alleles, one encodes an ω-gliadin with AMW of about 50 kDa, and another is null. The peculiarity of the *Gli-B3* is that there are indications that some variants of this locus may produce glutenin-like molecules. In hexaploid wheat, variants of the *Gli-B3* produce, in contrast to *Gli-A3*, the D-type LMW glutenins. Therefore, this locus is called *Glu-B2* in some publications.

Both *Gli-A3* and *Gli-B3* are mapped (genetic distance) just between the *Glu-1* and *Gli-1* on the chromosomes 1A and 1B, respectively. The recombination frequency between *Gli-A3* and *Gli-A1* varied considerably (about 10-35%) in different crosses. This difference can be explained by the divergence of homologous chromosomes known among common wheat genotypes.

In some publication, the designation *Gli-3* was erroneously used to name an ω-gliadin-encoding DNA sequences tightly linked to *Gli-1* loci although the designation *Gli-A3* and *Gli-B3* [Payne et al 1988b] has long been used in genetic studies on gliadin. The existence, in common wheat, of the *Gli-D3* locus homoeologous to *Gli-A3* and *Gli-B3* is not proved.

The loci *Gli-A5* and *Gli-B5* loci are tightly linked to *Gli-A1* and *Gli-B1*, respectively. There are two known alleles at each *Gli-5* locus, one of them is null (*Gli-A5a* and *Gli-B5a*, respectively). There are two contrast families of alleles at the *Gli-B1*, first, linked to *Gli-B5b*, and the second, linked to *Gli-B5a*.

There are three known alleles at the *Gli-A6* locus, one of them (*Gli-A6a*) is null. Thus, the chromosome 1A carries, in addition to other genes, at least six storage-protein-encoding loci situated in the order: *Glu-1* - Centromere - *Gli-A3* - *Glu-A3*/*Gli-A1* - *Gli-A5* - *Gli-A6* - *Hg-1*. The only possible arrangement of some loci in the chromosome 1B of common wheat is *Glu-B1* - Centromere - *Gli-A3* (*Glu-B2*) - *Glu-B3*/*Gli-B1* - *Gli-B5* (the locus *Rg-1* is distal).

The existence of the *Gli-D5* locus homoeologous to *Gli-A5* and *Gli-B5* in common wheat is still to be confirmed. The apparent molecular weight of ω-gliadins encoded by different minor *Gli* loci is in the range of 40-50 kDa.

7. MUTATIONS AT THE *GLI* LOCI, ALLELE FAMILIES AND HETEROALLELES IN COMMON WHEAT

7.1. Nulls are the most easily detected mutations at the *Gli* loci in wheat

Since the introduction of gel electrophoresis, wheat grain scientists have been intrigued by the enormous polymorphism of the gliadin/glutenin family of wheat-grain storage proteins. Why are they so diverse if their sole function is storage?

Genetic variability originates from spontaneous mutations. Mutations in genes encoding functional proteins, such as enzymes, often prove lethal (and they thus disappear) due to consequent loss of enzymes' activity. Without the restriction of loss of specific function, the endosperm proteins have retained the consequences of repeated mutations, thus leaving the great polymorphism of wheat germplasm. It was shown long ago that different seeds of the same wheat cultivar might differ in their gliadin electrophoregrams, and therefore, in their genotypes [Wrigley & Shepherd 1974; Appleyard et al 1979; etc.]. In early publications, nearly all cases of non-uniformity among seeds of the same cultivar were considered as "biotypes", or "electrophoretic types", especially if an off-type for gliadin was morphologically identical to the main type of the cultivar [Appleyard et al 1979; Pogna et al 1982b; etc.].

There are three main types of intra-varietal non-uniformity, quite different in their origin and significance, that could be detected by a seed-by-seed EP analysis of a given grain sample, firstly, authentic multiple biotypes of the cultivar (the genotype of each biotype may be traced to the progeny of the cross of parental genotypes), secondly, foreign seeds (in contrast to authentic biotypes, it is not a genuine part of a cultivar being casual admixtures or off-types), and thirdly, intra-varietal non-uniformity caused by assumed spontaneous mutations [Metakovsky et al 2019c].

There are several types of changes in the APAGE electrophoregrams which can be considered as being caused by spontaneous mutations [Metakovsky et al 2020] the most frequent of them is a disappearance of a whole block of EP bands controlled by one *Gli* locus in a given genotype (null allele at this locus) [Mecham et al 1978; Sozinov & Kopus 1983; Lafiandra et al 1987a; 1989; etc.]. A modification for the APAGE procedure has been recently proposed especially for detecting mutations causing changes in gliadin pattern [Watry et al 2020]. However, there is no difficulty in revealing a simultaneous lack of several gliadin bands (a whole block) using any gliadin electrophoregram, including SDS-EP (see **1.4.**). Any APAGE procedure allows to observe the decrease in staining intensity of several EP bands simultaneously, thereby revealing a null mutation in heterozygous state.

The frequency of null alleles was found to be strongly disproportional among six *Gli* loci; thus, the *Gli-B1null* occurred 20 times more often than the *Gli-B2null* [Metakovsky et al 2020]. Null mutation at any *Gli* locus is frequently (but not always) accompanied by the over-production of gliadins encoded by some other (as a rule, homoeologous), *Gli* locus (loci). For example, nulls at the *Gli-B1* locus (in different genotypes) were, as a rule, accompanied by the over-production of ω-gliadins encoded by the *Gli-D1* [Metakovsky et al 2020]. A mutation ceasing the synthesis of the HMW glutenin subunit 1Dy12 (see **4.3.**) causes an increased production of gliadins and LMW glutenins [Chen et al 2021]. Deletion of all HMW glutenin polypeptides in the mutant genotype LH-11 provokes the increase of the ratio of ω-gliadins among gliadins [Zhang et al 2018].

Taking into account the size of a *Gli* locus and distances between transcribed genes in this locus (see **5.6.**, **5.7.**), it is difficult to offer some other mechanism of appearance of a null mutation besides of some kind of aneuploidy. The suggestion [Mecham et al 1978] that an aneuploidy (the deletion of the satellite) was a reason for null mutants registered at the *Gli-B1* locus was confirmed [Payne et al 1984c; Felix et al 1996].

Although a null mutation identified through the absence of a block of EP bands may not be detected by cytological approach [Lafiandra et al 1987b; Iakobashvili 1989; Chernakov & Metakovsky 1993; etc.], it was always accompanied by the loss of some DNA sequences in the mutant genotype [D'Ovidio et al 1991; Sabelli et al 1992; Martin & Carrillo 1999; Chernakov & Metakovsky 1993; etc.], For example, it was proved that the genotype of the cv Alpe carrying *Gli-D1null* [Pogna et al 1995] and of the cv Xiaoyan-81 carrying *Gli-D2null* [Li et al 2018] lacked corresponding DNA sequences.

It is well-known from classical studies [Sears 1954; 1974; MacKey 1960; 1968; Morris & Sears 1967] that common wheat easily acquires (with the frequency of appearance of about 1%) and survives aneuploidy and chromosomal aberrations. Therefore, it is not a surprise that genotypes carrying any spontaneous null mutations at any *Gli* locus occurred in wheat grain material studied with a frequency of 0.44%-0.55%. It means that, using seed-by-seed analysis of any thousand common-wheat individuals (seeds or plants), several null mutants at the *Gli-1* loci (mainly *Gli-B1* and *Gli-D1*) would be found with a high level of probability. The frequency of null alleles was found to be strongly disproportional among six *Gli* loci; thus, the *Gli-B1null* occurred 20 times more often than the *Gli-B2null* [Metakovsky et al 2020].

Moreover, among grains of the common wheat cv Tselinogradka, a genotype was discovered lacking all gliadins controlled by the *Gli-B1* and *Gli-D1* loci (this loss was accompanied by the over-production of all gliadins encoded by the *Gli-A1* and of some APAGE bands controlled by *Gli-2* loci) and of all HMW glutenins encoded by *Glu-B1* and *Glu-D1* loci, simultaneously. However, this genotype, being apparently a double nullisomic for chromosomes 1B and 1D, produced normal good seeds characteristic for cv Tselinogradka [Chernakov & Metakovsky 1993; Metakovsky et al 2019c]. Registered cultivars carrying null alleles at their *Gli* loci, such as cv Darius with the *Gli-D1null,* and cv Touzelle with null at both *Gli-B1* and *Gli-D1* loci simultaneously (a double-null mutant), were found [Metakovsky & Branlard 1998].

Null mutations at the *Gli* loci might inherit as an ordinary co-dominant Mendelian trait (Metakovsky & Baboev 1992; Chernakov & Metakovsky 1993; Pogna et al 1995). However, some homozygous null mutation showed a lower level of inheritance by the progeny. For example, only eight F2 seeds among 61 studied were homozygous for *Gli-B1null* in an inter-varietal cross between the mutant genotype of the cv Bez1, carrying the *Gli-B1null*, and a genotype having the allele *Gli-B1g* [Metakovsky et al 2020].

Besides null mutations, several cases the lack of more than one, but not all gliadin bands of a block encoding by the *Gli-B1*, was registered in different common wheat cultivars. It was suggested that, in this type of spontaneous mutation at the *Gli-B1*, in contrast to the null at this locus only some portion of the satellite of the chromosome 1B was deleted [Metakovsky et al 2020].

Null mutations at the *Gli* loci are caused by the loss of genetic material. Therefore, they are not included, unlike *Glu-1* loci, in any catalogue of alleles. Other types of spontaneous mutations observed in common wheat (see **7.4.**) actually produce new variants of the *Gli* loci encoding new variants of APAGE blocks. However, we assume that a new variant of a given *Gli* locus will be recognized as a new allele only being found in a registered cultivar.

Null mutants at a *Gli* locus were discovered in tetraploid durum wheat (Lafiandra et al 1987a) and even in a diploid wheat *T.boeoticum* (Metakovsky and Baboev 1992). Aneuploids in semi-dwarf common wheat cultivars caused the appearance of tall mutant off-type plants [Storlie & Talbot 1993; Storlie et al 1996].

7.2. Families of similar alleles at one *Gli* locus: APAGE studies

There is a similarity between some blocks encoded by alleles of one *Gli* locus. Some blocks differ in EM of only one band, while other alleles of the same locus produce blocks that differ significantly in their APAGE band's composition [Metakovsky 1991]. We assume that similar blocks could arise as a result of mutations in individual genes located in a given *Gli* locus (see **7.4.**). Therefore, such alleles relate closely and represent a "family". On the other hand, some blocks controlled by alleles at the same locus can differ so much in their band composition that it is impossible to accept that they could arise from one another.

The largest family of similar blocks in common wheat is encoded by alleles *Gli-A2a*, *f*, *h*, *m*, *n*, *o*, *r*, etc. (in total, 10 alleles). Each of them have a band in the β-zone of APAGE, some slow-moving α-gliadins and two fast-moving α-gliadins. Some pairs of these alleles differ in the EM of only one band. For example, alleles *Gli-A2f* and *Gli-A2r* differ only in the EM of the fastest α-gliadin [Blocks 2018c]. A group of other alleles (*Gli-A2b*, *c*, *g*, *l*, *p*, etc.) also encode similar blocks. For example, alleles *Gli-A2b* and *Gli-A2p* differed only in the EM of one (the fastest) α-gliadin (higher EM, lower AMW) [Metakovsky et al 2018b]. The two main families of blocks share no one band.

Five very similar alleles (*Gli-A2i*, *j*, *k*, *ai*, *ak*) encoding a characteristic fast-moving α-gliadin represent one more family. The pairs of blocks encoded by alleles *Gli-A2aa* and *Gli-A2ab*, as well as *Gli-A2d* and *Gli-A2ad*, have a special (unusual) compositions of bands and differ (between members of one pair) only in the EM of one band. In addition, there are up to 10 different alleles encoding blocks whose band composition have no apparent similarity with any block encoded by other alleles.

No one *Gli-A2*-encoded block was registered that could be considered as an intermediate variant between different families of alleles at the *Gli-A2* locus.

Each of blocks produced by alleles *Gli-A1e*, *m* and *n* has the strongly-stained band γ50 [Bushuk & Zillman 1978] in its APAGE composition. Alleles *Gli-A1p* and *Gli-A1r* encode blocks which include a strong band with EM slightly higher than γ50. Therefore, the *Gli-A1m* and *Gli-A1r*, for example, differ only in the EM of one band of the block. These five alleles represent a well-discernible family because there is no one other allele produce a band similar to γ50 in its EM and staining intensity.

Other family includes alleles *Gli-A1h*, *i* and *j*. Each of them is composed of three characteristic bands, the EM of one of them (γ-gliadin) slightly differ among these alleles. Alleles *Gli-A1a*, *c* and *k*, also differ only in EM of one γ-gliadin [Blocks 2018c].

Obvious similarity is between blocks produced by alleles *Gli-B1i*, *k*, *m*, *o* and *p*. Each of these alleles encodes a characteristic γ-gliadin, the slowest of all known γ-gliadins produced by common wheat [Blocks 2018c]. Also, all alleles at the *Gli-B1* may be divided into two families, linked to *Gli-B5a* + *rg-1*, and linked to *Gli-B1b* + *Rg-1* (see **6.9.**; **6.10.**).

There are two well-known families of alleles at *Gli-D1* differing in the number of encoded ω-gliadins, ChS-type (alleles *Gli-D1a*, *b*, *f*, *h*, *i*, *k*) and Cheyenne-type (alleles *Gli-D1g*, *j*, *o*) (see **2.5.**). Alleles of the same family could arise as a consequence of mutations. For example, allele *Gli-D1c* found in two related cultivars obviously has come from the *Gli-D1a* as a result of recent spontaneous mutation that increase the AMW and decrease the EM of the major γ-gliadin encoded by this locus [Metakovsky et al 2018a]. Some alleles at the *Gli-D1* might originate as a result of genetic recombination between two allelic variants of this locus (see **7.4.**).

Families of alleles, can be noticed among alleles at the locus *Gli-D2*. Most of blocks encoded by alleles at this locus are composed of five APAGE EP bands (24 blocks of 30 in the catalogue), and three blocks are composed of four bands. All these alleles are similar in their band composition and might originate one from another due to spontaneous mutations changing the EM of individual gliadins or ceasing of activity of one of genes in the locus. For example, blocks encoded by *Gli-D2v* and *Gli-D2w* differ only in the EM of one band of encoding block: being encoded by the *Gli-D2v*, this band is located in the β-zone of APAGE electrophoregram while in the *Gli-D2w*, the band has considerably lower EM and occurs in the γ-zone. The allele *Gli-D2w* was found in a few Australian cultivars [Metakovsky et al 2019b] and probably has a recent mutational origin. The *Gli-D2a*- and *Gli-D2b*-encoded blocks can by distinguished due to the fine differences in the EM of two bands in the β-zone of the APAGE [Metakovsky & Sozinov 1990; Metakovsky et al 2018b]. Alleles *Gli-D2a* and *Gli-D2l* differ only in the absence, in the *Gli-D2l,* of a fast gliadin in the β-zone of APAGE [Blocks 2018c].

However, there are three alleles at the *Gli-D2* (*Gli-D2f*, *h* and *s*) encoding blocks of only three APAGE bands. Each of these alleles encodes a block composed of unique bands (characteristic only for this allele) whose EM differ from EM of any band producing by any other allele at the *Gli-D2*. Each of these three particular alleles may be considered as a unique representative of its own family, or all three may represent a family of alleles encoding only three gliadin polypeptides.

7.3. Families of similar alleles at one *Gli* locus: DNA sequence study

The RFLP seems to be especially useful technique for the comparison of allelic versions of a *Gli* locus, because sites of restriction for a given endonuclease may occur anywhere along the whole long (see **5.6.**) gliadin locus, including active genes and non-coding sequences [Castagna et al 1994]. Fragments of genomic DNA carrying the analysed sequences can be specifically recognized by corresponding probes thereby revealing a general pattern of restriction sites in the DNA region [Reddy et al 1993].

Application of this approach (the restrictase *Hae*III, the probe K32 (pTag1436) that is a cDNA clone specifically recognizing γ-gliadin sequences [Bartels et al 1986]) to the analysis of nulli-tetrasomic lines of ChS (see **2.1.**), made it possible to identify two alternative patterns of RFLP bands carrying sequences belonging to the locus *Gli-A1*. The first pattern was produced by alleles *Gli-A1a* and *Gli-A1l*, both encoding an apparently identical slow-moving γ-gliadin (APAGE). Another pattern was shown by several alleles including, surprisingly, both *Gli-A1m* and *Gli-A1o* [Vaccino & Metakovsky 1995]. Their occurrence in the same group could not be predicted because these two alleles produce blocks sharing no one band neither in the γ-zone of APAGE, nor in its other zones [Blocks 2018c].

The polymorphism of the γ-gliadin sequences of locus *Gli-A1* was also studied in 52 Australian cultivars using the allele-specific PCR markers based on SNPs. Thus, all common wheat genotypes studied were divided into two groups [Zhang et al 2003]. Our identification of alleles at the *Gli-A1* in many cultivars mentioned in this publication showed that these two groups were composed of cultivars, first, having the allele *Gli-A1a*, and, second, some other alleles including *Gli-A1m* and *Gli-A1o*. Therefore, in the two different studies [Vaccino & Metakovsky 1995; Zhang et al 2003], similar results were achieved.

The γ-gliadin-encoding sequence of the locus *Gli-B1* in the cv Cheyenne (*Gli-B1b* [Alleles 2018c]) is significantly more different from the alternative sequence in cvs ChS (*Gli-B1a*), Katepwa (*Gli-B1d*), Neepawa (*Gli-B1d*), Kanzler (*Gli-B1f*) and Galaxie (*Gli-B1f*) than these cultivars differ from each other [Zhang et al 2003]. Up to seven different families of alleles at the *Gli-B1* locus in common wheat were revealed by comparing DNA of allelic variants of this locus (see **7.5.**).

Microsatellite MW1B002 is a part of the γ-gliadin-encoding gene situated within the *Gli-B1* locus [Devos et al 1995]. One of alleles of this microsatellite (244 bp) was found to specifically differentiate red-spike from white-spike cultivars in Bulgarian common wheat germplasm [Landjeva et al 2006a]. Therefore, this allelic variant of the microsatellite reveals the difference between "red" and "white" alleles at the *Gli-B1* locus (see **6.10.**).

The comparison of the RFLP patterns (γ-gliadin probe K32) of allelic variants of the *Gli-D1* locus showed that alleles *Gli-D1k* and *Gli-D1*a belong to the same family and that the lack of the fast-moving γ-gliadin in *Gli-D1k* was not caused by a deletion of genetic material, but, probably, by a mutational transformation of the gene encoding that γ-gliadin into a pseudogene [Vaccino & Metakovsky 1995].

The gene *GAG56D* of common wheat was discovered using the PCR approach, cloned and sequenced. The gene was located within locus *Gli-D1*, and was shown to have two allelic states, "*ok*" (encodes unusual γ-gliadin due to a deletion of nine bp in the DNA sequence) and "*fo*" [von Büren et al 2000; von Büren 2001]. Identification of alleles at the *Gli-D1* in cultivars studied in these publications permitted to assert that the allele "*ok*" of the *GAG56D* is responsible for the synthesis of the *Gli-D1*-encoded fast-moving γ-gliadin, the faster (APAGE) of its two variants (alleles *Gli-D1a, c, h, o*). Another group of alleles (*Gli-D1b, f, g, j, l*, etc.) do not encode this variant of the fast-moving γ-gliadin in their blocks. Therefore, using the *GAG56D*, it is possible to divide all alleles at the *Gli-D1* into two groups (families) depending on the EM of the fast-moving γ-gliadin.

The use of RFLP technique and the probe pTU1 recognizing the α-type sequences [D'Ovidio et al 1992] made it possible to differentiate two contrast RFLP patterns belonging to genotypes with ChS-type or Bez1-type alleles at the *Gli-A2* (see **2.5.**), respectively. The same technique and probe could distinguish alleles *Gli-A2f* and *Gli-A2o* (both of the ChS-type). Also, some RFLP bands belonging to allelic variants *Gli-A2b*, *Gli-A2c* and *Gli-A2g* of Bez1-type were identified. However, it was not possible to differentiate alleles at loci *Gli-B2* or *Gli-D2* using this approach [Vaccino & Metakovsky 1995].

Thus, each technique that was used for the analysis of DNA of any *Gli* locus (RFLP, allele-specific PCR markers based on SNPs, PCR-revealed sequence) separated all alleles at a given locus into two groups, while APAGE reveals several groups (families) of alleles at each *Gli* locus (see **7.2.**).

7.4. Spontaneous mutations at *Gli* loci cause enormous polymorphisms of gliadins

There are at least four different types of changes in the APAGE electrophoregram which can be considered as being caused by presumed spontaneous mutations at the *Gli* loci: first, is a disappearance of all bands controlled by one *Gli* locus (null mutants) (see **7.1.**), second, lack of more than one, but not all gliadin bands of a block encoding by the *Gli-B1* (see **7.1.**), third, lack of only one EP band, and fourth, change of the EM of one EP band [Metakovsky et al 2020]. The third and fourth types of alteration of the gliadin EP pattern of a given cultivar produce a new variant of the block, encoded by a new variant of the corresponding *Gli* locus. It should be noted that the method of APAGE designed for revealing mutants in gliadin EP pattern [Watry et al 2020] would reveal null mutants (see **7.1.**), but would hardly detect a change of the EM of one EP band because of a too short distance between the top of gel and the fastest gliadin band.

Several cases of the lack of one gliadin EP band were registered in the analysis of more than 12,000 single seeds of world-wide common wheat cultivars. For example, the loss of the major γ-gliadin in the *Gli-D1a*-encoded block, was observed in the cv Saratovskaya-29. This mutant occurred, as a single seed, in each of four independent grain samples of this cultivar. There were three independent cases (cvs Albidum-3700, Saratovskaya-29, Tselinogradka) of the lack of the *Gli-B1e*-encoded slow-moving ω-gliadin, while three other bands of the encoded block remained in the APAGE pattern [Metakovsky et al 2020, Fig. 2]. Similar lack of the ω-gliadin encoded by the allele *Gli-B1g* can be seen in the APAGE electrophoregram of the cv Autonomia-B [Pogna et al 1982, Fig. 1]. The cessation of the synthesis of a single gliadin polypeptide may be the result of transformation of the corresponding gene into a pseudogene. That might be caused by different mechanisms, such as movements of TE (see **5.2.**), gene conversion [Morrell et al 2006], point mutation. For example, silencing of the gene encoding 1Dy12 HMW gluttenin subunit was caused by appearace of a stop codon produced by a base C/T mutation leading to truncated translation [Chen et al 2021]. A few alleles in the catalogue differ only in the presence or absence of one band in the encoded block. For example, a pair of alleles *Gli-B1e* and *Gli-B1t* which differ in the presence or absence, respectively, of one ω-gliadin [Blocks 2018c], so that the allele *Gli-B1t* repeats the mutant variant of the *Gli-B1e* found, for example, in cv Tselinogradka.

Changes of the EP mobility of one APAGE band caused by presumed spontaneous mutations occurred in grain samples of common wheat cultivars several times less frequently than nulls, but more frequently than loss of one band [Metakovsky et al 1993a; 2020] (see **7.5.**). In contrast to null mutations (see **7.1.**), mutations altering the EP mobility of one gliadin polypeptide occurred mainly at *Gli-2* loci, particularly at the *Gli-B2* [Metakovsky et al 2020]. Within families of alleles at the same *Gli* locus (see **7.2.**), there are pairs of alleles encoding blocks differing in the EM of only one band. Such pairs in which one allele could originate from another as a result of single mutation changing the EM of one gliadin polypeptide, are, for example, *Gli-A1i* and *Gli-A1j*, *Gli-B1e* and *Gli-B1f*, *Gli-D1a* and *Gli-D1c*, *Gli-A2f* and *Gli-A2r*, *Gli-D2k* and *Gli-D2u*, and so on. Each two alleles at the *Gli-B2* differed in EM of at least two bands of encoded block [Blocks 2018c], although in no one case two or more simultaneous changes in EM of different EP bands in one mutant genotype was registered [Metakovsky et al 2020]. The use of the 2D APAGE/SDS-EP permitted to established that, among related pairs of alleles at the same *Gli* locus, the polypeptide with higher EM had lower MW [Metakovsky et al 2018a; b].

An appearance of a new gliadin polypeptide (for example, with altered EM, caused by a spontaneous mutation of the length of a microsatellite within gliadin gene) gives rise to a new gliadin block controlled by mutant *Gli* allele. The genotype carrying this mutation may be occasionally involved in the breeding process. As a result, a new (daughter) cultivar may obtain a new (mutant) variant of the *Gli* locus (new *Gli* allele) differing from the "parental" one by the EM of one polypeptide. So, families of gliadin alleles at the same locus may appear with very similar band composition of controlled blocks increasing wheat polymorphism with time.

It is important to note the difference between blocks encoded by *Gli-D* loci and blocks produced by *Gli* loci of genomes A and B. The comparison of band composition of blocks encoded by *Gli* loci of the genome D suggests that some alleles at these loci could appear not as a result of mutations, but rather as a result of genetic recombination between genes of the locus. For example, the allele *Gli-D1j* (five ω-gliadins) could appear by acquiring the fifth ω-gliadin in a cross between genotypes having alleles *Gli-D1g* (four ω-gliadins) and *Gli-D1i*. Also there are four blocks in the catalogue representing four possible combinations of either ChS- or Cheyenne-types of ω-gliadins and two variants of EM (either fast or slow) of the fast-moving γ-gliadin (*Gli-D1a*, *Gli-D1f*, *Gli-D1o*, *Gli-*

D1g, respectively) [Blocks 2018c]. In the distal part of the chromosome 1DS of diploid *T.tauschii*, duplications of DNA sequences are found, which can contribute to recombination events at the *Gli-D1* locus [Spielmeyer et al 2000]. At the *Gli-D2* locus, some allelic variants could also arise as a result of recombination. For example, the allele *Gli-D2d* could have its origin in the cross of genotypes having alleles *Gli-D2a* and *Gli-D2b* [Blocks 2018c]. The recombination frequency between genes located within the *Gli-D2* can reach 2.73±1.04% [Kolyuchii1989, ref. in Kozub et al 2003].

The natural explanation for the existence of families of alleles, encoding blocks that are very similar in composition of bands (see **7.2.**), is that each family had its own original ("parental") version of a given *Gli* locus. Each parental variant obtained from diploid donor (see **8.1.**) gave rise to the family as a result of a mutation (and, in the case of genome D, also probable recombination) process.

It is suggested that an enormous polymorphism of polyploid wheat is generally based on its tolerance to mutations due to the presence of the homoeologous copies of mutant gene [Krasileva et al 2017]. Storage protein gene, to tolerate any mutation, even do not need in any "help" of homoeologous gene. In the absence of any restrictions for changes, new polypeptides with an altered length (EM) would constantly accumulate in wheat germplasm increasing its polymorphism.

7.5. Mutations of microsatellites within gliadin genes changes the EM of one EP band

Microsatellites are tandem repetitive DNA sequences of 1-6 nucleotides (the most frequent of them di- and trinucleotides) which present in abundance in the wheat genome [Li et al 2002] (see **9.3.**).

A characteristic feature of microsatellites is their tendency to spontaneous mutations Different mutational mechanisms of microsatellite variation were offered and discussed [Li et al 2002]. Generally, significantly higher mutation rates were observed in longer microsatellites [Li et al 2002; Yamada et al 2002]. The most polymorphic, longest and most often mutating microsatellites are located in the distal part of common wheat chromosomes with presumably high recombination rates (where the *Gli* loci are located, see **5.7.**) [Huang XQ et al 2002; Thuillet et al 2004].

The microsatellite mutation frequency was studied in durum wheat lines during 11 generations, and 12 mutational events were found at three out of 10 microsatellite loci observed. The frequency of emergence of new variants was up to 2.5 x 10^{-3}, and the average frequency was 0.24 x 10^{-3} per allele, per generation. Out of new 12 mutant variants, eight acquired one additional repeat, and four lost one repeat [Thuillet et al 2002]. In experimental population of common wheat, 13 cases of the lengthening and 10 cases of shortening were observed in nine microsatellite loci during 15 generations. Frequency of mutations for different loci varied from zero up to 5 x 10^{-3} per locus, per generation. Thus, the polymorphism of the studied population was constantly increasing [Raquin et al 2008].

Microsatellites occur relatively rare in the coding portion of genes [Li et al 2004], but gliadins genes are an exception: the poly-glutamine domains are an essential parts of genes encoding gliadins, the longest ones are inside α-gliadin genes [Devos et al 1995; Lee SJ et al 1995; Anderson & Greene 1997; etc.] (see **3.1.**). It can be assumed that the mutation of the microsatellite which is a part of the gliadin gene, should cause an alteration in the length of the gliadin polypeptide encoded by this mutant gene. So, at least some of spontaneous alterations of the EM of one gliadin band (see **7.4.**) may be explained through some mutational change of the structure of the corresponding gliadin gene such as increase or decrease of the length of the microsatellite (poly-glutamine domain) of the gliadin-coding sequence.

The frequency of appearance of new spontaneous mutations causing alteration of the EM of one APAGE gliadin band was studied using spike material of different cultivars, was identified as occurrence of a mutation in the heterozygous state in only one seed of a spike, and was calculated to be about 10^{-4} per locus, per generation [Metakovsky et al 1993a]. Taking into account the fact that up to 10 transcribed genes can be present in one *Gli* locus (see **2.3.**, **5.1.**), it is possible to tentatively assume a frequency of 10^{-5} per gliadin-coding gene, per generation.

It should be noted that the absence of gliadin EP bands, especially null mutations at *Gli* loci (including being in the heterozygous state) (see **7.1.**), are relatively easy-identified in the gliadin APAGE electrophoregram of a given genotype. On the contrary, multiple overlap of gliadin bands in electrophoregram, especially faint bands controlled by *Gli-A2* or *Gli-B2* loci in β- and α-zones [Redaelli et al 1994, Metakovsky et al 2018b] (see **2.5.**) can mask a slight change of the EM of one gliadin band. As a result, some alterations of the EM, especially being in the heterozygous state, escape from notice in the registration process. Therefore, the calculations of the frequency of appearance of new spontaneous mutations in wheat germplasm represent the lowest likely level [Metakovsky et al 2020].

Spontaneous insertion or deletion of no triplet sequence, when microsatellite is located in coding sequence, is expected to introduce frameshift loss of function mutations at high frequency (Moxon et al 1994). In this case, the transformation of the transcribed gliadin gene into a pseudogene would be expected accompanied by the lack of encoding gliadin EP band in the electrophoregram of a given genotype. One can also suggest that spontaneous mutation in the poly-glutamine domain can restore, in some cases, of the reading frame in a gliadin pseudogene.

In higher eukaryotes, the frequency of mutations was found to be 10^{-5} - 10^{-6} per gene, per generation (Neel 1983) while in plants, the frequency of spontaneous mutagenesis is 10^{-5} - 10^{-6} per nucleotide per generation (Feldman & Kislev 2007).

7.6. Heteroalleles and homoalleles at the *Gli-B1* locus of common wheat

Variants of the same locus, which are similar (identical) in their transcribed part, may differ in their non-transcribed regions [Wicker et al 2003; Gu et al 2004a; 2006; Isidore et al 2005; Gao et al 2007; Anderson et al 2009; etc.]. Therefore, transcribed and non-transcribed regions of different alleles at the *Gli-B1* locus were compared.

About 1000 common wheat cultivars bred in the 20th century was studied, 25 alleles at the *Gli-B1* locus were registered [Blocks 2018c], and 19 the most frequent and interesting of them were used in the analysis.

The PCR sequence length (SNP primers GliB1.1 and GliB1.2 [Zhang et al 2003]) for the γ-gliadin pseudogene were studied in 46 cultivars from 15 countries. As it was claimed [Zhang et al 2003], all alleles at the *Gli-B1* locus were divided into two big groups, and the allele *Gli-B1l* (translocation 1BL.1RS, see **2.6.**) did not give any product. One group, giving a PCR product with the GliB1.1 primers, was represented by three families of alleles differing in the length of their product, first, alleles *Gli-B1i*, *k*, *m*, *o*, *p*, *r* (was called family I); second, alleles *Gli-B1n*, *q*, *s* (family II); and third, a single allele *Gli-B1b* (family III). Another big group (PCR product with the GliB1.2) was represented by a family of alleles *Gli-B1c*, *e*, *f*, *g* (family IV) and three distinct alleles *Gli-B1h* (family V), *Gli-B1d* (family VI), and *Gli-B1a* (family VII), each having its own length of the product. Any allele occurred in distinct genotypes, for example, *Gli-B1b* in cvs Bezostaya-1, Gabo, Marquis and Mironovskaya-808, produced the PCR sequence of identical length. Nine cultivars with allele *Gli-B1d* bred in four countries also had the PCR sequence of identical length [Popovych et al 2020].

A few cultivars with identical allele at *Gli-B1* provided PCR sequences differing slightly in length. For example, the *Gli-B1h* allele produced, among three cultivars studied, three PCR (GliB1.2) sequences of slightly different length that were not, however, produced by any other allelic variant of the locus. Generally, a strict congruity was revealed between variants of the EM (APAGE) of the *Gli-B1*-encoded γ-gliadin and the polymorphism of the PCR sequence length variants of a γ-gliadin pseudogene tested. Thus, the genetic diversity of the region of chromosome 1B marked by alleles at the *Gli-B1* is limited to only several (7-8) principal variants [Popovych et al 2020].

It was found that each of groups of "red" and "white" alleles at the *Gli-B1* (see **6.10.**) was composed of several families, so that families of alleles I, II and V were "red" whereas III, IV, VI and VII were "white".

Several clear-cut families of alleles at the *Gli-B1* locus, distinct in their restriction site positions, were revealed using RFLP technique (K32 probe for γ-gliadin sequences and three different restrictases). The families revealed by RFLP coincided with groups of alleles discovered in the analysis of a γ-gliadin pseudogene tested using SNP primers. For example, this technique could not distinguish between *Gli-B1k* and *Gli-B1m* (family I), between *Gli-B1n* and *Gli-B1q* (family II), and between *Gli-B1e* and *Gli-B1g* (family IV). Alleles *Gli-B1c* and *Gli-B1f* (family IV) only slightly differed in their RFLP pattern from other two members of the family IV, while alleles *Gli-B1b* (family III), *Gli-B1d* (family VI) and *Gli-B1a* (family VII) differed strongly from each other and from other alleles. No intermediate forms between alleles of different families were registered [Vaccino & Metakovsky 1995].

The DNA sequence of the γ-gliadin pseudogene *GAG56B* was compared in 15 different allelic variants (revealed using APAGE) of the common wheat locus *Gli-B1*. To reveal this pseudogene in the genotype of cultivar, primers GAG29 and GAG30 [von Büren et al 2000] were used, and the resulting products were sequenced. In total, 30 cultivars from 15 countries were studied [Metakovsky et al 2021b].

Comparison of all obtained sequences of the *GAG56B* permitted to identify 14 positions with SNP, one of them with three different possible nucleotides. Furthermore, the sequence included a single sequence repeat (SSR) CAA for which 10 possible variants of the length were detected. All these polymorphisms were combined in 11 different haplotypes.

In the most cases, any allele at the *Gli-B1* present in different cultivars had identical sequence of the pseudogene studied. For example, each of the alleles *Gli-B1n* and *Gli-B1g* was registered in two unrelated cultivars from two countries, allele *Gli-B1d* was found in three cultivars from three countries, allele *Gli-B1h* occurred in four cultivars from four countries, and so on.

All 11 haplotypes of the pseudogene *GAG56B* were divided into seven deeply diverged groups (families) of alleles, *Gli-B1i, k, m, o* (haplotypes 1, 8, 9, family I); *Gli-B1n, q, s* (haplotype 2, family II); *Gli-B1b* (haplotype 3, family III); *Gli-B1e, g, f, c* (haplotypes 4, 10, 11, family IV); *Gli-B1h* (haplotype 5, family V); *Gli-B1d* (haplotype 6, VI); and *Gli-B1a* (haplotype 7, family VII), The most similar were haplotypes 2 and 3 (substitution at position 182 and a minimal difference in the length of the SSR sequence) and haplotypes 4 and 7 (a difference in the length of the SSR sequence). The largest difference (substitution in 10 positions and in the length of the SSR) was registered between haplotypes 1 and 5.

There was three alleles (APAGE) for which the *GAG56B* sequence differed among wheat genotypes: *Gli-B1m* (haplotypes 1, 8 and 9), *Gli-B1e* (4 and 11) and *Gli-B1f* (4 and 10). The differences between haplotypes belonging to the same group are minimal and easy to be explained through a mutational change in a given allelic variant of the *Gli-B1* locus causing the appearance of a new haplotype of the pseudogene studied. For example, an increase of the number of CAA repeats present inside the *GAG56B* pseudogene converts allelic variant *Gli-B1e* (cv Glenlea, haplotype 4 usual for alleles of the variant IV) into the new variant of the same allele (cv Saratovskaya-36, haplotype 11). Three variants of the allele *Gli-B1m* differed in the number of the CAA repeats (18, 17 and 20, respectively), while the only substitution in position 348 distinguished haplotype 4 from 10 [Metakovsky et al 2021b].

Two pairs of allelic variants, II and III, and IV and VII, were more closely related than the others. Nevertheless, even similar allelic variants, for example, VII and IV, differed from each other by the EM (APAGE) of the encoded γ-gliadin, by the RFLP-*Taq*I test, by the length of the PCR product, and by the sequence of the *GAG56B*. The allele *Gli-B1c* was identical to alleles of the group IV in the RFLP-*Taq*I test, RFLP-*Rsa*I test, SNP and *GAG56B* tests, but differed from them in RFLP-*Hae*III test and in the APAGE. Therefore, the allele *Gli-B1c* may be considered as one more variant (VIII) related to the group IV [Metakovsky et al 2021b].

In general, the comparison, by different techniques, of coding and non-coding parts of most of allelic variants of the *Gli-B1* locus present in common wheat germplasm of the 20th century revealed at least seven families of alleles at this locus. Alleles belonging to distinct families differ strongly from each other, without intermediate forms. Such alleles may be called "heteroalleles", while alleles of the same family would be "homoalleles". Heteroalleles differ deeply from each other both in their coding and non-coding sequences whereas homoalleles may differ only by single nucleotide substitutions and other small differences of mutational origin. The data attest in favour of the independent origin of at least seven "parental" variants of the *Gli-B1* locus that might initially come to common wheat from deeply diverged genotypes of the donor(s) of the genome B in hexaploid wheat (see **7.4.**). Due to a rather low level of the cross-pollination in wheat (although its frequency depends on genotype and external conditions [Pogna et al 1986; Kozub et al 2008a]), this chromosomal region could be maintained relatively intact with time being different and even contrast between actual genotypes of common wheat. The particularities of the "parental" (donor's) variants of the agriculturally important region of chromosome 1B marked by heteroalleles at the *Gli-B1* locus might be conserved since the origin of polyploid wheat.

Also, it can be tentatively suggested that heteroalleles are an accessory (markers) of heterohomologous chromosomes (see **6.3.**).

7.7. Summary. Spontaneous mutations produce families of homoalleles at *Gli* loci

The most frequent alteration in the APAGE electrophoregram of a given wheat genotype is a disappearance of a whole block of EP bands controlled by one *Gli* locus (null allele) obviously caused by some kind of aneuploidy. The lack of one EP band may be caused by switching off its controlling gene. Alteration of the EP mobility of one APAGE band (occurs mainly at the *Gli-2* loci, particularly at the *Gli-B2*) was observed several times less frequently than nulls, but more frequently than loss of one band.

The appearance of new mutations identified in the heterozygous state in only one seed in a spike was calculated to be about 10^{-4} per *Gli* locus, per generation, or about 10^{-5} per gliadin gene, per generation. Because of overlapping of gliadin bands in APAGE electrophoregram, some changes of the EM of one gliadin band escape from notice. So, the calculation of the frequency of appearance of new spontaneous mutations in wheat germplasm represent the lowest likely level. Due to spontaneous mutations, the polymorphism of wheat evaluated through the analysis of gliadins would slowly but constantly increase with time.

Microsatellites (poly-glutamine sequences), an essential part of genes encoding polypeptides of gliadin (the longest ones present inside α-gliadin genes), mutate frequently, changing their length. Spontaneous alteration of the length of a microsatellite located inside *Gli* gene may be registered as a mutation causing the change of the EM of encoded gliadin polypeptide thereby producing a new variant of the block encoded by mutant version of a given *Gli* locus. The genotype carrying this mutation may be occasionally involved in the breeding process. As a result, a new cultivar may obtain the mutant allelic variant of the *Gli* locus differing from the initial one by the EM of one encoded gliadin polypeptide. Thus, a family of gliadin alleles may appear with similar band composition of controlled blocks. On the other hand, some blocks controlled by alleles at the same locus can differ strongly in their band composition. Such different parental alleles and their families could originate from variants of a given *Gli* locus obtained from deeply diverged genotypes of diploid donor of genome. With time, it could give rise to the family as a result of mutations (in the case of genome D, recombination is also probable).

The comparison of coding and non-coding parts of most of allelic variants of the *Gli-B1* locus present in common wheat germplasm of the 20th century revealed at least seven families of alleles at this locus. Alleles belonging to distinct families differ deeply from each other both in their coding and non-coding sequences, without intermediate forms. Such alleles may be called "heteroalleles". Alleles of the same family would be "homoalleles" differing only by single nucleotide substitutions and other small differences of mutational origin.

8. ORIGIN OF POLYPLOID WHEAT AND RELATIONSHIPS BETWEEN WHEAT SPECIES

8.1. The pathways of origin of hexaploid wheat

The pathways of origin of hexaploid wheat through successive crosses between diploid wheats with the genomes A and B, and later, between tetraploid AABB and diploid with the genome D are known in detail [Dvorak et al 2011; Matsuoka 2011].

There are three species of diploid wheat with the genome A [Kihara 1954; Gill & Friebe 2002]. Wild species *T.urartu* and *T boeoticum* do not interbreed [Huang S. et al 2002]. *T.monococcum*, which is the cultivated form of *T.boeoticum*, began to be grown about 9,000 years ago, at the same time as tetraploid wheat and in roughly the same geographic region of the Middle East [Luo et al 2007]. Wheat species *T.urartu* and *T.monococcum* are not distinguishable by cytological methods [Gill & Friebe 2002]. However, the results of many studies show that it was *T.urartu* that was the donor of the genome A in common wheat [Dvořák et al 1988; 2006; Kerby et al 1988; Dubcovsky et al 1995; Allaby et al 1999; DuPont et al 2004; Petersen et al 2006; etc.]. The use of the AFLP DNA markers showed that the genome of *T.urartu* had 20% more similarity with the genome A of polyploid wheats *T.aestivum*, *T.durum* and *T.dicoccoides* than *T.boeoticum* [Brandolini et al 20.6]. The divergence of diploid wheat with the genome A, which led to the emergence of tree diploid species with this genome, and that gave rise to the genome A of polyploids and up-to-day's genotypes of *T.urartu*, began about half a million years ago [Huang S. et al 2002].

The donor of the genome B was probably *T.speltoides* or some diploid species closely related to it [Gill & Friebe 2002]. Two diploids, *T.urartu* and a species related to *T.speltoides*, hybridized to form the cultivated allotetraploid emmer wheat 0.36-0.50 million years ago [Huo et al 2018a]. Possible participation of *T.speltoides* was confirmed by comparing of sequences of two DNA regions (DMC-1 and EF-G) present in the genomes of hexaploid, tetraploid and several diploid wheats [Petersen et al 2006].

It was firmly established that *T.tauschii* was a donor of the genome D [McFadden & Sears 1946; Autran et al 1979b; Petersen et al 2006; etc.].

The place of origin of hexaploid wheat with genome AABBDD is considered to be the north-western regions of the present-day Iran and south regions of Caucasus [Flaksberger 1930, ref. in Dvorak et al 2012]. The distribution of genetic variation was studied in 480 different genotypes *T.aestivum* and *T.tauschii* usung the RFLP technique (probes specifically recognizing DNA sequences of the *Nor-3* and *Glu-1* loci, as well as 53 other unique sequences were used). It was found that the probable place of origin of hexaploid wheat is the territory of Transcaucasia, Armenia in particular, and north-western Iran [Dvořák et al 1998b]. Single nucleotide polymorphism (SNP) was studied in 402 accessions of *T.tauschii* and 75 cultivars of common wheat and it was concluded that a population of *T.tauschii* growing in the south-western and southern Caspian was the main source of the hexaploid wheat genome D. As a result of analysis of the gene within *Glu-D1* locus which encodes y-type subunit of HMW glutenin, the southeast of Turkey or Syria were suggested to be the place of origin of common wheat, and its reappearance is associated with the territory of Iran [Giles & Brown 2006]. Analysis of more than 120 genes of genome D showed that among the modern wheat genotypes, the closest to the young (recently emerged) hexaploid wheat are the wheat of China, in particular, the local landrace ChS [Dvorak et al 2012].

It was hypothesized that the geographical centre of diversity of a crop coincides with their geographical centres of origin [Vavilov 1923; 1992]. However, the currently established centres of genetic diversity of polyploid wheat do not coincide with their supposed areas of origin. This discrepancy is explained by the peculiarities of domestication and subsequent evolution of polyploid wheat, which were discovered relatively recently [Dvorak et al 2011].

It is assumed that diploid wheats with elemental genomes A, B, and D diverged from a common ancestor about 5-7 [Allaby et al 1999], or about 4 [Dvořák et al 2006], or 2.5-4.5 [Huang S. et al 2002] million years ago. After their emergence, the three diploid species (in future, donors of genomes in polyploid wheat) evolved independently diverging from each other and increasing its own intra-species polymorphism. Analysis of nucleotide sequences that encode the synthesis of some enzymes indicated a rapid inter-species and intra-species differentiation and divergence of genomes [Huang S. et al 2002] which was assisted by the activity of transposable elements [Charles et al 2008; Senerchia et al 2013] (see **5.2.**).

The hexaploid wheat appeared through crossing between tetraploid (genome AABB) and diploid (genome DD) wheat about 8-10 thousand years ago [Nesbitt & Samuel 1996; Allaby et al 1999; Gill & Friebe 2002; Dubcovsky & Dvorak 2007; Huo et al]. Summarizing the results of many studies, it can be said that, most probably, *T.aestivum* originated in the south and west of the Caspean Sea about nine thousand years ago [Peng et al 2011].

So, at the time of origin of allopolyploid forms of wheat, each diploid species (carrying genomes A, B, or D) was already highly polymorphic and represented by a huge number of genotypes differing by allelic variants of their polymorphic loci. In repeated spontaneous inter-species crosses on the way to polyploid wheat, different already deeply diverged genotypes of a polymorphic donor of a given elemental genome could participate and pass their own variants of polymorphic loci to allopolyploid. Later, each such putative "parental variant" of each locus of a given elemental genome might be transformed into a family of homoalleles (see **7.2.**) as a result of spontaneous mutations (see **7.5.**).

Moreover, during the expansion of young hexaploid wheat to new cultivation areas, its polymorphism increased as a result of spontaneous hybridization with local wildly-growing tetraploid wheat. Besides, hexaploid and tetraploid wheat were often cultivated together and could interbreed with each other father increasing the polymorphism of hexaploid wheat [Dvorak et al 2006].

The use of RFLP technique made it possible to show that polymorphism of domesticated tetraploid (AABB) wheat is twice the polymorphism of genomes A and B of common wheat. Thus, a significant part of the polymorphism of tetraploid wheat is also present in common wheat. It is assumed that the enrichment of hexaploid wheat with alleles from tetraploids was, ultimately, one of the main sources of young hexaploid. At the same time, there was practically no increase in polymorphism in the genome D of hexaploid wheat due to the impossibility of its crossing with diploid wheat having genome D [Dvorak et al 2006; 2011].

8.2. Origins of the polymorphism of polyploid wheat: storage-proteins studies

Due to the high level of polymorphism, grain storage proteins are successfully used in studies of the origin and evolution of wheat species.

The nucleotide sequences of the gene encoding x-subunit of the *Glu-B1* locus, and pseudogene of the y-subunit located at the *Glu-A1* locus were studied. The analysis of 58 (in total) genotypes of tetraploid and hexaploid wheat permitted to discover five, and six contrast variants (alleles) at the *Glu-A1*, and *Glu-B1* loci, respectively. None of these variants could arise from the other as a result of an elementary mutational event. Consequently, at least five different genotypes of the donor of genome A and six of the donor of genome B took part in the emergence of the tetraploid AABB (later, it gave rise to the hexaploid wheat AABBDD). It was concluded that contrasting variants (alleles) of the *Glu-A1* and *Glu-B1* loci already existed in polymorphic diploid donors of genomes A and B [Blatter et al 2004]. Judging from the number of distinct heteroalleles at the *Gli-B1* locus, there are at least seven diverged genotypes of donor(s) of the genome B in common wheat (see **7.5.**) and at least three-five donor genotypes of the genome A (see **7.2.**).

Considerable differences in the DNA sequence among variants of the *Glu-D1* locus of common wheat also indicated that hexaploid wheat appeared as a result of more than one act of hybridization between tetraploid (AABB) and diploid (DD) genotypes [Allaby et al 1999]. Different genotypes of donor of the genome D could participate in crossing with different genotypes of tetraploid AABB [Dvorak et al 1998b].

The oldest grains of easily threshed wheat which were found during excavations date back to the period 6-7 thousand years ago. The nucleotide sequences of genes *Acc-1* and *Pgk-1* studied in common wheat and in *T.tauschii* were found to be almost identical. That indicates a rather recent acquisition of genome D by common wheat [Huang S. et al 2002].

Different genotypes of *T.tauschii*, donor of the genome D, could participate in crossing with different genotypes of tetraploid AABB [Dvorak et al 1998b]. At least two independent crosses of tetraploid AABB and diploid having the genome D were evidenced from the results of analysis of polymorphism of a small (0.5 Kb) low-copy DNA sequence located on the chromosome 1D [Talbert et al 1998]. Also, the DNA sequences of genetic loci encoding enzymes *Xwye838* (sequence 247 bp) and *Gss* (326 bp) were compared in common wheat and *T.tauschii*. The results obtained were interpreted as an indication that there were at least two acts of the inter-species hybridization producing hexaploid wheat [Caldwell et al 2004].

Two allelic variants of the γ-gliadin sequence *GAG56D* (chromosome 1D) were found in hexaploid wheats as well as in *T.tauschii*. It was concluded that hexaploid wheat originated as a consequence of at least two independent crosses between tetraploid AABB and *T.tauschii* [von Büren 2001].

The polymorphism was studied of the DNA sequence (284 bp) located in the promoter region of the gene encoding the y-subunit of HMW glutenin (locus *Glu-D1*). In 100 genotypes of *T.tauschii*, and in 160 common wheat cultivars studied, nine, and seven alternative (allelic) variants of the sequence were found, respectively, two of them were identical in the two species. It was concluded that the alleles sharing by two species were transferred from *T.tauschii* to hexaploid wheat as a consequence of at least two independent inter-species crosses [Giles & Brown 2006].

The polymorphism was studied of single nucleotide substitutions (SNP) in 7185 DNA markers scattered throughout the genome (in total, about 10Kb of sequences) in 402, and 75 genotypes of *T.tauschii*, and hexaploid wheat, respectively. It was discovered that genotypes studied of diploid wheat represent two groups having little genetic contact and that both groups contributed to the genetic polymorphism of common wheat [Wang et al 2013].

Generally, the old hypothesis that hexaploid wheat arose as a result of repeated acts of spontaneous hybridization between different genotypes of donors of genomes [Kuckuck 1959; 1964; Dekaprelevich 1964; etc.] may be considered as being reliably and firmly confirmed.

The polymorphism of common wheat at least at the *Gli-B1* locus (see **7.5.**), is not continuous ("gradual"), but interrupted, composed of several rather diverged (discrete) components. Any two heteroalleles mark a chromosome region that may differ in many characters including allelic variants of other genes located in this region. Therefore, genetic diversity (polymorphism) of common wheat germplasm is based mainly on the differences between deeply diverged variants of genotypes inherited by hexaploid wheat from its diploid donor(s). The donor's genotypes supplied common wheat with their own genetic diversity which is observed now being marked by heteroallelic variants of the *Gli* loci. Therefore, a considerable downfall of the level of genetic diversity of common wheat for the chromosome region marked, for example, by the *Gli-B1* locus would be caused, only by a complete disappearance from wheat germplasm (impossible) of one of these variants.

On the other hand, the chromosomal region marked by different homoalleles could carry similar allelic combination of genes located in the region. Consequently, the contribution of a given chromosomal region marked by different homoalleles, to characteristics important for the breeder will not differ significantly. Although the loss of rare alleles (probably, homoalleles having recent mutational origin) with time could be also considered as a display of the genetic erosion, the disappearance of a rare allele would have only a minor influence on the level of genetic diversity [Metakovsky et al 2021b].

8.3. Origins of durum wheat and its difference from other tetraploid wheats

Tetraploid cultivated wheat *T.durum* Desf. (subspecies of tetraploid wheat *T.turgidum* L.), like common hexaploid wheat, is easily threshed. It is one of the most important agricultural crops, second among wheat both in terms of its economic importance and sown area [Luo et al 2007; Dvorak et al 2011; Matsuoka 2011; etc.]. Origin and evolution of tetraploid wheat with the genome AABB, including the question of when, where and how many times wild emmer wheat was domesticated, is known in details [Feldman & Kislev 2007].

The cultivation of tetraploid wheat with the genome AABB began about 9000-9600 [Luo et al 2007], or 9500-10300 [Feldman & Kislev 2007], or about 11000 [Ozkan et 2005] years ago. The existence of contrasting variants of the DNA sequence in the region of the *Glu-B1* locus [Allaby et al 1999], as well as chloroplast DNA fingerprinting [Mori et al 2003] indicated that the cultivation started at least twice. Due to the wide genetic base, the area of cultivation of tetraploid wheat expanded rapidly and soon covered, in particular, the whole of Europe [Feldman & Kislev 2007]. In each new area of cultivation, the tetraploid genotypes (AABB) most adapted to the given eco-climatic conditions were naturally selected [Matsuoka 2011]. So, natural selection as well as numerous spontaneous crosses with local wild-growing tetraploids caused an intensive regional differentiation of the genotypes of domesticated (cultivated) tetraploid wheat [Huang S. et al 2002; Dvorak et al 2006; Luo et al 2007].

Durum wheat differentiated from the cultivated tetraploid AABB and emerged as a completely independent form in the eastern Mediterranean region of Europe [Feldman & Kislev 2007; Luo et al 2007; Dvorak et al 2011]. The oldest grains corresponding to modern durum wheat were found during excavations and dated back to 5500-4200 BC [Feldman & Kislev 2007].

The use of DNA markers (AFLP) made it possible to establish that durum progenitor genotypes appeared shortly after the beginning of wheat cultivation and therefore the genotypes of modern *T.durum* and wild-growing tetraploid *T.dicoccoides* differ significantly [Ozkan et al 2005]. Alleles characteristic for durum are only rarely occur in wild-growing tetraploid wheat. For example, gliadin APAGE bands ω35 + γ45, as well as some *Glu-B3*-encoded LMW glutenins specific for durum could be observed in single accessions of *T.dicoccoides* [Ciaffi et al 1993].

So, durum and common wheat originated in geographic regions remote from each other and significantly different in their eco-climatic conditions. As a result, genotypes of domesticated tetraploid (AABB) wheat grown in eastern Mediterranean region of Europe (place of origin of durum) and region of northern Iran and Transcaucasia (place of origin of common wheat, see **8.1.**) must have been different before the appearance of two modern wheat species, *T.durum* and *T.aestivum*. Indeed, the use of DNA genetic markers (RFLP) made it possible to relate the genotypes of modern durum wheat to tetraploid wheat (AABB) cultivated in south-eastern Europe. Genotypes of *T.durum* species resulted to be perfectly adapted to the local arid areas of cultivation [Feldman & Kislev 2007; Matsuoka 2011].

Comparison of more than 130 DNA sequences showed that the genomes A and B in durum wheat are genetically closer to the corresponding genomes of wild tetraploid wheat *T.dicoccoides* (subspecies of *T.turgidum*) than of common wheat [Dvorak et al 2012]. In similar study, nine common wheat and 14 durum wheat cultivars of distant origin were compared, using DNA markers AFLP, with four genotypes of cultivated tetraploid wheat *T.dicoccum*. It was established that the genome A of *T.dicoccum* (subspecies of *T.turgidum*) is relatively closer to the genome A of common wheat than to durum wheat [Brandolini et al 2006].

8.4. Comparison of orthologous loci of common and durum wheats

Hexaploid common wheat (genome AABBDD) and tetraploid durum wheat (genome AABB) have orthologous genomes A and B. The level of difference between two orthologous chromosomes (for example, between chromosomes 1A in common and durum wheat) is greater than between heterohomologous (variants of the chromosomes 1A in different genotypes of the same wheat species see **6.3.**). For example, the difference between the chromosome 1A of common wheat and diploid wheat *T.monococcum*, detected using RFLP markers, is five times greater than the difference between the chromosome 1A of common wheat cvs ChS and Cheyenne [Dubcovsky et al 1995]. Homoeologous genomes differ even more. The difference between the ω-gliadins-encoding DNA sequences located in the A and B genomes of common wheat was found to be so great, as if they represented different classes of storage proteins [Anderson et al 2009].

Genes of the *Glu-A3* locus were compared in several wheat species of different ploidy. It was shown, in particular, that there are different kinds of amino acid sequences of LMW glutenins encoded at this locus, those occurring in both common and durum wheat, in common wheat only, and in durum wheat only [Lee et al 1999b].

The distance (bp) among genetic markers, for example, between a group of three closely linked genes (kinase receptor 2, globulin and the HMW-y-subunit-encoding gene) and the gene encoding the x-subunit of the HMW glutenin differed among orthologous genomes A in wheat species [Gu et al 2006; Dong et al 2012].

Not more than 33% of 1the 30-Kb DNA sequence located on the chromosome 1AS of common wheat (cv Renan) was similar to the orthologous sequence in durum (cv Langdon). Moreover, within this sequence, durum wheat harboured genes *Lr10* and *RGA2*, which were absent in this region of common wheat [Isidore et al 2005].

The DNA sequences of the chromosome region which includes locus *Glu-1* differed in durum and common wheat, and this difference is significantly higher for the genome A than for the genome B. It was concluded that diploid donor of genome A was more polymorphic than donor of genome B at the time of repeated spontaneous crosses between these two diploid donors [Gu et al 2004a; 2006]. Later, crosses between tetraploid AABB and donor of the genome D (see **8.1.**) led to the origin of two lines (groups) of genotypes of common wheat differing in the degree of similarity of their genome A to genome A of durum: one line is more similar (for example, ChS), and another much more different (cvs Renan, Cheyenne) [Gu et al 2006].

Differences were found between orthologous intergenic regions of the *Glu-1* loci in the number and composition of TEs located in these regions (see **5.2.**) and in the type of LMW glutenin polypeptides encoded at the *Gli-B3/Glu-B2* locus (see **6.4.**). In addition, durum cv Edmore has the *Glu-B4* locus which is absent in common wheat [Liu 1995].

It was discovered that the 42-kDa LMW glutenin of common wheat is almost identical to the B-type LMW glutenin encoded by the *Glu-B3* locus of durum wheat. A strong similarity between these two polypeptides, produced by orthologous genomes B, was observed in their MW (SDS-EP), amino acid sequence (among the 369 amino acids that compose this polypeptide, only eight differences between common and durum wheat were found) and some other criteria [Masci et al 2000]. In durum, this molecule is a defining part of the so-called LMW-2 complex which is produced exclusively by high quality cultivars while the alternative LMW-1 complex is present in poor-quality cultivars [Payne et al 1984a; Masci et al 2000]. The LMW-2 differs from LMW-1 precisely in the presence of the 42-kDa LMW glutenin polypeptide [Masci et al 1995; D'Ovidio & Masci 2004]. Identification of gliadin alleles in common wheat cultivars with and without the 42-kDa LMW glutenin permitted to establish a full compliance between the presence of this polypeptide and the allele *Gli-B1d* in the same genotype (see **4.9.**). This allele is rather frequent in common wheat, especially in some regions (Canada, Mexico, see **9.1.**), and its presence in the genotype may relate to a better flour quality (see **4.7.**). In the catalogue of gliadin alleles in durum, there is a block called *Gli-B1d* [Melnikova et al 2012] which has APAGE band composition apparently identical to that encoded by *Gli-B1d* in common wheat. Marker genotype for the *Gli-B1d* in durum is "Candeal Select la Prevision" [Melnikova et al 2012], which is probably Spanish landrace. A block similar to that controlled by this allele may exist in two of 49 durum wheat cultivars studied [Upelniek et al 2013]. The presence of the *Glu-B3*-ecoded 42-kDa glutenin subunit correlated with improved dough quality of durum wheat also. The presence of the allele *Gli-B1d* in those durum cultivars producing the 42-kDa LMW glutenin was not studied, however.

8.5. Comparison of sets of gliadin alleles of common and durum wheat

More information about the similarities and differences of orthologous genomes A and B was obtained by analysis of the sets of alleles at the *Gli* loci identified in the two species.

There is a catalogue of blocks of jointly inherited EP components of gliadin based on the analysis of 700 durum varieties from 45 countries [Autran et al 1995; Melnikova et al 2012]. The catalogue includes 16, 19, 41, and 55 allelic variants of blocks at the *Gli-A1*, *Gli-B1*, *Gli-A2*, and *Gli-B2* loci of durum wheat, respectively [Melnikova et al 2012]. There are few data on the comparison of blocks of gliadin EP (APAGE) bands in durum and common wheat.

There is an allele at the *Gli-A1* locus in durum wheat (for example, in cv Kharkovskaya-7) encoding the same block of EP (APAGE) bands that is produced by the allele *Gli-A1o* [Blocks 2018c], fairly frequent among common wheat cultivars from different countries (for example, cvs Albatross, Ducat, Longbow, etc.). This allele occurred in up to 40% of durum wheat cultivars studied. Alleles identical to *Gli-A1p* (cv Pyrotrix-28), and *Gli-A1r* (cv Laura) were found in durum cvs Melanopus-7, and Langdon and Wells, respectively [Metakovsky et al 1989].

Due to the absence of the genome D and, therefore, gliadins encoded by the *Gli-D2* locus in the α-zone of the APAGE electrophoregram (see **2.5.**), all α-gliadins of durum are encoded only by *Gli-A2* locus located on the chromosome 6A [Joppa et al 1982] and are inherited together as one block. There are some blocks in durum similar (identical) to those encoded by alleles at the orthological *Gli-A2* locus in common wheat, for example, *Gli-A2b* (Bez1 in common wheat vs. Hordeiforme-4 in durum), alleles of CS-type (CS, Marquis, etc. vs. Bezenchukskaya-105) and *Gli-A2u* (Aragon-03 vs. Narodnaya). The close similarity (identity) of component composition of blocks encoded by orthological *Gli-A2* loci in common and durum wheat were confirmed using 2D APAGE/SDS procedure (see **1.4.**) [Kudryavtsev et al 1988], Thus, the polymorphism of the A genomes of common and durum wheat partially coincides. However, there are alleles that are characteristic only for common wheat, or only for durum wheat.

There are two main groups of genotypes of *T.durum* that differ in allelic states of loci of the genome B. For example, there are two variants of the DNA sequences of the *Wx-B1* locus, so that cvs Mexicali and Langdon have distinct variants of the sequence [Guzmán et al 2012]. Anyone who analysed intra-varietal polymorphism of gliadins knows that the vast majority of durum APAGE electrophoregrams represent two contrast groups differing in the EM of the slow-moving strongly-stained *Gli-B1*-encoded γ-gliadin, so-called γ42 and γ45 [Damidaux et al 1978; Zillman & Bushuk 1979b; Kosmolak et al 1980; Joppa et al 1982; etc.].

Using a probe pTDA16 that specifically recognizes the γ-gliadin DNA sequences, alleles at the *Gli-B1* of durum wheat were differentiated into two alternative groups, depending on whether the cultivar produces the γ42 or γ45 band [D'Ovidio et al 1992]. Moreover, the DNA sequence of the γ-gliadin pseudogene *GAG56B* differ in the γ45- and γ42-producing genotypes [von Büren 2001]. It was claimed that he common wheat genotypes studied lacked the sequences both encoding gliadins γ45 and γ42, which would be recognized by specific probes (PCR) [Masci et al 2000]. Durum cultivars with the gliadin γ45 provide significantly better dough quality than genotypes with γ42 (see **4.7.**).

Indeed, no one of more than 1000 common wheat genotypes studied produced gliadin APAGE band γ-42 [Blocks 2018c]. However, one of several durum-produced blocks, which include the band γ45, is probably identical to the block produced by the allele *Gli-B1h* of common wheat [Blocks 2018c]. The *Gli-B1h* (family V, see **7.5.**) is not an occasional "durum allele" in common wheat: this allele was observed in registered cultivars from different countries, although everywhere as single cases (for example, cvs Tincurrin, Viking, Pepital, Barbilla-Leon, Maris-Ranger,

Krasodonka, and Newcaster were bred in Australia, France, Netherlands, Spain, UK, Ukraine, and USA, respectively. The frequency of the occurrence of *Gli-B1h* in common wheat germplasm studied does not exceed 1% [Alleles 2018c].

It is important to note that, in the catalogue of blocks encoded by the *Gli-B1* locus in durum [Melnikova et al 2012], at least six alternative positions (EM) of the main γ-gliadin (including γ45 and γ42) were mentioned. In many other publications on durum, always only the γ45- and γ42-producing genotypes were considered. The discrepancy can be explained by assuming that, to create the catalogue of gliadin blocks in durum, many non-registered genotypes and landraces were examined extending the gliadin polymorphism. Besides, durum genotypes having common wheat in their pedigree and therefore carrying some alleles characteristic of common wheat in their genotypes might be involved in the study.

The data obtained by comparing sets of gliadin alleles identified in durum and common wheat indicate that the similarity in polymorphism between the orthologous genomes A are obviously higher than between the genomes B.

8.6. Polymorphism of orthologous genomes in common wheat and *T. carthlicum*

Local Georgian tetraploid wheat *T.carthlicum* (Nevski) (another name *T.persicum*) is strikingly similar to *T.aestivum* in morphology. Probably, it originated from hybridization between a cultivated tetraploid wheat with the genome AABB and common wheat (Matsuoka 2011). Some collection accessions of *T.carthlicum* as well as some authentic spikes kept without reproduction since 1930s were studied [Iakobashvili 1989].

Three varieties studied of *T.carthlicum* (*dekaprilevichi*, *rubiginosum*, *stramineum*) were found to have α-gliadins similar to the block encoded by the allele *Gli-A2b* of common wheat. Fourth variety of *T. carthlicum* studied, *var. fuliginosum,* was found to be identical to the block encoded by the allele *Gli-A2x* in common wheat (cv Solo) [Metakovsky et al 1989]. This allele is not characteristic to common wheat being found in unique common wheat genotypes of more than 1000 cultivars studied [Alleles 2018c]. Judging from the pedigree of cv Solo, this allele came to this cultivar from one of its parents, *T.carthlicum*. It was found to be allelic to other variants of the *Gli-A2* locus of common wheat [Koval & Metakovsky 1985]. In durum wheat, there is a long family of obviously related blocks similar to that encoded to the allele *Gli-A2x* in common wheat [Kudryavtsev et al 1988] (see **7.2.**, **8.5.**).

Two varieties of *T.carthlicum* had the allele *Gli-A1p* present in the genotypes of durum and common wheat [Metakovsky et al 1989]. Thus, some allelic variants of the common wheat *Gli-A1* and *Gli-A2* loci were found not only in durum wheat (see **7.7.**) but also in another tetraploid wheat, *T.carthlicum.*

In contrast to durum wheat, however, the set of alleles at the *Gli-B1* locus had a marked similarity between common wheat and *T.carthlicum.* Thus, In the genotypes of varieties *rubiginosum* and *dekaprelevichi* alleles *Gli-B1k*, (family 1) and *Gli-B1n* (family 2) (see **7.3.**) were observed, respectively, both were red-spike alleles (see **6.10.**).

8.7. Summary. Origins and polymorphism of polyploid wheat species

The pathways of origin of hexaploid wheat through successive crosses between diploid wheats with the genomes A and B, and later, between tetraploid AABB and diploid DD are known in detail. The hexaploid wheat appeared about 8-10 thousand years ago.

It was proved that donors of the elemental genomes A and D to hexaploid wheat were diploid wheat *T.urartu* and *T.tauschii*, respectively. The donor of the genome B was probably *T.speltoides* or some diploid species closely related to it. It is assumed that diploids with genomes A, B, and D diverged from a common ancestor several million years ago. After their emergence, they evolved independently diverging from each other and increasing their own intra-species polymorphisms. The divergence was assisted by the activity of transposable elements.

Polymorphism of modern common wheat can be traced to at least seven diverged genotypes of donor(s) of the genome B and at least three-five donor genotypes of the genome A. Different genotypes of donor of the genome D could participate in crossing with different genotypes of tetraploid AABB. The donor's genotypes supplied common wheat with their own genetic diversity which is observed now being marked by heteroallelic variants of the *Gli* loci. Therefore, the polymorphism of common wheat, at least at the *Gli-B1* locus, is not continuous ("gradual"), but interrupted, composed of several rather diverged (discrete) components. A considerable downfall of the level of genetic diversity of common wheat would be caused only by a complete disappearance from wheat germplasm (impossible) of one of heteroallelic variants.

The cultivation of tetraploid wheat with the genome AABB began about 9000-11000 years ago, starting at least twice. Natural selection, as well as numerous spontaneous crosses with local wild-growing tetraploids, caused an intensive regional differentiation of the genotypes of domesticated (cultivated) tetraploid wheat. Durum wheat differentiated from the cultivated tetraploid and emerged as a completely independent form in the eastern Mediterranean region of Europe. So, durum and common wheat originated in geographic regions remote from each other and significantly different in their eco-climatic conditions. As a result, genotypes of domesticated tetraploid (AABB) wheat grown in eastern Mediterranean region of Europe (place of origin of durum) and region of northern Iran and Transcaucasia (place of origin of common wheat) must have been different before the appearance of two modern wheat species, *T.durum* and *T.aestivum*. The data obtained by comparing sets of gliadin alleles identified in durum and common wheat indicate that the similarity in polymorphism between the orthologous genomes A of these two species are obviously higher than between their genomes B. In contrast to durum wheat, however, the set of alleles at the *Gli-B1* locus had a marked similarity between common wheat and *T.carthlicum.*

9. WHEAT GENETIC DIVERSITY OVER TIME AND NATURAL SELECTION

9.1. The world-wide polymorphism of cultivated common wheat is highly structured

It was discovered that world-wide polymorphism of spring common wheat studied is differentiated and strongly structured between countries and also among geographical regions of the same country. Some *Gli* alleles frequent in one region scarcely occurred in unrelated cultivars from other countries or regions. For example, the alleles *Gli-B1b, Gli-B1d*, and *Gli-B1e* (see **7.5.**), were the most frequent in spring common wheat germplasm of the 20th century in Australia, Canada, and the former USSR, respectively. The same world-wide unevenness was observed of the distribution of alleles known for other *Gli* loci. For example, alleles *Gli-A1m*, *Gli-A1af*, and *Gli-A1a* were the most frequent in spring cultivars bred in Canada, France, and Italy, respectively [Metakovsky et al 2019b]; allele *Gli-A1a* was also frequent in winter cultivars bred in Serbia. Allele *Gli-B2c* ruled in Canada and Mexico but occurred in many other countries also, while alleles *Gli-D1j*, *Gli-A2m*, and *Gli-D2h*, dominated in Canadian wheat germplasm for much of the 20th century being rare in the world-wide germplasm [Alleles 2018c; Metakovsky et al 2019b]. Alleles *Gli-A1af*, *Gli-1Bf*, *Gli-D1b*, *Gli-B2g*, *Gli-B2l*, *Gli-D2a* and *Gli-D2g* occurred with the high frequency (from 33.0% to 73.9%) in the group of 44 UK winter wheat cultivars studied registered in the 20th century.

Several of the 25 cultivars grown in Greece had the same genotype (allelic variation at the *Glu-1* and *Gli-1* loci was studied). The coincidence could not be explained from the pedigree of these cultivars [Xynias et al 2006].

Some alleles were endemics or nearly endemics occurring only in one region and nowhere else. For example, *Gli-B2d* occurred only in Kazakhstan, but in 12 of 22 spring common wheat cultivars studied of this country [Metakovsky et al 2019b], while *Gli-D1d* might be found exclusively in Italian wheats. The genotype of every second cultivar bred in Croatia include the allele *Gli-A2e*, which is rare in cultivars from other countries [Alleles 2018c].

The genetic distances between wheat germplasms (cultivars bred in the 20th century) from different countries were evaluated using cluster analysis based on the frequencies of each allele at each locus. As a result, some pairs of closely related groups of cultivars, considering any *Gli* locus, were found, such as Russia-winter and Ukraine, Russia-spring and Kazakhstan, Germany and UK. However, genotypes of two groups of cultivars might be similar for one or two *Gli* loci differing considerably for other loci. For example, cultivars of Croatia and Serbia, France-winter and Spain-winter, Hungary and Russia-winter, as well as (surprisingly) Spain-spring and Nordic countries (Finland+Norway+Sweden) were similar only at the *Gli-A1* locus, and Canada and Mexico at the *Gli-B1* locus, differing strongly at any one of the five other major *Gli* loci. Australian and Mexican cultivars were similar at the *Gli-A1* and *Gli-D1* but differed in the polymorphism at four other *Gli* loci. Bulgaria and Hungary were most similar at the *Gli-D1* and *Gli-D2*, and so on (Metakovsky & Melnik, unpublished).

The cultivars bred in Bulgaria and Italy differed significantly from each other and from the cultivars of the UK and France if polymorphism is taken into account for all six main *Gli* loci. However, Bulgarian, or Italian cultivars become similar to those of France and the UK, if polymorphism at the *Gli-D1* locus, or *Gli-B2* and *Gli-D2* loci were considered [Metakovsky & Branlard 1998]. Distinct relationships among groups of wheat cultivars studied for different *Gli* loci firmly attest against the possibility that the use (even worldwide use) of a limited number of "the best" genotypes, would reduce the genetic diversity of wheat germplasm as a whole (see **9.7.**).

Cultivars from different regions of one country with a large territory may have a set of alleles that are characteristic only for this region. For example, each Australian state has its own the most frequent allele at five major *Gli* loci (besides *Gli-B1b* which is the most frequent everywhere in Australia), including some "endemic" alleles such as the *Gli-B2an* which is characteristic only for cultivars bred in the state of Western Australia. Alleles *Gli-A1f*, *Gli-B1e* and *Gli-B2s*, and *Gli-A1k*, *Gli-B1b* and *Gli-B2k*, are the most frequent in the Saratov, and Eastern Siberia regions of Russia, respectively. The genotypes of cultivars grown in the north and south of France differ significantly [Metakovsky & Branlard 1998]. On the contrary, cultivars grown in the west and east of Canada (with the exception of Quebec) have similar composition of the most frequent gliadin alleles [Metakovsky et al 2019b].

The cultivars bred at the same period (1970-88) in two regions of the former Yugoslavia (now Croatia and Serbia) were similar only in sharing the frequent allele *Gli-A1a*, but differed strongly in the frequency of some alleles at other *Gli* loci. For example, the frequency of *Gli-B1b, Gli-B1e, Gli-D1b, Gli-A2g, Gli-A2e,* and *Gli-D2m* were (%%) 6.7 vs. 56.4; 50.0 vs. 0.0; 10.0 vs. 48.9; 3.3 vs. 50.0; 76.7 vs. 23.1; 36.6 vs. 0.0 in groups of Croatian and Serbian cultivars, respectively [calculated from Metakovsky et al 1991].

Cultivars bred in 1968-83 in two breeding centres of Hungary, Martonvásár and Szeged, had contrasting genetic diversity *H* ($H = 1 - \Sigma p^2$ where *p* is the frequency of a given allele [Nei 1972;1973]), $H = 0.36$ and $H = 0.74$, respectively. In Martonvásár, allele *b* at *Gli* locus (obviously obtained from Bez1 and its relatives) presented in 70-90% of cultivars studied. Therefore, cultivars of this centre had only 18 alleles identified, in sum, at six loci in 10 cultivars studied. On the contrary, in Szeged, a good allelic variation (30 alleles) was observed, and the alleles *b* at *Gli* loci, being also relatively frequent, were found only in 20-50% of cultivars studied.

Polymorphism at the *Gli-1* and *Glu-1* loci was compared in 128 Ukrainian and 48 cultivars bred in seven European countries. Significant differences in allelic frequencies between these two groups of cultivars, both grown in Ukraine in 2018, were registered [Sobko et al 2019].

Differences in allelic variation at the *Glu-1* loci between groups of cultivars bred in different countries [Carrillo et al 1988; Ng et al 1989; Rodriguez-Quijano et al 1998; Branlard et al 2003; Shan et al 2007; Lerner et al 2009; Kozub et al 2017; etc. (see **4.1.**)] indicate that the global polymorphism of HMW glutenins is structured also. Four groups of China's winter wheat cultivars representing the main cultivation areas of the country differed strongly in the frequency of different alleles at the *Glu-1* and *Glu-3* loci [Liu et al 2005].

The polymorphism of microsatellites was studied in common wheat cultivated in Austria, Germany and Hungary representing three different agro-ecological regions. As a result, all cultivars studied were divided into three isolated groups of genotypes according to their country of origin. The difference between groups was statistically significant. It is assumed that it is natural selection that causes the differentiation of genotypes as a result of plant adaptation to specific eco-climatic conditions [Stachel et al 2000]. Analogously, the frequency of alleles at the *Gli* loci studied in 98 local genotypes of spring durum wheat differed in regions of Bulgaria [Melnikova et al 2010].

Wheat cultivars having phenotypes *albidum*, *graecum*, and *lutescens* occurred with the frequency of 40.6%, 28.1%, and 0.0% in Australian cultivars, of 0.0%, 2.8%, and 66.7% in Canadian ones, and of 40.5%,

0.0%, and 33.3% in Saratov region of the former USSR. Moreover, variety *alborubrum* occurred almost exclusively in cultivars bred in the south-eastern regions of Australia, South Australia (37.5% of all cultivars of this state), New South Wales (15.4%), and Victoria (53.8%). Only three non-Australian cultivars out of more than 1000 studied were *alborubrum*, one of them, Canaleja from Spain, being a descendant of the *alborubrum* parent from Australia [Metakovsky et al 2019b].

At the country scale, groups of wheat cultivars differed considerably in the level of their genetic diversity, from *H*= 0.61 (Canada) to *H*= 0.84 (Spain). The genetic diversity may also differ between regions of one country, for example, between two states of Australia, Queensland (*H*= 0.53) and South Australia (*H*= 0.67) [Metakovsky et al 2019b]. Groups of cultivars bred in some regions might have very low genetic diversity, about *H*= 0.30 and less (see **9.8.**).

Differentiation of wheat genotypes by countries and regions, revealed by analysis of genetic markers, is most easily explained by the adaptive value of polymorphism of these markers (due to natural selection), as it was done for wild wheat species (see **9.9.**). The difference of common wheat genotypes bred even in the geographically close regions can be also used for impotant practical purposes, for example. to establish, in a controversial situation and in the case of alleged theft of grain, the origin of a given wheat genotype.

9.2. The difference of genotypes of spring and winter common-wheat cultivars

Theoretically, gliadin allelic composition cannot be linked to wheat growth habit (winter or spring), as the genes controlling these characters are located on different chromosomes [McIntosh et al 2003]. In practice, however, winter and spring wheat genotypes represent two different *Gli* gene pools separated not only by breeding programs in different countries, but probably by different adaptability requirements for spring and winter wheat genotypes also. Therefore, a significant difference can be expected between the genotypes (sets of alleles favourable in a given eco-climatic conditions) of spring and winter wheat cultivars.

Groups of winter and spring cultivars bred in the former USSR have different sets of the most frequent alleles at the *Gli-1* loci. It was shown that alleles *Gli-A1b*, *Gli-B1b*, *Gli-D1b* and *Gli-D1g*, predominated in winter cultivars, while alleles *Gli-A1f*, *Gli-A1a*, *Gli-B1e* and *Gli-D1a* were the most frequent in spring ones [Sozinov et al 1999; etc.]. Analysis of the USA cultivars using AFLP molecular markers showed that the difference of genotypes between cultivars of the same habit is less than between groups of winter and spring cultivars [Barrett & Kidwell 1998]. The same result was obtained in the study of allelic variants of the *Glu-1* loci (see **4.3.**) and of the D-zone of the SDS-EP (see **4.12.**) in German cultivars [Brönneke et al 2000] and in a large selection of Swedish wheat genotypes [Johansson 1996].

The genetic diversity of winter common wheat may be either higher [Barrett & Kidwell 1998], or lower [Johansson 1996] in comparison with spring wheat. To make such a comparison, it is necessary to have equivalent groups of cultivars. In one publication [Siedler et al 1994], besides other results, the genetic diversity of spring and common cultivars was compared. The observed higher diversity of spring cultivars the authors of the publication explained by the inequality of the two groups of cultivars (seven world-wide spring cultivars vs 52 European winter cultivars, including some relatives). The best material for comparison of genotypes of cultivars of different habits is represented by cultivars bred in one country.

Cluster analysis of world-wide common wheat germplasm [Alleles 2018c] showed that there were two groups of winter genotypes. The first group includes cultivars bred in France, Germany, Netherlands and UK frequently having alleles *Gli-A1o*, *Gli-A1f*, *Gli-B1f*, *Gli-B1g*, *Gli-D1b*, *Gli-A2g*, *Gli-A2l*, *Gli-B2o*, *Gli-B2g* and *Gli-D2a*. On the contrary, the second group was composed of cultivars bred in Bulgaria, Hungary, Romania, Russia, Ukraine and former Yugoslavia with the most frequent alleles *Gli-A1b*, *Gli-B1b*, *Gli-D1b*, *Gli-A2g*, *Gli-A2b*, *Gli-B2b* and *Gli-D2b*. Gliadin allelic pool of the second group is obviously based on the cv Bez1 with its alleles *b* at each mayor *Gli* locus (see **2.3.**). There are four distinct groups of spring genotypes each of them being represented by cultivars of one country, Australia, Canada, Mexico and Russia [Metakovsky et al 2019b].

Genetic distances between groups of cultivars bread in different countries were studied using the allelic frequencies at individual *Gli* loci. It was discovered that, when allelic variation at the *Gli-A2* locus was considered, the cluster analysis separated the groups of world-wide wheat cultivars into two quite different groups (branches of the tree), the group of spring genotypes and the group of winter genotypes. The allelic frequency at any other *Gli* locus could not separate spring from winter cultivars.

Only few of the countries studied produced both types of cultivars (spring and winter) in sufficient quantities for comparison. It was found that Spanish spring and winter cultivars were similar at the *Gli-A2* (but not at the *Gli-A1* or *Gli-D1*) and both occurred in the "spring" part of the *Gli-*

A2-made tree, while Italian spring and winter cultivars were nearly identical at each of the six mayor *Gli* loci and occurred among winter groups of the *A2*-made tree. Spring and winter cultivars bred in France or in Russia had quite different genotypes (frequent alleles) at each of the six mayor *Gli* loci including *Gli-A2* [Metakovsky & Melnik, unpublished].

The high frequency of the allele *Gli-B2c* (for APAGE band composition of the encoded block, see **2.5.**) is a common characteristic of spring genotypes and differentiates them from winter ones. In Canadian, Australian and Mexican spring wheat (99 cultivars considered), in Western European winter wheat (four countries, 167 cultivars), and in Eastern European winter wheat (five countries, 153 cultivars) the frequency of this allele was 58.1%, 6.0%, and 1.0%, respectively. In 127 Russian and Kazakh spring cultivars, the frequency of the *Gli-B2c* (11.6%) is somewhat higher than in winter wheat due to use, in breeding programs of some regions of Russia, of Canadian and Mexican cultivars (Kitchener, Marquis, Nadadores-63). The allele *Gli-B2c* is of interest because of its possible relationship with grain quality in common wheat (see **4.10.**).

9.3. Using DNA markers to detect wheat polymorphism and its erosion over time

During the 20th century, the common way for wheat improvement was based on crosses between different genotypes. However, the repeated use in breeding of a restricted number of parental genotypes, successful in producing new cultivars, may cause, theoretically, a decrease of the level of genetic diversity of a crop and thereby restrict the potential of its improvement [Harlan 1972; Porceddu et al 1988; Thormann & Engels 2015; etc.].

There are various interpretations of the term "erosion" and different approaches for its evaluation (discussed in Fu & Somers; 2009; Brown and Hodgkin 2015; etc.). Sometimes, the number of alleles identified in a given group of cultivars was accepted to be one of the indicators of the genetic diversity. For example, in an examination of genotypes of 75 Canadian cultivars (370 microsatellite markers were applied) released in 1845-2004, a considerable allelic reduction (up to 17%) with time in every part of wheat genome was detected. Therefore, it was suggested that modern breeding reduces the genetic diversity of wheat [Fu & Somers 2009; 2011]. At the same time, genetic diversity among cultivars released from different breeding periods, breeding programs, and ancestral families did not change significantly during the 20th century (Fu et al., 2006). In most publications, index *H* was used for evaluation of genetic diversity in a given group of cultivars.

Molecular markers, especially microsatellites (see **7.5.**), are frequently used, as genetic markers, for identification and differentiation of wheat genotypes (including closely related ones) and for comparison of genetic diversity in different time-groups of cultivars. For these purposes, microsatellites are more informative compared to RFLP or RAPD [Lee SJ et al 1995; Röder et al 1995; Huang XQ et al 2002]. It was shown that blocks of dinucleotides $(GA)^n$, and $(GT)^n$ occur in haploid genome of common wheat 36000, and 23000 times, respectively and the average distance between two microsatellites is about 270 Kb [Röder et al 1995].

The average length of microsatellite in common wheat is up to 40 dinucleotides [Röder et al 1995]. Allelic variants of the same microsatellite differ from one another, as a rule, by two or three nucleotide pairs [Huang XQ et al 2002]. There may be, on average, up to 10 and more allelic variants of one microsatellite [Plaschke et al 1995; Stachel et al 2000; Prasad et al 2000; Christiansen et al 2002; etc.]. The $(GT)^n$ microsatellites were shown to be the most polymorphic [Prasad et al 2000].

For use as genetic markers, microsatellites have an important advantage over alleles of *Gli* loci in that they may be found on each wheat chromosome while *Gli* loci are located on six of the 21 chromosomes of the common wheat haploid set. Using only 20 microsatellites for analysis of 55 elite genotypes of wheat permitted to analyse 155 alleles assigned to 13 chromosomes in common wheat [Prasad et al 2000]. The use of 84 lines with deletions of chromosomal material made it possible to cytogenetically map 725 microsatellites [Sourdile et al 2004]. In common wheat, at least 93, 115, and 71 microsatellite loci were mapped in the genomes A, B, and D, respectively [Röder et al 1998b].

Spontaneous mutations generate new allelic variants of microsatellites with a high frequency (see **7.5.**). A high mutation rate of microsatellites may explain, at least in part, the dramatic changes in the set of their allelic variants over time discovered in some studies. For example, 24 microsatellites were used to evaluate the diversity of genotypes in local common wheats cultivated in several regions of Albania, Austria, India and Nepal. It was shown that, in two sets of samples separated by a cultivation time interval of 40-50 years, the level of genetic diversity did not change in each of the studied regions. However, only two thirds of all alleles present in the first sample of genotypes were retained in the second sample [Khlestkina et al 2004a]. Also, there were no difference in the genetic diversity between groups of cultivars bred before (H= 0.60) and after 1964 (H= 0.58) in three regions of Russia. Again, a strong difference in the set of microsatellite alleles were discovered between genotypes bred before 1960 and in 1960-90: 34, and 29 alleles were found only in group of old and newer genotypes, respectively, while the frequency of occurrence of 79 of 88 alleles found in both groups differed considerably [Khlestkina et al 2004b].

The analysis of genotypes of 497 French cultivars bred in 1840-2000 using microsatellite showed that their genetic diversity *H* varied from 0.615 to 0.620 in groups of cultivars bred in 1930-45, 1960-70 and 1990-2000, and was slightly higher in 1840-1930 (*H*= 0.645) and in 1980-90 (*H*= 0.630). In addition, the number of rare alleles in the cultivars grown in 1840-1930 was found to be 44% more than in cultivars created in 1970-2000 [Roussel et al 2004]. Some minimal reduction of genetic diversity with time during 20th century was registered as a result of the analysis of microsatellites in 480 cultivars (20 countries) that represented 15 European eco-geographic regions (*H*= 0.690, *H*= 0.695, and *H*= 0.655, in cultivars bred in 1960-69, 1970-89, and 1990-2000, respectively). The number of microsatellites decreased by 18% in 1960-2000 [Roussel et al 2005].

In these publications, the genetic diversity in the group of cultivars grown in 1840-1929 was reported to be the same as in the second part of the 20th century (*H*= 0.685) [Roussel et al 2004; 2005]. This result, to our opinion, should be taken with caution: there was a risk of loss of some part of genetic diversity by samples of old cultivars maintained in collections due to errors of various kind [Ribeiro-Carvalho et al 2004; Metakovsky & Graybosch 2006; Caballero et al 2010; etc.]. It is known that landraces, local and old varieties have a high level of genetic diversity, higher than that of the modern cultivars (see **10.1.**).

A total of 47 microsatellite markers were assessed 45 Swedish, 17 Norwegian, 9 Finnish, and 4 Danish 75 spring cultivars bred during 20th century. It was discovered that the index *H* varied from *H*= 0.34 (1900) up to *H*= 0.64 (1940), being *H*= 0.49 in the 1990. Thus, the level of the *H* could vary significantly over the century, but there was no clear tendency towards genetic erosion of polymorphism in Scandinavian wheat studied [Christiansen et al 2002].

In Bulgaria, 19 microsatellites were used to analyse polymorphism in three groups of winter common wheat cultivars, grown before 1960, in 1971-80, and after 1990, and no signs of genetic erosion were observed (*H*= 0.64, *H*= 0.57 and *H*= 0.71, respectively) [Landjeva et al 2006].

Microsatellites were also used to evaluate genetic diversity in 253 spring common wheat cultivars bred in CIMMYT (Mexico) in 1950-97. It was shown that some decrease of index *H* in cultivars bred in 1950-89 was compensated by its increase in 1990-97 due to a new breeding program adopted in the late 1879s aimed at expanding the genetic base of the cultivars being created [Reif et al 2005].

Microsatellites and AFLP genetic markers were used to compare genotypes of Argentinian common wheat cultivars bred in 1932-95. It was shown that microsatellites did not reveal any change of the genetic diversity with time, while AFLP showed a statistically significant increase of the index *H* in 1980s (*H*= 0.34) compared to 1970s (*H*= 0.28) [Manifesto et al 2001]. Three systems of molecular markers (sites of integration of retrotransposones, RFLP, microsatellites) as well as some morphological traits (in total, 200 loci) were used to analyse the genotypes of the leading winter wheat cultivars grown in Europe in 1950-2000. As a result, no decrease of the genetic diversity was registered. Moreover, involvement of the CIMMYT's genotypes caused a significant increase in the genetic base of European cultivars created in 1970-80s [Orford et al 2005].

The use of microsatellite markers did not reveal differences in genetic diversity between three groups of spring durum wheats, 25 cultivars of different origin grown in Bulgaria (*H*= 0.53±0.17), nine modern Bulgarian cultivars (*H*= 0.51±0.24) and 102 local varieties cultivated in Bulgaria since 19th century hasta 1920s (*H*= 0.53±0.19) [Ganeva et al 2010].

Certain decrease of genetic diversity at *Gli* loci in Russian durum cultivars was registered during the 20th century (*H*= 0.68, *H*= 0.70 and *H*= 0.58 in 1929-50, 1951-80 and after 1980, respectively) [Melnikova & Kudryavtsev 2009]. On the contrary, a progressive widening of the genetic basis (for its evaluation, polymorphism of microsatellites was used) in the elite durum wheat germplasm was discovered instead of genetic erosion. It was shown that 45 modern cultivars were based on ten genetically very remoted foundation-genotypes [Maccaferri et al 2003].

The experimental data obtained in studies of wheat polymorphism by using various DNA markers with the subsequent calculation of the index *H* in different time-groups of common wheat germplasm, permits the conclusion that there are no convincing proofs of noticeable genetic erosion in wheat at the germplasm or country scales [Landjeva et al 2007; van de Wouw et al 2010].

9.4. Using gliadin alleles to assess wheat genetic diversity over time

Microsatellites are present in the coding sequence (in the polyglutamine domains, see **3.1.**) of α- and, to a lesser extent, γ-gliadin-encoding genes. Therefore, some part of the polymorphism of gliadin detected by APAGE may be due to microsatellite variation (see **9.3.**).

A high level of genetic diversity of worldwide spring common wheat (290 registered cultivars from nine countries, four continents) was revealed using alleles at the *Gli* loci as genetic markers. At a germplasm-studied scale, the genetic diversity varied during the 20th century between *H*= 0.80±0.04 (1986-93, 35 cultivars) and *H*= 0.85±0.02 (1963-69, 40 cultivars) with an insignificant decrease with time not exceeding 0.04 units of *H* during the whole century [Metakovsky et al 2019a].

Also, no indications of the narrowing of polymorphism with time throughout the 20th century (genetic erosion) in any country studied were found. In Canada, genetic diversity of cultivars bred in 1907-52 (eight cultivars studied) and in 1986-88 (six) was *H*= 0.55 and *H*= 0.54, respectively. In Italy, spring cultivars bred in 1913-55 (ten cultivars) and 1966-87 (eight) had the same level of diversity, *H*= 0.73. In the former USSR, six time-based groups (16-20 cultivars per group) of cultivars bred in 1923-93 were studied. It was found that genetic diversity in this germplasm varied, over time, within narrow limits, from *H*= 0.65 (1976-80) to *H*= 0.71 (1987-93) [Metakovsky et al 2019a].

Bulgarian cultivars studied were divided into two time-groups, 17, and 15 cultivars bred in 1966-79, and 1981-91, respectively. The average genetic diversity calculated for six major *Gli* loci was higher in newer cultivars (*H*= 0.71 vs. *H*= 0.66) with the biggest contribution of *Gli-A1* (*H*= 0.83 vs. *H*= 0.68) and *Gli-A2* (*H*= 0.78 vs. *H*= 0.69) loci. No considerable differences were registered between two contrast groups of eight cultivars each (bred in 1948-73, and 1984-91), either in allelic frequencies or in the genetic diversity (*H*= 0.72 vs. *H*= 0.67). On the contrary, in Croatia, the genetic diversity of 10 older cultivars (1967-81) was higher than in 11 newer ones (1983-88) *H*= 0.77 vs. *H*= 0.73). The biggest contribution to the total decrease in the genetic diversity of newer cultivars was made, as in Bulgarian wheat germplasm, by loci *Gli-A1* (*H*= 0.71 vs. *H*= 0.79) and *Gli-A2* (*H*= 0.61 vs. *H*= 0.76) [calculated from Alleles 2018c].

Alleles at the *Gli* loci were identified in two groups of Spanish common wheat genotypes, 79 accessions of old and local wheat varieties grown in Spain before 1966 and 72 cultivars registered after 1966. It was found that both groups of genotypes had very high and similar level of genetic diversity (*H*= 0.828 and *H*=0.804, respectively) [Metakovsky et al 2000; Ruiz et al 2002b].

Allelic frequencies in two time-groups of Ukrainian cultivars (18, and 19 cultivars bred in 1979-92, and 1996-98, respectively) were compared [calculated from Sobko & Sozinov 1999]. It was found that genetic diversity calculated for six major *Gli* loci was higher in the group of older cultivars (*H*= 0.55 vs. *H*= 0.46). However, if to take into account the results for only three *Gli-1* loci, or five *Gli* loci (without *Gli-A2*), then the genetic diversity in the two groups of cultivars becomes statistically indistinguishable, *H*= 0.573 vs. *H*= 0.590, or *H*= 0.564 vs. *H*= 0.557, respectively [calculated from Sobko & Sozinov 1999] The result is explained.by the fact that the EP technique used in this work did not allow to distinguish any allelic variation at the *Gli-A2* locus in newer cultivars. The complete absence of the polymorphism at one of *Gli* loci (*H*= 0.00) greatly reduced the mean *H* value calculated for six *Gli* loci.

In Serbia, 94 cultivars were studied distributed into three time-groups (1964-1973, 1974-1988, and 1989-2004) [Novoselskaya-Dragovich et al 2007]. Our calculation showed that there was no sign of genetic erosion in Serbian cultivars (*H*= 0.63, *H*= 0.61, and *H*= 0.63, respectively) [calculated from Metakovsky et al 1991; Novoselskaya-Dragovich et al 2007]. It should be noted that the data for Italian cultivars mentioned in this publication are at least doubtful. Italian cultivars were not studied in this work, and the results of analysis of others [Metakovsky et al 1994] were used. The same three time-groups were organised for Italian cultivars [Novoselskaya-Dragovich et al 2007]. However, nothing about Italian cultivars bred in 1989-2004 (the third time-group) could be published in 1994.

In Portuguese common wheat, sets of ω-gliadins (assumed to be markers of allelic variants of the *Gli-1* loci) and alleles at the *Glu-1* and *Glu-3* loci were studied. The genetic diversity calculated using data for the nine storage-protein-coding loci studied were similar in the group of 52 "new" (*H*= 0.592) and 82 "old" (*H*= 0.574) cultivars. It was shown that genotypes (allelic frequency) of cultivars in these two groups differed considerably, and the average genetic diversity for three *Gli-1* loci was slightly higher in the new genotypes (*H*= 0.497 vs. *H*= 0.460) [Ribeiro et al 2011].

There was no one country studied for which any clear proof in favour of genetic erosion of wheat polymorphism in time course in the 20th century was registered. The main set of alleles frequently occurring in a given country could be maintained in groups of cultivars bred throughout the 20th century, with only some minor differences between time-groups of cultivars as it was, for example, in wheat of Bulgaria, Croatia, the former USSR, France, Serbia, Spain, UK and Ukraine.

9.5. Using coefficients of parentage to evaluate the genetic diversity over time

One of the usual methods for determining the genetic relationships between cultivars and the level of genetic diversity in a group of cultivars is to analyse the published pedigrees of these cultivars [Martynov et al 1992; internet site wheatpedigree.net; etc.]. An indication of genetic erosion in a given country or breeding centre would be an increase with time of the relatedness between newly registered cultivars.

Using this approach, erosion of genetic diversity was not found in the study of two groups of the USA winter common wheat cultivars bred during the 20th century [Cox et al 1986]. Coefficients of parentage were used to examine genetic diversity and compare trends in spring wheat cultivars grown in 1970-80s in two remoted regions (in Mexico and Pakistan). A small trend towards increased genetic diversity was found in Pakistan's data while in Mexico, large oscillations in diversity occurred without significant trends. It was concluded that cultivar improvement programs did not erode genetic diversity [Souza et al 1994].

Pedigrees of spring cultivars bred in Canada (47 cultivars studied), Mexico (90) and USA (133) were studied in three time-groups, those which were released before 1950 (group I), in 1950-75 (II), and after 1975 (III). It was shown that pedigrees of 270 cultivars studied included 124 parental forms from 32 countries (five continents), although only a few of these parental forms contributed to the genotypes of modern USA spring cultivars [van Beuningen & Busch 1997b]. The greatest, and the least diversity of parental forms in their pedigrees had cultivars of the group III, and I, respectively, from Mexico and USA. In the group III of Canadian cultivars (but not in cultivars of the USA or Mexico), there were some small decrease of genetic base, due to the presence of cv Thatcher in most of pedigrees. It is suggested that the observed decrease was caused by stringent requirements to grain quality in Canadian breeding programs [van Beuningen & Busch 1997a, b].

It was shown that the pool of original ancestors of 340 spring cultivars produced on the territory of Russia in 1929-2003, numbered 255 local and ancient varieties, including those from Europe (63), Asia (40), America (34) and Africa (12). A significant differentiation of ancient wheat was found according to their contribution to cultivars created in different regions. For example, Japanese wheat Akakomugi often participated in the pedigrees of cultivars of the North Caucasus and Western Siberia, although its contribution is insignificant in cultivars bred in other regions. The main contribution to the genotypes of cultivars bred in the Far East was made by Krymka and Iumillo, in cultivars from Volga region, it was English Goldendrop, and in the North of Russia, it was the local Indian wheat Gehun. Obviously, the set of parental forms used for crosses in a given region is determined by the environmental conditions of this region (see **9.8.**). Some qualitative alterations in the genetic base of cultivars over time accompanied by its widening were revealed. Since 1970s, the rate of expansion of the genetic base has increased significantly. The genetic diversity, judging by the number of original ancestors per cultivar, increased due to involvement in crosses of genotypes from different countries and regions [Dobrotvorskaya et al 2004].

It should be noted that the use of coefficients of parentage to the study of genetic diversity has disadvantages associated with the absence, incompleteness and inaccuracy of many pedigrees [Huang et al 1996; Cao et al 2000; Metakovsky, unpublished].

9.6. New breeding program: its impact on allelic frequency and genetic diversity

The direction of selection in breeding programs is subject to change. Do these changes affect the frequency of alleles at the *Gli* loci and the genetic diversity calculated from this frequency? In general, does the structure of common wheat polymorphism (see **9.1.**) alter with time?

At least ten statistically significant changes in the allelic frequencies were found in the group of 24 Australian cultivars released after 1970 compared to 16 cultivars studied registered before 1951: the diffusion of alleles *Gli-A1a* (%%) (33.3 vs. 0.0), *Gli-D1b* (37.5 vs. 6.3), *Gli-D1f* (41.7 vs. 6.3), *Gli-A2c* (41.7 vs. 12.5) and *Gli-B2c* (62.5 vs. 12.5) was accompanied by the drop of *Gli-A1m* (8.3 vs. 43.8), *Gli-B1i* (8.3 vs 37.5), *Gli-A2a* (14.6 vs. 62.5), *Gli-B2an* (4.2 vs. 31.3) and *Gli-D2w* (8.3 vs. 50.0) [Alleles 2018c]. Changes in genotypes of Australian cultivars could be a consequence of the transition from using local cultivars (Gluyas-Early, Nabawa and Ranee) to Mexican (CIMMYT's) wheats [O'Brien et al 2001]. Indeed, all alleles that became frequent in the group of newer Australian cultivars were found in Mexican cultivars, especially the *Gli-B2c* positively affecting wheat quality (see **4.10.**). Despite the dramatic qualitative changes in allelic frequencies over 20th century, each time-group of Australian cultivars kept the same genetic diversity. Thus, in 1901-41 (16 cultivars studied; 31 gliadin alleles were identified), 1951-66 (16; 31), 1967-73 (16; 30), and 1974-85 (15; 31), the diversity was found to be *H*= 0.66, *H*= 0.68, *H*= 0.69, and *H*= 0.65, respectively [Metakovsky et al 2019a].

In Canadian spring common wheat breeding programs, there has been a transition from using cv Marquis, as the main local genotype in crosses to cvs Thatcher and Neepawa. This shift caused, like in Australia, a dramatic alteration in allelic frequencies in the group of 11 cultivars released in 1981-88 as compared with 11 cultivars bred in 1907-69. In particular, *Gli-B1d* became much more frequent, and new frequent alleles *Gli-A1a*, *Gli-D1b* and *Gli-A2f* appeared in the genotypes of newer cultivars. At the same time, some alleles frequent in old cultivars were eliminated (*Gli-D1a*, *Gli-B2w*). or significantly decreased their frequency (*Gli-A1m*, *Gli-B1b*, *Gli-A2m*). A total number of alleles at six *Gli* loci increased in the group of newer cultivars (23 vs. 18). The average genetic diversity calculated for six *Gli* loci slightly increased (*H*= 0.494 vs. *H*= 0.466), especially at the *Gli-A1*(*H*= 0.69 vs. *H*= 0.32), *Gli-A2* and *Gli-D2* loci [Alleles 2018c]. A big variation in the generic diversity between time-groups of Canadian cultivars throughout the 20th century was observed (*H*= 0.55, *H*= 0.38, and *H*= 0.54 in groups of cultivars released in 1907-52, 1953-85, and 1986-88, respectively) [Metakovsky et al 2019a].

In Serbian wheat (breeding centres Novi Sad and Kragujevac), the frequency of the allele *Gli-B1l* increased significantly in the group of 21 cultivars registered in 1988-91 compared to 13 cultivars bred in 1960-76 (40.5% vs. 7.7%). As a result, the *Gli-B1b* allele that dominated (69.2%) in genotypes of older cultivars was partially displaced by *Gli-B1l* in the genotypes of newer cultivars increasing the genetic diversity at *Gli-B1* locus (*H*= 0.62 vs. *H*= 0.50). This alteration in allelic frequency in the group of new cultivar was associated with a new breeding strategy aimed at improving the resistance of cultivars to plant diseases. Therefore, cvs Avrora and Kavkaz having translocation 1BL.1RS in their genotypes (see **2.6.**), began to be frequently involved in crosses [Javornik et al 1991]. Genetic diversity calculated for six major *Gli* loci was found to be slightly higher in the group of new cultivars than in the group of old ones (*H*= 0.65 vs. *H*= 0.61) [calculated from Metakovsky et al 1991; Alleles 2018c].

Analogously, each fourth of 24 UK winter wheat cultivars studied released after 1980 had the allele *Gli-B1l* in its genotype whereas this allele was absent in 19 cultivars bred in 1936-80 (25.0% vs.0.0%). The appearance of the *Gli-B1l* increased the *H* at this locus (*H*= 0.70 vs. *H*= 0.64). The allele *Gli-B1l* was introduced in English wheat germplasm by cv Hedgehog descended from genotypes with translocation 1BL.1RS created in the 1930s. There were more differences in allelic frequencies between two time-groups, for example, a significant increase in the frequency of the allele *Gli-A2r* (25.0% vs.2.6%). The presence of this allele in the genotype may relate to increased cold resistance [Metakovsky

& Branlard 1998].

The level of genetic diversity of UK cultivars was found to be similar (about *H*= 0.63) in the two time-groups of cultivars studied [calculated from Chernakov & Metakovsky 1994]. Polymorphism of the HMW glutenin and gliadin [Cooke & Law 1998] and DNA polymorphism (RFLP, probes specific to grain storage proteins) [Donini et al 2000] were studied in 55 English winter cultivars bred in 1934-94. Some considerable qualitative changes over time were registered but no evidence for any reduction or erosion of diversity over time was found.

Also, alleles *Gli-B1l* and *Gli-A2r*, which were absent in German cultivars registered in 1952-68, appeared in the group of cultivars bred in Germany in 1980-89. The two groups of cultivars differed also in that the genotypes of newer cultivars lacked the homoallele *Gli-B1o* (family I, see **7.6.**). Allele of this family may negatively relate to the quality (see **4.7.**). The genetic diversity in the group of cultivars bred in 1980-89 was slightly higher (*H*= 0.67 vs. *H*= 0.62).

In spring wheat of Spain, three main differences were observed between two time-groups, 1927-66 and 1981-92, two substitutions of alleles, *Gli-D1i* for *Gli-D1b* and *Gli-D2a* for *Gli-D2j*, as well as the appearance of the allele *Gli-B2c*, whose presence in the genotype was positively associated with quality (see **4.10.**), in the second group. The genetic diversity did not differ (*H*= 0.77 vs. *H*= 0.79, respectively) between the two groups of cultivars studied.

In Romania, two time-groups of cultivars were studied (1962-71 and 1984-87) and two substitutions of alleles, *Gli-B1d* for *Gli-B1l* and *Gli-A2b* for *Gli-A2t*, were observed. In addition, the *Gli-D1b* became dominant at the *Gli-D1* locus. The genetic diversity did not differ in the two groups of cultivars (*H*= 0.64 vs. *H*= 0.62, respectively) [Alleles 2018c].

Among winter wheat cultivars grown in the south of Ukraine in 1910-60 (an average yield of 8-15 c/h), the most frequent alleles were *Gli-A1m*, *Gli-B1b*, *Gli-B1e*, *Gli-D1a*, *Gli-D1g* and *Gli-D1j*, while in cultivars released in 1961-95 (yield 30-40 c/h), such alleles were *Gli-A1b*, *Gli-B1b*, *Gli-D1b* and *Gli-D1g* (the designations of alleles used in the publication have been tentatively transformed, see **2.3.**). The alterations in genotypes were apparently associated with a new breeding strategy aimed mainly at increasing yield [Sozinov et al 1999].

Omsk is a large region of Russia having a specific eco-climatic conditions. In this region, spring cultivars bred in Saratov (Volga region) and other sources began to be widely used in breeding programs in the second part of the 20th century with the aim to combine a good flour/dough quality and a high adaptive properties of plants to the environments of West Siberia. So, about a half of Omsk's spring cultivars bred in the period from 1972 to 1993 involved winter cultivars (Bez1, Kavkaz, etc.) as a principal parent. As a result, noticeable qualitative changes in allelic frequencies, for example, the appearance of the allele *Gli-A1o* in the genotypes of newer cultivars (32.1% vs. zero), increase in the frequency of alleles *Gli-A1k*, *Gli-B1b*, *Gli-A2k* and *Gli-D2a*, and the disappearance of the *Gli-B1q* (zero vs. 28.6%) were registered. The genetic diversity of cultivars bred in Omsk throughout the 20th century varied from *H*= 0.63 (five cultivars, 1990-93) and *H*= 0.64 (six and seven cultivars, 1929-40, 1989-93, respectively) to *H*= 0.70 (eight cultivars, 1980-88) [calculated from Chernakov & Metakovsky 1994; Metakovsky et al 2019b].

The use of a new breeding program in a given country could affect, theoretically, the allelic frequencies and the genetic diversity of the released cultivars. Indeed, different degrees of qualitative changes in allelic frequencies might be observed. For example, Australian cultivars bred in 1901-49 and 1967-85 have only one frequent allele *Gli-B1b* in common [Metakovsky et al 2019b] while only a few significant changes in allelic frequencies were provoked in wheat genotypes bred in Serbia or UK. However, in all cases, no narrowing of polymorphism (genetic erosion) was observed at the country/big-region level even if large quantitative changes caused by the use of new wheat germplasm in breeding occurred in the genotypes of cultivars produced in a given country.

9.7. The effect of selection for end-use quality on genetic diversity

One of the main factors which may influence negatively on wheat genetic diversity is the selection, common for many countries during the second part of the 20th century, for improved end-use (dough) quality. Allelic variation at the *Glu* loci directly influences the dough quality (see **4.2.**, **4.4.**). A strict selection for high flour/dough quality, which was carried out for decades in many countries, caused an increase in the frequency of occurrence of certain alleles of *Glu* loci in the genotypes of cultivars of these countries [Morgunov et al 1993; Branlard et al 2003; Ribeiro et al 2011; etc.], as well as a narrowing of the genetic base for this trait in new varieties [van Beuningen & Busch 1997a]. Some significant differences between old and newer cultivars of Portugal in the frequency of variants of the *Glu-1* loci were revealed, in particular, increasing the frequency of alleles having positive relation to quality, *Glu-B1b* (from 3.7% to 40.4%) and *Glu-D1d* (from 9.8% to 67.3%). It is assumed that these alleles come into new genotypes due to new breeding programs aimed at increasing the flour/dough quality of wheat cultivars. However, there was no evidence of a negative effect of changes in the allelic frequencies at the *Glu* loci on wheat genetic diversity. For example, the genetic erosion was not found in the study of the polymorphism of the *Glu-1* loci in the collection of USA winter common wheat cultivars [Graybosch 1992].

Gliadin alleles may serve as an indirect targets of breeders' selection for the dough quality. The increase in frequency, in genotypes of new cultivars, of gliadin alleles positively influencing dough strength (for example, *Gli-B1b*, *Gli-B1d*, or *Gli-B2c* [Metakovsky et al 2019a]) accompanied by elimination of homoalleles at the *Gli-B1* belonging to the family I (see **7.6.**) whose presence in the genotype negatively correlates with quality (see **4.6.**, **4.9.**) might cause changes in the genetic diversity at the corresponding *Gli* loci.

The first "green revolution" took place in 1920-30 in Italy, where famous breeders N.Strampelli, C.Strampelli, G.Venturoli and C.Orlandi created new, outstanding for their time, wheat genotypes. New varieties combined a number of advantages (yield up to 80 c/h, resistance to lodging and rust, earliness), which not only allowed Italy to fully provide itself with wheat, but had a decisive influence on wheat breeding throughout the world [Boggini 1978; D'Amato 1989; Worland 1999]. Later, however, special requirements began to be imposed on the quality of flour of cultivars being bred [Boggini 1978; Borghi et al 1991]. To implement the strategy, several Mexican and Canadian cultivars (donors of *Gli-B2c*) as well as Bez1 (donor of *Gli-B1b* and *Gli-A2b*) were involved in crosses.

As a result, significant differences in the frequency of gliadin alleles between the group of old cultivars created in the 1930s and new genotypes registered after 1986 have been revealed in Italian common wheat germplasm. It was found that, in the group of 18 new cultivars, dominated alleles were *Gli-B1b*, *Gli-A2b* and *Gli-B2c* (the frequency of each about 30%), while none of these alleles were found even once in the group of 19 old cultivars ($P<0.02$). At the same time, frequency of alleles *Gli-B1k, Gli-A2p*, and *Gli-B2h* dropped statistically significantly (37% vs. 6%, 26% vs. zero, and 32% vs. 11%, respectively) [Metakovsky et al 1994; 1919a]. The exclusion from germplasm of the allele *Gli-B1k* was obviously a consequence of new breeding aims because the presence of this allele in common wheat genotype related to the low quality. In fact, *Gli-B1k* underwent hidden negative selection which, of course, none of the breeders knew or controlled. Despite the differences in the frequency of some alleles, the groups of old and new Italian cultivars have identical genetic diversity (about *H*= 0.73) and the number of identified alleles [Metakovsky et al 2019a]. Analogously, the frequency of homoalleles *Gli-B1k* and *Gli-B1o* was reduced in the group of 38 Spanish cultivars grown after 1982 (2.6%) as compared with group of 28 cultivars bred before 1966 (41.1%) [calculated from Metakovsky et al 2000]. The genetic diversity of newer Spanish cultivars was slightly higher (*H*= 0.73) as compared with older ones (*H*= 0.70) [Metakovsky et al 2019a].

Analogously, in northern Kazakhstan, there was a transition, in the breeding programs, from using local accessions to the wide use of successful cultivars from other regions. The first local cultivar in this region was a poor-quality Akmolinka-1 which did not inherit any *Gli* allele from cv Marquis which was claimed to be one of its two parents. To improve quality, alien successful cultivars bred in Saratov (Volga region) and other geographic and eco-climatic regions (Canada, West Europe, South America) were involved in crosses [Movchan 1988]. This caused the dramatic qualitative changes in the genotypes of produced cultivars, comparable in scope to those registered in Australian wheat (see **9.4.**). Comparison of two time-groups, each composed of 10 cultivars (1945-75 and 1979-93), showed that the Saratov alleles, but not the alleles of other foreign cultivars that were used, that began to dominate in the genotypes of Kazakh wheat germplasm at each *Gli* locus, *Gli-A1f*, *Gli-D1a*, *Gli-A2q*, *Gli-A2q*, *Gli-D2e* [Metakovsky et al 1988]. At the *Gli-B1* locus, the homoallele *Gli-B1m* of the "poor-quality" family I was completely eliminated (from 40% to zero) from newer cultivars. The local alleles dominated in genotypes of older cultivars, *Gli-A2ak* (this allele occurred with the frequency 70% of cultivars bred in 1945-75), *Gli-A1q*, *Gli-D1g*, *Gli-D1i* were eliminated also. The frequency of the last two alleles of Akmolinka-1, *Gli-B2d* and *Gli-D2s*, dropped strongly. The average genetic diversity calculated for six major *Gli* loci was similar, *H*= 0.623, and *H*= 0.595 in older, and newer cultivars, respectively. At individual *Gli* loci, *H* decreased at the *Gli-B1*, *Gli-D1* and *Gli-D2* loci and increased at the *Gli-A1*, *Gli-A2* and *Gli-B2* [Alleles 2018c].

Allelic frequencies and genetic diversity were studied in French winter cultivars bred in 1945-92, and changes in gliadin allelic frequencies were documented caused by the use of new alien genotypes in crosses prompted by Increased requirements for the quality of cultivars. Significant changes were registered at the *Gli-B1* locus where the allele *Gli-B1f* dominated in 13 cultivars grown in 1945-70 was partially replaced (the frequency dropped from 80.8% to 46.7%) by the *Gli-B1b* in 15 cultivars registered in 1988-96 (increased from 3.8% to 46.7%). As a result, the genetic diversity calculated for the *Gli-B1* locus was found to be much higher in the group of newer cultivars (*H*= 0.56 vs. *H*= 0.33). The average genetic diversity at six *Gli* loci was also higher in newer cultivars (*H*= 0.64) than in older ones (*H*= 0.56) [calculated from Metakovsky & Branlard 1998; Alleles 2018c].

In Ukraine, gliadin polymorphism was studied for the *Gli-1* and *Glu-1* loci in 28 winter cultivars bred before 1996 and 49 cultivars registered in 1996-2007 and some significant differences in the set of alleles between these two groups were found. Thus, in newer cultivars, alleles *Gli-A1w* and *Gli-B1h* appeared, the frequency of alleles *Gli-B1l* and *Gli-D1b* increased, and alleles *Gli-A1f*, *Gli-B1b* and *Gli-D1g* decreased. The genetic diversity for *Gli-1* loci was similar in two groups of cultivars studied (*H*= 0.658 and *H*= 0.641, respectively), while for six loci (*Gli-1+Glu-1*) it was noticeably higher in the group of newer cultivars (*H*= 0.554) in comparison with older ones (*H*= 0.461). The difference was due to the appearance of new alleles at the *Glu-B1* locus in the more recently bread group due to a new strategy of breeding aimed at improving quality of Ukrainian common wheat cultivars [Kozub et al 2009].

Thus, selection aimed at improving the flour/dough quality did not cause a decrease of wheat polymorphism. On the contrary, the use of successful high-quality alien germplasm in new breeding programs may increase, as studied by gliadin alleles, the genetic diversity of new cultivars.

9.8. Regional "specialization" over time of wheat polymorphism

In contrast to world germplasm and country scales (see **9.1.**, **9.4**), a small, but clear time-based narrowing of genetic diversity at the regional scale could be observed using gliadin alleles. For example, in groups of cultivars bred in the Australian State New South Wales, the genetic diversity in cultivars bred in 1901-60 (eight cultivars), 1945-72 (eight), and 1973-83 (nine) was *H*= 0.63, *H*= 0.61, and *H*= 0.53, respectively. Similar decrease was found for cultivars bred in the state of Victoria, from *H*= 0.60 (seven cultivars bred in 1922-56) to *H*= 0.50 (six cultivars, 1962-85) [Metakovsky et al 2019b].

In two Eastern regions of Canada (Manitoba + Ontario), the decrease in genetic diversity was especially pronounced, from *H*= 0.506 (1907-59, seven cultivars) to *H*= 0.284 (1965-86, eight cultivars) ($p < 0.05$). This decline was accompanied by the propagation of alleles *Gli-B1d* ($p < 0.02$) and *Gli-B2c* ($p < 0.05$), both related to better dough quality (see **4.7.**, **4.10.**)

and, conversely, by the disappearance of the *Gli-D1a* and *Gli-B2w* in the genotypes of more recently bred cultivars [Metakovsky et al 2019b].

In the Saratov breeding centre (Volga region) of the former USSR, genotypes selected from local landraces Selivanovskii-Rusak and Poltavka (see **10.1.**) in the beginning of the 20th century were exploited in breeding for many years. Besides, some "foreign" cultivars (Kitchener, Selkirk, etc.) were used in crosses. As a result, several high-quality cultivars well-adapted to difficult eco-climatic conditions of the Saratov region were produced. In the genotypes of these cultivars, however, only local alleles (only a small part of their original diversity) were present. That resulted in a significant ($p < 0.05$) loss of genetic diversity of Saratov cultivars to a record-low level: each of ten cultivars registered in 1963-76 carried alleles *Gli-B1e* and *Gli-D1a* (no polymorphism at these loci), and at least seven of them had alleles *Gli-A1f*, *Gli-A2q* and *Gli-D2e*. The genetic diversity in this group was $H= 0.30$, while in 17 cultivars bred in 1923-47, it was $H= 0.59$. Alleles from "foreign" wheat germplasm could scarcely compete with local alleles of a good, but too "specialized" Saratov wheat genotype. So, only a few alien (non-Saratov) alleles (*Gli-A1m, Gli-A2m*, *Gli-B2w*) occurred in the genotypes of seven cultivars released in 1977-87 increasing the genetic diversity of the group up to $H= 0.46$ [Metakovsky et al 2019b]. The average index of relationship for pedigrees of cultivars bred and registered in Saratov during the 20th century was as high as half-siblings and higher (Dobrotvorskaya et al 2004).

It can be assumed that the regionally-specific natural selection may cause some kind of temporal specialization of polymorphism for rather small agricultural regions of one country [Perronner et al 2017; Metakovsky et al 2019b]. Breeders hesitate to involve remoted genotypes for improving cultivars bred for a given region [Chen et al 1994]. The appearance of new alleles in an already differentiated genotype, well-adapted to specific natural conditions, should almost always have a negative effect on the adaptive properties of the plant [Thomas & Schaalje 1997]. To breed cultivars with improved adaptive properties is the aim of many breeding programs [Ehdaie & Waines 1989].

Cultivars produced in regional breeding centres of the same country can vary greatly in their genetic diversity, for example, Omsk ($H= 0.75$) and Saratov ($H= 0.51$) of Russia [Metakovsky et al 2019b], or Martonvasar and Szeged of Hungary (see **9.1.**). Groups of cultivars produced by French breeding centres INRA or Rustica have a high level of diversity ($H= 0.720$ or $H= 0.724$, respectively), while the cultivars released by breeding group Mennesson are similar to each other [Metakovsky & Branlard 1998].

There are several groups (classes) of the USA common wheat cultivars (HRS, HRW, SWS, SWW, WC) differing in technological parameters and adapted to certain environmental conditions. The use of molecular markers (PCR, RFLP) showed that genotypes of elite cultivars of the same class have up to 88% similarity to each other [Chen et al 1994] and that genetic distances between cultivars of the same group were much shorter than between cultivars from different classes [Barrett & Kidwell 1998]. So, there is obvious specialization of wheat genotypes belonging to the same group.

The specialization of genotypes grown in different regions of one country may increase differences between genotypes bred in these regions. As a result, the index *H* calculated for a country composed of distinct regions may unexpectedly show an elevated genetic diversity of wheat at a country level. For example, the group of common wheat cultivars registered in former Yugoslavia during 30 years starting from mid-1960s, has a relatively high level of genetic diversity calculated based on allelic frequency at the *Gli* loci, *H*= 0.73. However, Yugoslav wheat germplasm consisted of two quite different parts, the cultivars produced in Croatia (breeding centres Osijec and Zagreb) and Serbia (breeding centres Kragujevac and Novi Sad). It was found that groups of cultivars bred at the same time in Croatia and Serbia differed strongly in the set of alleles at the *Gli* loci (see **9.1.**) but had similar level of the genetic diversity, (*H*= 0.67 and *H*= 0.66, respectively). So, the genetic diversity in each of these countries is much lower than in the former Yugoslavia as a whole [calculated from Metakovsky et al 1991].

The same mechanism can operate at the species level as well. Indeed, cultivars bred in different countries and regions may have quite different sets of alleles in their genotypes showing, all together, increased level of generic diversity *H* in comparison with each group of cultivars studied. separately.

9.9. Both the structure of polymorphism and genetic diversity depend on natural selection

The interaction between organism and environment is central in evolution of wild cereals. Environmental factors of stress, abiotic (solar radiation, temperature, drought, excess rainfall, soil properties, mineral poverty, etc.) and biotic (pathogens, parasites) select genotypes mostly adapted to specific (the given) eco-climatic conditions decreasing the genetic diversity [Allard 1988; Nevo 1988; 1994; 2011; Nevo et al 2012]. Moreover, the available laboratory tests are of limited value in screening cultivars because of the high genotype x environment interactions and of the weak association with the performance under field stress conditions [Borghi et al 2014].

Obviously, different environments will select the different most adapted genotypes. There is a significant difference between populations of wild-growing self-pollinating plants detected using allelic frequencies of polymorphic genetic markers [Worland et al 1994; Hamrick & Godt 1997]. Using analysis of polymorphism of three enzyme systems in populations of several species of wild wheat, it was shown that, in each species studied, the inter-population genetic diversity was always greater than intra-population one [Asins & Carbonell 1986]. It is possible to say that populations of the same species are different because they are "specialized".

Analysis of morphological characters, enzymes, grain storage proteins [Nevo et al 1995] and DNA molecular markers [Pagnotta et al 1995] showed that populations of wild tetraploid (AABB) wheat *T.dicoccoides* differed substantially in the frequency of their alleles and genotypes [Nevo 1994]. There are weighty arguments that a number of environmental eco-climatic factors (climate, peculiarities of soil, plant diseases) might cause a divergence and differentiation of polymorphism observed between these populations [Nevo et al 1982; 1988; 1991; 1995]. The difference between populations was observed both for individual genes and multilocus associations [Golenberg & Nevo 1987]. Due to polyploidy, wheat has a hidden phenotypic variation that promotes the plasticity of wheat and its high ability to adapt to new or changed environmental conditions of growth [Krasileva et al 2017].

Adaptive changes were discovered in microsatellite allelic composition and flowering time comparing the genotypes of 10 independent populations of *T.dicoccoides*, which were sampled firstly in 1980 and again in 2008. It was found that each of populations studied displayed, with time, earliness and a highly significant reduction of the number of microsatellite alleles. The higher environmental correlations in 2008 than in 1980 were interpreted as an indication of stronger adaptation of the populations. Climate change is suggested to be a major environmental stress threatening biodiversity [Nevo et al 2012].

Thus, the use of *T.dicoccoides* as an example made it possible to demonstrate that genetic diversity is structured according to ecological and geographic conditions, is related to the influence of the external environment. Finally, the structure of polymorphism of a wild plant species is a product of natural selection [Nevo & Beiles 1989; Nevo 1994].

There seems to be a logical analogy between a group of cultivars (genotypes) bred in a given region and a group of plants (genotypes) composing a population of wild wheat: in both cases, larger genetic distances were observed among regions (populations) and less inside each region (population). Breeders' selections for end-use wheat quality, more or less similar in different countries, would not cause the world-wide structuring of polymorphism observed in common wheat germplasm. Therefore, the differentiation of the polymorphism studied of common wheat among countries and regions (see **9.1.**) may have adaptive significance being the result of natural selection, and the environment, in the form of natural selection, could participate in the creation of wheat genotype acceptable for a given region. It was shown that the environmental factors, including climate changes (temperature regime), strongly influence the allelic frequencies of storage protein loci in time-groups of common wheat cultivars [Kozub et al 2020], analogously to that described [Nevo et al 2012] for wild populations of *T.dicoccoides*. The genotype and environment is of similar importance for the determination of the content and composition of the compounds within the wheat grain determining quality of cultivated wheat [Johansson et al 2020]. Effects of dwarfing genes can be modified by genetic background and year conditions [Chebotar et al 2012b].

The decisive influence of eco-climatic conditions and natural selection on the level of genetic diversity in groups of cultivars is evidenced by the fact that the highest level of common wheat genetic diversity in Europe is observed in the south-eastern region, which is most favourable for wheat cultivation [Huang XQ et al 2002]. On the contrary, in the relatively less favourable conditions of Canadian region of Manitoba or in the region of Saratov (Russia), much lower level of genetic diversity of cultivars was observed throughout the 20th century (H= 0.41, H= 0.51, respectively) [Metakovsky et al 2019b]. Significant ($P<0.001$) differences in the genetic diversity of groups of spring cultivars depending on the cultivation region were found when analysing coefficients of parentage of cultivars (see **9.5.**) bred in different regions within Russia [Dobrotvorskaya et al 2004].

In western Canada (province Alberta), the elite genotypes of winter common wheat that display a survival ability were tall, low in yield and weak strawed. The role of environment in the selection of genotypes having a high cold resistance was confirmed using a mixture of cultivars. As a result, no improved type with good survival ability was found to be as hardy as the elite lines [Thomas & Schaalje 1997]. Consequently, in hard climatic conditions, natural selection, which operates throughout the breeding process, will limit the breeder's options and contribute to narrowing the genetic diversity of cultivars adapted for these conditions. Thus, groups of common wheat genotypes (cultivars) produced in regions with more favourable eco-climatic conditions would have higher genetic diversity than cultivars grown in more difficult conditions.

9.10. Some *Gli* alleles may relate to wheat adaptability

Photoperiod response genes play a mayor role in determining the climatic adaptability of wheat cultivars. In European winter wheat, time of flowering and subsequent grain's maturity determine the yield in a given eco-climatic conditions. Photoperiod insensitivity in European wheat is probably determines by a *Ppd1* allele. Besides accelerating ear emergence time, there are pleiotropic effects of this allele such as the reduction of plant height, tillering, and spikelet numbers, but also increase of spikelet fertility providing an increased numbers of grains per spike [Worland et al 1998]. It was shown that genotypes of modern cultivars bred in different countries carried genes predicted to promote the best levels of adaptability to precise environmental conditions. For example, all cultivars bred in UK are late flowering due to the presence of allele *ppd1* [Worland et al 1994]. In southern Europe, early flowering *Ppd1* genotypes produce larger grain and grater yields. In England and Germany, the pleiotropic effects of the *Ppd1* on yield vary annually depending on weather conditions. In Central European countries, breeding of photoperiod sensitive or photoperiod insensitive cultivars have a clear adaptive significance, and the annual variations in climate make it extremely difficult for breeders to produce varieties with good adaptability to changing environmental conditions [Worland et al 1998].

One of genes influencing the heading date is located on the chromosome 6A [Zemetra et al 1986]. Spontaneous null mutation at the *Gli-A2* located on this chromosome revealed by analysis of gliadin electrophoregrams affected earing time so that it was 6-7 days late compared to control [Lafiandra et al 1987b]. It was also discovered that allelic variation at the *Gli-A2*, *Gli-B1* and *Gli-D2* correlated with time of flowering of French cultivars. Indeed, alleles *Gli-B1b*, *Gli-A2f*, *Gli-A2t* and *Gli-D2m* were significantly more frequent in group of the earliest cultivars that were grown in the south of the country. The alleles *Gli-D2g* and *Gli-D2n*, and *Gli-D2m* and *Gli-D2j*, occurred more frequently ($P<0.05$) among cultivars grown in the north (late flowering), and south (earliness), areas of France, respectively. The allele *Gli-B1l* (1BL.1RS) occur only in late cultivars that are grown only in the north of the country [Metakovsky & Branlard 1998].

The presence of some alleles at the *Gli* and other storage-protein-coding loci in the genotype may mark the ability of wheat plants to adapt to specific eco-climatic conditions and stress factors. For example, the presence of the allele *Gli-B1b* (family III, see **7.6.**) in the genotype of Ukrainian winter wheat plant provided better survive rate compared to *Gli-B1k* (family I) [Poperelya et al 1980]. In French germplasm, alleles *Gli-B1l*, *Gli-A2r* and *Gli-A2g* were the most frequent among cultivars having the highest cold resistance [Metakovsky & Branlard 1998]. In Polish cultivars, the correlation was noted between the presence of the allele *Glu-A1c* in the genotype and high frost resistance. This finding explains the high frequency of the allele *Glu-A1c* among Polish cultivars, despite its negative impact (see **4.3.**) on flour quality [Witkowski et al 2008]. The relationship between gliadin and glutenin alleles and cold tolerance is unexpected because, in common wheat, the major genes controlling frost resistance are located on the chromosomes 5A and 5D [Sutka 2001].

Eco-climatic conditions differ in northern and southern zones of Ukraine. Differences in the genotypes of cultivars (sets of alleles at the *Gli* loci were studied) bred for growing in the south or the north of Ukraine were documented [Sozinov et al 1999]. Polymorphism at the *Gli-1*, *Glu-1*, *Gli-A3*, *-A6* and *-B5* loci was studied in 362 winter common wheat cultivars grown in Ukraine, and significant differences in frequencies of alleles between groups of cultivars bred in different soil and climatic zones were revealed. Alleles characteristic for a given zone were reproduced in different time-groups of cultivars bred in this zone [Kozub et al 2017].

Significant differences were found in the frequencies of gliadin alleles between genotypes of spring cultivars grown in in Western and Eastern Siberian regions of Russia. It was shown that some genotypes occurred in the Eastern Siberia only [Nikolaev et al 2009].

Considerable temporal changes in the assemblage of genotypes were documented in studies of the hybrid experimental populations of winter [Kolyuchii & Sozinov 2000; Raquin et al 2008] and spring [Koval & Metakovsky 1985] common wheat, and they were interpreted to be caused by natural selection in a specific environment. Acting through adaptively valuable genes and their associations, natural selection could change the frequency of occurrence of polymorphic genetic markers (including alleles at the *Gli* loci) in breeding material in a given region. For example, in the population created by crossing five parental cultivars of winter wheat, a considerable increase (in 2.4 times) was observed after 10 years of reproduction (under the action of natural selection in one of Ukrainian regions) in proportion of plants having the allele *Gli-A2n* [Kolyuchii & Sozinov 2000]. The presence of this allele in the genotype was assumed to be associated with a better survival of plant in winter. In Siberia (region of Novosibirsk), the presence of the alleles *Gli-B2b* and, especially, *Gli-B2o* in the genotype led to relative increase in the yield and weight of 1000 grains [Koval & Metakovsky 1985; Metakovsky et al 2006].

9.11. Summary. Natural selection acts against wheat genetic erosion

Finally, it can be suggested that each newly-bred cultivar is always the fruit of the joint efforts of a breeder pursuing breeding aims and natural selection acting in a given region at all stages of the breeding process, from the choice of parental genotypes (only a short list of genotypes successful for a given area or region is available which is mainly determined by the eco-climatic conditions of the region) and producing crosses to propagation of the finally selected genotype. Natural selection acts in different ways in different regions, determining and limiting the level of genetic diversity of wheat grown in that region. Therefore, successful genotypes may perhaps contribute to a narrowing of polymorphism. Some results do indicate a temporal genetic erosion, but in a few cases and at the regional level only. It is difficult to imagine a super-successful wheat genotype which would fit any combination of eco-climatic conditions of growing at the level of world-wheat-germplasm or even at a country level. Therefore, unavoidable erosion of genetic diversity of cultivated *T.aestivum* at a species level would occur only if eco-climatic conditions became identical everywhere where wheat is grown.

The genetic base for the released cultivars broaden when new donors of useful traits are used in breeding programs, for example, increasing resistance to plant pathogens. In this case, both natural selection and the breeder work towards increasing plant adaptability. Local varieties and landraces have a particular value in breeding as a source of new alleles and their associations improving adaptation of wheat cultivars for the conditions of growth in a given region.

The high level of genetic diversity maintaining by common wheat over time indicates that the movement of alleles from country to country, as well as breeder's selection, does not significantly affect the structure of polymorphism of wheat as a species. The erosion of genetic diversity that could be caused by the frequent use of a limited number of "the best" successful genotypes, is resisted by a powerful force, natural selection in the given specific eco-climatic conditions of a country or region. It is natural selection that determines which alleles will be successful in this particular conditions, and which will be rejected. It is impossible to imagine that any wheat "super-genotype" includes alleles that will be successful in any eco-climatic conditions, will be picked up by natural selection and inherited by the new cultivar. A high level of diversity of common wheat germplasm of the 20th century was, is, and will be maintained over time due to diversity, on a world-wide scale, of natural selection in different eco-climatic environments of wheat cultivation and, in addition, due to distinct breeders' aims in their breeding programs.

It can be concluded that natural selection contributes to the maintenance of the genetic diversity of cultivated wheat. The use of outstanding donor cultivars in breeding programs does not cause erosion of genetic diversity either at the country level or in wheat as a species.

The difference of common wheat genotypes bred even in the geographically close regions can be used for important practical purposes, for example. to establish, in the case of alleged theft of grains, the origin of these grains.

10. WHEAT GENETIC RESOURCES AND THEIR CONSERVATION AND MAINTAINING

10.1. High levels of polymorphism in old local varieties and landraces

Ancient local varieties, landraces and populations of common wheat grown in a particular area for decades, if not centuries, are highly adaptive and resistant to the natural conditions of a particular region of cultivation. This germplasm is of particular breeding value as a source of alleles and their associations to improve the adaptability of cultivars produced in a given region [Allard 1988; Ehdaie & Waines 1989; Reif et al 2005; etc.].

Typically, genetic diversity of the landraces of common wheat is considerably higher than that of modern cultivars bred in the same region [Hao et al 2006; etc.]. An accession of landrace or local variety is frequently composed of many genotypes distinguishable using suitable genetic markers. Old Australian heterogeneous varieties Bobin and Gular [Wrigley & Shepherd 1977] are a good example of such polymorphism. Hungarian common wheat old variety Bànkùti-1201 had the nature of population containing several genotypes with different gluten protein alleles [Juhàsz et al 2003]. The oldest cultivar of the north of Kazakhstan, Akmolinka-1, was bred using local wheat germplasm. It consists of 16 genotypes differing in the set of alleles at four *Gli* loci [Metakovsky et al 1988].

Every second of 158 accessions of old Italian durum wheat studied using the analysis of gliadin and microsatellites was found to be heterogeneous [Figliuolo et al 2007]. Gliadins of 98 accessions were studied of local durum wheat cultivated in Bulgaria in the 19th to the first quarter of the 20th century. It was found that 61 accessions were composed of two or more genotypes, and the genetic diversity of the material studied was H= 0.70 [Melnikova et al 2010].

Local common wheat Barbela which is still grown by farmers in several northern regions of Portugal was studied. As a result, polymorphism of microsatellites was detected both between and within 59 accessions studied. The average index of the genetic diversity H of genotypes present in one sample was H= 0.52. High genetic diversity both within each accession and in the group of genotypes as a whole has been preserved for decades [Ribeiro-Carvalho et al 2004].

Accessions of seven groups (agrotypes) of Spanish local common wheat varieties were studied. The classification of agrotypes historically developed in farms is based on the agro-morphological characteristics of plants. Accessions were maintained in the collection of CRF-INIA (Spain) for over 50 years with a few reproductions during this time. Each accession was accompanied by a "control spike" which was kept without reproduction. As a result, all six accessions of the agrotype Barbilla originated from different locations of Spain differed among themselves, and their genotypes included some alleles at the *Gli* loci not previously registered. In sum, from four to six alleles at one *Gli* locus were identified in Barbilla. General genetic diversity for the six accessions studied was *H*= 0.70. In another agrotype, Candeal, 11 accessions were studied grown in different regions. This germplasm was found to include 17 genotypes, 39 alleles at the *Gli* loci (10 of them were not observed in world-wide registered cultivars studied [Alleles 2018c]) and genetic diversity as high as *H*= 0.71. Similar results were obtained in analysis of other agrotypes (Blanquillo, Hembrilla, Jeja, Negrete, Rojo): there was a difference both between agrotypes and between genotypes of different accessions of the same agrotype [calculated from Ruiz et al 2002a; Metakovsky, unpublished].

In 44 individual seeds studied of the authentic sample (grains were stored without reproduction since 1930s) of Russian landrace Selivanovskii-Rusak grown in Volga region in the 19th and beginning of the 20th centuries, at least 20 alleles at six major *Gli* loci were identified, from two to six alleles per locus. Some of these alleles occur frequently in modern cultivars bred in Saratov (*Gli-A1i*, *Gli-A2q*, *Gli-A2s* and *Gli-B2q*). Genotypes of the four (of 44) single seeds studied included alleles *Gli-A1i*, *Gli-A2s* and *Gli-B2q* simultaneously. These genotypes were probably used by breeders at the beginning of the 20th century when producing the first cultivars in the Saratov breeding centre. An authentic sample of another important spring landrace grown in Russia, Poltavka, was less heterogeneous. Among 45 studied seeds, only 13 different genotypes were identified. However, 18 seeds of the sample studied were identical having alleles *Gli-A1f*, *-B1e*, *-D1a*, *-A2q, -B2s* and *-D2e* in their genotypes. These alleles are typical for all generations of Saratov's cultivars bred throughout the 20th century. Moreover, about 70% of modern spring cultivars bred in Russia were bred with the participation of Saratov genotypes [Dobrotvorskaya et al 2004]. Some gliadin genotypes found in Poltavka or Selivanovskii-Rusak included alleles not found in common wheat germplasm [Alleles 2018c], as well as, more interestingly, alleles that are very common among modern world-wide cultivars, but scarcely occur in Saratov's cultivars, for example, *Gli-B1b* (20 of 44 seeds of Selivanovsii-Rusak studied) and *Gli-D1i* (22 seeds) [Sozinov et al 1986; Metakovsky et al 1987a; Metakovsky unpublished].

Analysis of polymorphism of gliadin in 87 accessions from eight countries and regions (collection of the All-Union Institute of Plant Growing, St-Petersburg, Russia), showed an exceptionally high level of heterogeneity of some accessions. For example, in eight samples from Afganistan, a total 64 variants of the APAGE electrophoregram were observed. Samples к-8200 from the Crimea, and к-38453 from the European part of Russia were mixtures consisting of 14, and 15 genotypes, respectively [Dragovich et al 2009]. In 158 Iranian landraces, 57 haplotypes of glutenin combinations including five new HMW glutenin subunits were found [Maryami et al 2020].

Allelic composition at the *Glu-1* loci in 100 lines (seven botanical varieties) based on 14 landraces of Spanish wheat cultivated in Mexico during 16th-18th centuries (currently only in collections) were studied. As a result, 21 allelic variants of the *Glu-1* loci (five of them were not observed in modern common wheat) were identified and the genetic diversity was calculated to be H= 0.547. Some of alleles were found in one line only. That indicates the risk of their loss and, therefore, erosion (see **9.8.**) of the genetic pool studied [Caballero et al 2010].

10.2. Monitoring of wheat polymorphism in collections

At the end of the 20th century, up to 750 000 samples of wheat were stored in various collections [Börner et al 2000]. It is believed that the maintenance of the genetic diversity of cultivated wheat as a species depends largely on the preservation of the gene pool of landraces and local old and not used cultivars in the collections. The polymorphism observed in wild wheat species (for example, in *T.dicoccoides*) can be used as a sources of valuable traits for improving genotypes of cultivated wheat [Feldman & Sears 1981; Ciaffi et al 1992; Cakmak et al 2004; etc.].

Landraces, local and old varieties are usually extremely heterogeneous and exhibit high levels of genetic polymorphism (see **10.1.**). Currently, the most of up-to-day's wheat germplasm polymorphism is preserved only in collections. Therefore, monitoring of wheat polymorphism in collections is necessary for its adequate preservation.

However, there are some observations that, in big collections, wheat accessions may lose its polymorphism. For example, analysis of the polymorphism of microsatellites and gliadin showed a partial loss of genetic diversity in the collection of old samples of durum wheat in Italy [Figliuolo et al 2007]. One of two initially heterogeneous (as it was established through microsatellites analysis) accessions has lost a part of its heterogeneity being stored, with regular reproductions, in the Gatersleben collection (Germany) [Börner et al 2000]. The cv Maris-Freeman in 1970s was found to consist of two gliadin biotypes [Ellis & Beminster 1977], but the grain sample of this cultivar obtained from PBI (Cambridge) in late 1980s was homogeneous [Chernakov & Metakovsky 1994].

There were several cases of the presence of some gliadin allele in the genotype of a daughter cultivar, which was absent in parental cultivars mentioned in the pedigree but present in the genotype of grandmother cultivar. The only explanation for this fact is the loss of some of polymorphism, including this allele, by one of the parental cultivars studied during the maintenance in collection after its use in the cross. For example, the allele *Gli-A2g* is present in the daughter cv Score and its grandparents Cappelle-Desprez and Splendeur, but not in its parental cvs Horace and Maris-Huntsman. Analogously, the sample studied of the cv Maris-Hobbit did not have allele *Gli-A1ab* (this allele was found to differ from *Gli-A1b* [Blocks 2018c]) although some of its parents and descendants had this rare allele [Metakovsky & Graybosch 2006].

The problem that may affect the use of germplasm store for agricultural purposes is that any wheat collection may contain a considerable quota of misclassified accessions, or, as a result of various kinds of errors, the sample may be labelled incorrectly [Hurka 1994; Ribeiro-Carvalho et al 2004; Caballero et al 2010; etc.]. For example, in the earliest works in which gliadin polymorphism was used for establishing parents for the cultivar under study, it was found that the grain sample of the cv Bobin-Selection-W39 was not authentic [Wrigley & Shepherd 1977]. The results of the identification of the wheat-rye translocation (see **2.6.**) in the cv Charodeika were contradictory. It was explained through the lack of authenticity of one of samples of this cultivar studied [Landjeva et al 2006b].

The usefulness of a germplasm collection depends on how well it represents the genetic diversity of a crop. However, in a big collection, some accessions are not authentic. More than 2000 accessions out of more than 11.6 thousand recorded as durum wheat (from 60 countries) from the ICARDA (International Centre for Agricultural Research in the Dry Area) collection were found to be hexaploids [Pecetti et al 1992]. One grain sample (к-49068) of common wheat out of a few studied from the collection of the All-Union Institute of Plant Growing (Russia) was a mixture of hexaploid and tetraploid wheat [Dragovich et al 2009].

Sixteen of 124 accessions (13%) from National Small Grains Collection USA, labelled as hexaploids *T.macha*, *T.spelta* and *T.vavilovii*, were found to be diploids and tetraploids. DNA markers (RAPD) were used for analysis of 12 accessions classified as hexaploids (*T.macha* or *T.vavilovii*) but suspected to be misclassified based on their morphological characters. Indeed, these 12 accession were 5 tetraploids, 6 diploids and one sample of *T.timopheevii* [Cao et al 1999]. Also, four out of 69 accession of *T.spelta* studied were misclassified [Cao et al 1998]. About 30% of old Spanish accessions of common wheat were not correctly labelled compared to authentic spikes stored since 1950s [Ruiz et al 2002b].

Up to 15% or more duplicating genotypes occurred in big wheat collections [Pecetti et al 1992; Cao et al 1999]. The RAPD genetic markers were used for analysis of *T.spelta*, and several genotypically identical samples of this wheat were revealed [Cao et al 1998]. It was suggested that the removal of duplications would optimize the genetic diversity of the collection [Ruiz et al 2011]. The genotypes of 50 durum wheat accessions selected from natural populations in the beginning of the 20th century and since that time maintained in the collection were examined using polymorphism of gliadin, high molecular weight glutenin and microsatellites. It was shown that less than half of the supposedly duplicating genotypes could not be distinguished using these genetic markers [Ruiz et al 2011]. Probably, duplicate samples are unlikely to pose a serious problem, since genotypes identical when given genetic markers are applied, may differ when using other markers.

Much more serious problem is the authenticity of accessions of cultivars used in scientific studies. For example, grain samples of the cv Brimstone, and of the cv Insignia, each being studied in two independent works [Jackson et al 1996 and Branlard et al 2003], and [Gupta & Shepherd 1990 and Branlard et al 2003], respectively, differed considerably in the set of alleles at the *Glu-3* loci. Thus, in our studies, only 43 of 48 cultivars showed the same *Gli* allelic composition in two or more grain samples obtained from different collections. For example, no common *Gli* alleles in genotypes of four samples of Canadian cv Selkirk obtained from Australia, Canada, France and Russia were found; gliadin genotypes of grains in these four samples were completely different. Subsequent pedigree analysis confirmed the authenticity of the sample from Canada. Gliadin genotypes of grains of French cv Marius (samples from France and Spain) had nothing in common. Grain samples of the Dutch cv Opal were obtained from breeding centres of Australia and Russia. As a result of the analysis, both these samples were recognised to be no-authentic because, firstly, their genotypes were completely different, secondly, they included *Gli* alleles characteristic (see **9.1.**) for Australia and Russia, respectively, and third, no correspondence to pedigree (parents, descendants) of cv Opal was observed [Metakovsky & Graybosch 2006].

The absence of the correspondence between gliadin genotype revealed in the grain sample under study and its pedigree is an another problem of grains in collections. For example, all cultivars participated in 194 pedigrees (a daughter cultivar, its parents and some descendants, if exist) were studied, and only 134 pedigrees appeared to be completely correct. On the contrary, grains (always obtained from the country of origin of the cultivar) of cvs Abbondanza, Forlany (Italy), Admiral (UK), Orepi, Scipion (France), Sayanogorsaya (Russia), Spear (Australia), GK Szeged (Hungary), etc. carried in their genotypes several gliadin alleles absent in their corresponding parents whose genotypes were confirmed in other pedigrees [Metakovsky & Graybosch 2006]. The lack of correspondence between some daughter and its parental cultivars in 30% of pedigrees was found in the analysis of glutenin polymorphism in French wheat [Picard et al 1992]. Similar results were obtained in the analysis of polymorphism of gliadin in 38 cultivars of Syria and 58 genotypes (lines) from different world collections [MirAli 2002]. The absence of the correspondence between genotype of the cultivar and its parents and descendants may be explained either through errors in pedigrees, or through careless maintenance of cultivars in collections, including small local ones harboured in the institutions of wheat genetics and breeding.

10.3. Admixtures in wheat-grain samples

One more problem of world-wide common wheat germplasm is an impurity of their grain samples. An off-type (admixture, contamination, foreign genotype) is defined as not being a genuine part of a given cultivar, although being found in the grain sample under study [Metakovsky et al 2020c]. There are three characteristics of off-types distinguishing them from authentic biotypes. Firstly, an impurity occurs in a given grain sample at a low frequency. Secondly, the genotype of the off-type differs markedly from that of the cultivar. Finally, alleles present in the off-type's genotype are absent in parents of the cultivar under study.

In some earlier publications, almost any difference observed between single-seed gliadin electrophoregrams of a cultivar was regarded as the presence of "biotypes" in this cultivar. However, the application of knowledge about the inheritance of gliadin polypeptides (see **2.2.**) allows to determine the nature of the inter-varietal non-uniformity found. For example, differences in gliadin electrophoregrams between single seeds in the grain samples of cvs Alpe and Verna were called "biotypes" [Pogna et al 1982b]. Later, the difference in the gliadin pattern between seeds in cv Alpe was found to be a null mutation [Pogna et al 1995] (see **7.1.**). Our analysis of gliadin electrophoregram of cv Verna showed that the observed off-type seed [Pogna et al 1982b] was an admixture differing from the cultivar at least at three *Gli* loci simultaneously [Metakovsky & Graybosch 2006]. Another grain sample of this cultivar was pure [Peruffo 1982].

The admixture may be present in the grain sample in a significant amount. The grain sample of Russian cv Milturum-553 was composed of two genotypes, A and B, differing at the three *Gli* loci simultaneously. It was established that Milturum-553-A could originate from the initial cross (Milturum-351 x Kitchener) and contribute its alleles to several its descendants. On the contrary, the genotype B was recognised as an admixture although both genotypes occurred in similar amounts in the grain sample studied [Metakovsky & Graybosch 2006].

Off-type may appear in a grain sample under study as a descendant of a casual out-crossing between wheat genotypes and subsequent genetic segregation of the hybrid. In this case, the genotype of off-type would include some alleles characteristic for the cultivar under study as well as foreign alleles which are not present in parental cultivars [Sobko & Sozinov 1999]. Heterozygotes at some loci may be observed also. For example, the sample of the cv Priazovskaya obtained from the breeder contained about 30% of off-types. Genotypes of admixture grains differed, and some of them were heterozygous at some *Gli* loci. Obviously, this impurity appeared as a result of genetic segregation of recent out-crossing of plant(s) of the cultivar [Metakovsky & Kopus 1991]. Ten out-crossed seeds were registered out of 205 studied among nine common wheat cultivars [Pogna et al 1986]. Some common wheat genotypes, for example those carrying the 1BL.1RS translocation, demonstrate an increased tendency to out-crossing, up to 5.1% [Kozub et al 2008a]. The Russian cv Ilichevka was stopped to be used for growing due to numerous impurities caused by out-crossing [Kolyuchii & Sozinov 2000].

Admixture may also appear as a result of handling with grain samples. In this case, genotypes of the cultivar and off-type seed would differ at several genetic loci simultaneously [Metakovsky et al 2019c]. For example, off-type grains in the cv Moskovskaya-35 considered to be "biotypes" [Nettevich et al 1981] were clear admixtures because their genotypes include some alleles absent in genotypes of parents and relatives of the cultivar and because of their low (2.2%) frequency in the grain sample studied. Most of off-types observed in durum cultivars could be contaminants from other cultivar sources [Kosmolak 1979].

The frequency of cultivars having off-type seeds in their grain samples varied significantly between sets of cultivars obtained from different countries and collections. For example, 11 of 50 accessions studied of durum wheat collection contained admixtures [Ruiz et al 2011]. In most of sets of cultivars, this frequency did not exceed 5% of all cultivars studied in the set, but in a few cases, nearly every second cultivar from the set contained some seed of admixture [Metakovsky et al 2019c].

10.4. The use of gliadins as genetic markers in studies of somaclonal variation

The presence of off-type seeds in a wheat sample may lead to incorrect interpretations of results in scientific experimentation. A significant case of "damage" caused by admixtures to experimental work was provided in studies of somaclonal variation in common wheat. Somaclonal variation (changes in the genotype induced by cultivation of immature wheat embryos in vitro) in the 1980s claimed to be a promising new way for increasing wheat polymorphism suitable for wheat breeding [Larkin & Scowcroft 1981].

In the experiment, many differences were observed between genotypes of regenerated and control plants. When gliadins were used as genetic markers, up to three (average) differences per one genotype (gliadin electrophoregram) were registered [Larkin et al 1984]. However, the analysis of gliadin electrophoregrams presented in this article showed that genotypes of some regenerated plants differed from control to the same extent as usually differ two unrelated wheat cultivars. Indeed, both regenerant and control electrophoregrams were composed of different, but known (catalogued, see **2.4.**) allelic variants of blocks encoded by major *Gli* loci [Metakovsky et al 1987b]. Judging from the EP band compositions of gliadin blocks [Blocks 2018c] (see **2.5.**), the one-step transformation of any known allele into another known allele seems entirely unlikely because it would mean that the same allelic variant of the complex *Gli* locus encoding several specific gliadins (see **5.1.**) may arise more than once. That is more true for such transformation of several *Gli* loci simultaneously. Instead, the differences in gliadin electrophoregrams can be readily explained through the presence of admixtures in the material used for the experiment. Alien genotypes might appear as descendants from an occasional out-crossing, or handling, or due to some type of error [Metakovsky et al 1987b; Metakovsky & Sozinov 1990] (see **10.3.**).

The same approach was used in another attempt to find somaclonal variation in wheat. Again, any difference between regenerant and control in their gliadin electrophoregrams were considered as a sign of somaclonal variation [Maddok et al 1985]. As a result, two regenerant genotypes were found differing from the control cv Bersee. Our analysis of the electrophoregrams presented in this publication showed that one of these two deviating genotypes had substitutions at least at three loci, *Gli-B1, Gli-D1* and *Gli-D2* simultaneously. At each of these loci, the allele from cv Bersee was substituted for other allele catalogued earlier. Another deviating genotype was represented by the complete electrophoregram of the cv ChS. Obviously, it was a contamination [Metakovsky et al 1987b].

Genotypes of three regenerants of the line ND-7532 differed from the control line. Our analysis of published electrophoregrams [Cooper et al 1986] showed that the gliadin genotype of the line ND-7532 differed from cv Marquis (general control in the experiment) by alleles at least at two loci, *Gli-A1* and *Gli-D2*. The comparison of the first of deviating patterns with the control showed the presence of substitution of the allele *Gli-A1f* (from ND 7532) for *Gli-A1m* of Marquis. The second electrophoregram represented a heterozygous genotype *Gli-A1f/m*, and the third was a double heterozygous genotype at both *Gli-A1* and *Glu-D2* [Metakovsky et al 1987b]. Obviously, all three deviating regenerants represented some segregating progenies from out-crossing between ND-7532 and Marquis.

Gliadin and glutenin EP patterns were studied in 75 regenerants of five different common wheat cultivars. It was found that genotypes of two of these regenerants differed from their corresponding control cultivars in six and seven of nine storage protein loci studied (six *Gli* loci and three *Glu-1* loci). These two deviating regenerants were assumed to be admixtures [Novoselskaya et al 1987].

Thus, the most natural and simplest explanation of strong differences between deviating regenerant and its control is that the sample of material used to obtain the regenerants contained impurities of foreign genotypes.

Mutations that would cause the emergence of new variants of alcohol dehydrogenase 1 were not found in the analysis of 551 regenerants of the common wheat cv Millewa. All mutant variants of this enzyme observed in the regenerants were caused by chromosomal abnormalities [Davies et al 1986]. Moreover, 38 of 129 regenerants of common wheat were found to be aneuploids having from 38 to 45 chromosomes [Karp & Maddok 1984].

Also, among 468 regenerants of ChS, 32 deviating plants, in total, were revealed, 28 of them were null mutants at one of the *Gli* loci (see **7.1.**), and four mutants without one EP band encoded by the *Gli-B1* (see

7.4.). Among 260 control plants, one mutant plant was found lacking the *Gli-B1*-encoding band [Upelniek et al 1995].

Thus, the use of *Gli* allele to monitor changes in genotypes of wheat regenerants caused by the cultivation of immature embryos in vitro showed that, in the majority of cases, the cultivation causes chromosomal aberrations, which manifest themselves, in particular, in the occurrence of null mutations at the *Gli* loci. Mutations causing appearance of new alleles at the *Gli* loci and thereby increasing wheat polymorphism were not observed in regenerant plants.

Generally, analysis of publications dealing with wheat regenerants confirmed a high value of gliadins (gliadin alleles) as genetic markers. Due to their enormous polymorphism, gliadins have a high resolving power to evaluate and distinguish various types of differences which can be observed between experimental material and its control.

On another hand, researches dealing with wheat cultivars need to pay more attention to the authenticity (see **10.2.**) as well as the possible genetic heterogeneity of plant (grain) material used for the experiment. It has been firmly established that many registered cultivars of common wheat consist of multiple authentic biotypes originated from the primary cross performed by breeder [Metakovsky et al 2019c; 2020]. Also, the cultivar under study may contain foreign genotypes (admixtures) of different origin (see **10.3.**). Genetic heterogeneity can lead to misinterpretation of the results obtained.

10.5. Summary. Wheat collections and their problems

The maintenance of the genetic diversity of cultivated wheat depends on the correct preservation of the gene pool in the collections. However, there are problems that may affect the use of germplasm store for agricultural crop improvement.

Accessions of landraces or local varieties are frequently composed of many genotypes distinguishable using suitable genetic markers. However, there are publications indicating that, in collection, wheat accession may lose its polymorphism. There are cases of the presence of a gliadin allele in the genotypes of a daughter and grandmother cultivars but not in parental ones. The only explanation for that is the loss of some of polymorphism, including that allele, by one of the parental cultivars after its use in the cross. Also, wheat collections may contain a considerable quota of misclassified and incorrectly labelled accessions.

A serious problem is the authenticity of samples of cultivars maintained in collection and used in different scientific studies. Genotypes of two or more grain samples of the same cultivar obtained from different collections may differ considerably. The striking example represents the Canadian cultivar Selkirk whose grain samples, obtained from four countries, had nothing in common. Genotype of a cultivar may include *Gli* alleles characteristic for the country of origin of the grain sample, not of the origin of this cultivar.

The absence of the correspondence between gliadin genotype of the cultivar and its pedigree is an another problem. All cultivars participated in 194 pedigrees (a daughter cultivar, its parents and some descendants) were studied, and only 134 pedigrees appeared to be completely correct. The fact may be explained either through errors in pedigrees, or not adequate maintenance of cultivars in collections, including small local ones harboured in the institutions of wheat genetics and breeding,

One more problem of world-wide common wheat germplasm is an impurity of grain samples. An off-type (admixture, contamination, foreign genotype) is a part of a given grain sample but not of a given cultivar. An off-type may appear as a descendant of casual out-crossing between wheat genotypes and subsequent genetic segregation of the hybrid. In this case, the genotype of the off-type would include some alleles of the cultivar studied and some foreign alleles. Heterozygotes at some loci may be observed. An off-type may also appear as a result of handling with grain samples. In this case, genotypes of the cultivar and the off-type would differ at several genetic loci simultaneously. The frequency of grain samples carrying off-type seeds varied between sets of cultivars but did not exceed 5%. In a few cases, nearly every second cultivar from the set contained some seed of admixture.

The presence of off-type seeds in a wheat sample may lead to incorrect interpretation of results in scientific experimentation. A significant case of "damage" caused by admixtures to experimental work was provided in studies of somaclonal variation in common wheat. The simplest explanation of a big difference between genotypes of deviating regenerant and its control (claimed to be somaclonal variation) is impurity of material used to obtain this regenerant.

Due to their enormous polymorphism, gliadins have a high resolving power to evaluate and distinguish various types of differences which can be observed between experimental material and its control. Monitoring of inter-varietal and intra-varietal polymorphism seems to be necessary for adequate maintenance and preservation of wheat genetic resources and conservation of its biological diversity.

SHORT SUMMARY OF THE CONTENT OF THIS BOOK

It is a critical review of more than 740 publications containing world knowledge about particularities of wheat seed storage proteins, gliadins and glutenins. The possible use of them in scientific research and in breeding is discussed.

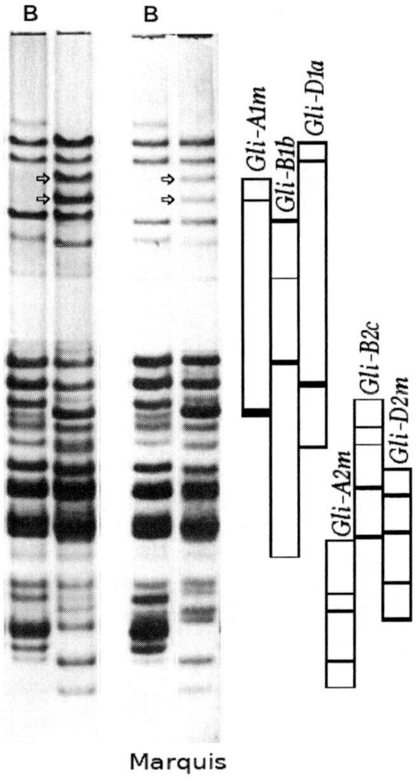

Figure 1. Gliadin electrophoregrams (APAGE) of common wheat cultivar Marquis. Blocks of jointly inherited electrophoretic bands encoded by alleles at the *Gli* loci (see **2.2.**) are shown schematically. Two different EP runs are shown. Arrows show *Gli-A5*-encoded bands (see **2.5.**, **6.7.**). B - a universal standard cultivar Bezostaya-1.

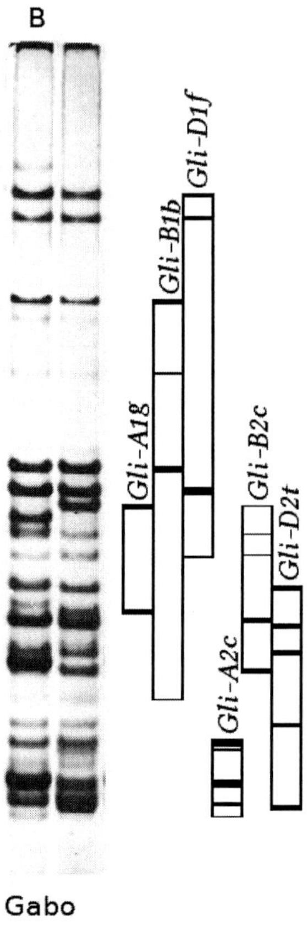

Figure 2. Gliadin electrophoregrams (APAGE) of common wheat cultivar Gabo. Blocks of jointly inherited electrophoretic bands encoded by alleles at the *Gli* loci of the cultivar Gabo are shown schematically. B - a universal standard cultivar Bezostaya-1.

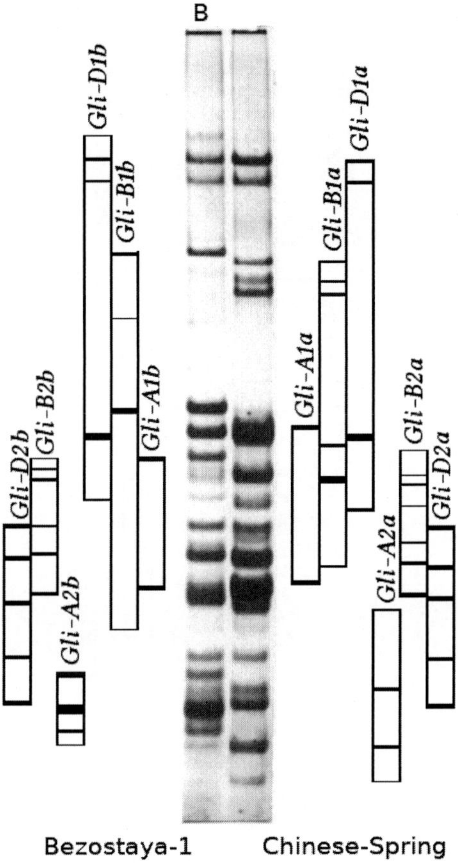

Figure 3. Gliadin electrophoregrams (APAGE) of common wheat cultivars Bezostaya-1 and Chinese-Spring. Blocks of jointly inherited electrophoretic bands encoded by alleles at the *Gli* loci are shown schematically. B - a universal standard cultivar Bezostaya-1.

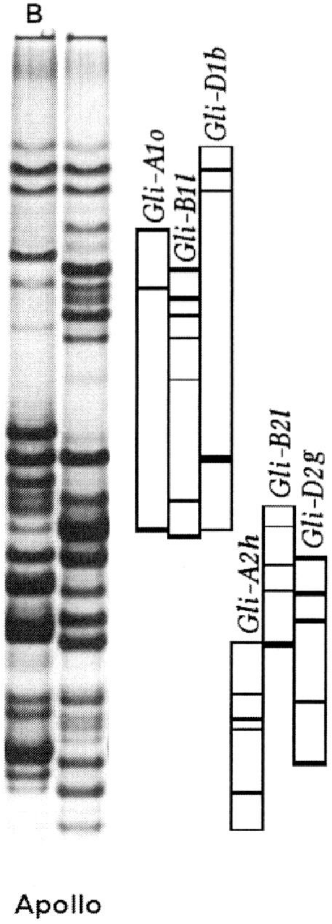

Figure 4. Gliadin electrophoregrams (APAGE) of common wheat cultivar Apollo. Blocks of jointly inherited electrophoretic bands encoded by alleles at the *Gli* loci of the cultivar Apollo are shown schematically. B - a universal standard cultivar Bezostaya-1.

REFERENCES

(Internet https addresses were taken from Google Scholar)

AACC 1995 AACC Approved method 56-70.

Akhmedov MG & Metakovsky E (1987) Inheritance of the gliadin component composition in hybrids from the cross between common wheat cultivars Bezostaya 1 and Chinese Spring. Russian J Genet 23(8): 1038-1046, https://agris.fao.org/agris-search/search.do?recordID=US8837764

Akhmedov MG, Metakovsky E, Sozinov AA (1988) Intergenomic transfer of genetic material in common wheat revealed by analysis of genetic markers. Doklady Acad Nauk USSR (Biol Sci) 299: 734-737 (in Russian), http://pascal-francis.inist.fr/vibad/index.php?action=getRecordDetail&idt=7160219

Akhunov ED, Akhunova AR, Linkiewicz AM, Dubcovsky J, Hummel D, ... Dvorak J (31 authors) (2003) Synteny perturbations between wheat homoeologous chromosomes caused by locus duplications and deletions correlate with recombination rates. PNAS 100 (19): 10836-10841, https://doi.org/10.1073/pnas.1934431100

Allaby RG, Banerjee M, Brown TA (1999) Evolution of the high molecular weight glutenin loci of the A, B, D, and G genomes of wheat. Genome 42: 296-307, https://doi.org/10.1139/g98-114

Allard RW (1988) Genetic changes associated with the evolution of adaptedness in cultivated plants and their wild progenitors. J Hered 79: 225-238, https://doi.org/10.1093/oxfordjoutnals.jhered.a110503

Altenbach SB, Chang HC, Simon-Buss A, Mohr T, Huo N, Gu YO (2020) Exploiting the reference genome sequence of hexaploid wheat: a proteomic study of flour proteins from the cultivar Chinese Sping. Funct Integr Genomics 20: 1-16, https://doi.org/10.1007/s.10142-019-00694-z

Álvarez ML, Guelman S, Halford NG, Lustig S, Reggiardo MI, Ryabushkina N, Shewry P, Stein J, Vallejos RH (2000) Silencing of HMW glutenins in transgenic wheat expressing extra HMW subunits. Theor Appl Genet 100: 319-327, https://doi.org/10.1007/s001220050042

Anderson OD (1991) Characterization of members of a pseudogene subfamily of the wheat α-gliadin storage protein genes. Plant Molec Biol 16: 335-337, https://doi.org/10.1007/BF00020565

Anderson OD & Greene FC (1997) The α-gliadin gene family. II. DNA and protein sequence variation, subfamily structure, and origins of pseudogenes. Theor Appl Genet 95: 59-65, https://doi.org/10.1007/s001220050532

Anderson OD, Litts JC, Gautier MF, Greene FC (1984) Nucleic acid sequence and chromosome assignment of a wheat storage-protein gene. Nucleic Acids Res 12: 8129-8144, https://doi.org/10.1093/nar/12.21.8129

Anderson OD, Litts JC, Greene FC (1997) The α-gliadin gene family. I. Characterization of ten new wheat α-gliadin genomic clones, evidence for limited sequence conservation of flanking DNA, and Southern analysis of the gene family. Theor Appl Genet 95: 50-58, https://doi.org/10.1007/s001220050531

Anderson OD, Hsia CC, Adelsteins AE, Lew EJ-L, Kasarda DD (2001a) Identification of several new classes of low-molecular-weight wheat gliadin-related proteins and genes. Theor Appl Genet 103: 307-315, https://doi.org/10.1007/s001220100576

Anderson OD, Hsia CC, Torres V (2001b) The wheat γ-gliadin genes: characterization of ten new sequences and futher understanding of γ-gliadin gene family structure. Theor Appl Genet 103: 323-330, https://doi.org/10.1007/s00122-001-0551-3

Anderson OD, Gu YQ, Lazo GR (2007) Structure, evolution, and expression of the wheat prolamin loci. Proc. 9th Annual Gluten Workshop (Sept. 2006, San Fransisco, Calif.), eds G.L. Lookhart & P.K.W. Ng; pp 90-93, https://www.cabdirect.org/cabdirect/abstract/20073176389

Anderson OD, Gu YQ, Kong X, Lazo GR, Wu J (2009) The wheat ω-gliadin genes: structure and EST analysis. Funct Integr Genomics 9: 397-410, https://doi.org/10.1007/s10142-009-0122-2

Anderson OD, Dong L, Huo N, Gu YQ (2012) A new class of wheat gliadin genes and proteins. PLoS ONE 7(12): e52139, https://doi.org/10.1371/journal.pone.0052139

Anderson OD, Huo N, Gu JQ (2013) The gene space in wheat: The complete γ-gliadin gene family from the wheat cultivar Chinese Spring. Funct Integr Genomics 13, 261–273, https://doi.org/10.1007/s10142-013-0321-8

Andrews JL, Blundell MJ, Skerritt (1996) Differentiation of wheat-rye translocation lines using antibody probes for *Glu-B1* and *Sec-1*. J Cereal Sci 33: 61-72, https://doi.org/10.1006/jcrs.1996.0006

Appelbee MJ, Mekuria GT, Nagasandra V, Bonneau JP, Eagles HA, Eastwood RF, Mather DE (2009) Novel allelic variants encoded at the *Glu-D3* locus in bread wheat. J Cereal Sci 49: 254-261, https://doi.org/10.1016/j.jcs.2008.10.011

Appleyard DB, McCausland J, Wrigley CW (1979) Checking of identity and origin of off-types in the propagation of pedigreed wheat seed. Seed Sci Technol 7: 459-466, http://pascal-francis.inist.fr/vibad/indexphp?action=getRecordDetail&idt=PASCALAGROLINEINRA8010185536

Appels R (1982) The molecular cytology of wheat-rye hybrids. Int Review Cytol 80: 93-132, https://doi.org/10.1016/S0074-7696(8)60368-X.

Asins MJ & Carbonell EA (1986) A comparative study on variability and phylogeny of Triticum species. 1. Intraspecific variability. Theor Appl Genet 72: 551-558, https://doi.org/10.1007/BF00289541

Autran JC & Bourdet A (1975) L'identification des variétés de ble: établissmaent d'un tableau general de déterminacion fondé sur le diagramme électrophorétique de gliadines du grain. Ann Amélior Plantes 25: 277-301 (in French), http://pascal-francis.inist.fr/vibad/index.php?action=getRecordDetail& idt=PASCAL7638001987

Autran JC, Bushuk W, Wrigley CW, Zillman RR (1979a) Wheat cultivar identification by gliadin electrophoregrams. IV. Comparison of international methods. Cereal Foods World 24: 471-475, http://jcautran.free.fr/archives_familiales/profession/publications/1979/79-01.pdf

Autran JC, Lew EJL, Nimmo CC, Kasarda DD (1979b) N-terminal amino acid sequencing of prolamins from wheat and related species. Nature 282: 527-529, https://doi.org/10.1038/282527a0

Autran JC, Pogna NE, Kudryavtsev AM (1995) Use of genetic variation in the improvement of quality in durum wheat. In: "Durum wheat quality in the Mediterranean region"; eds N.DiFonzo, F.Kaan, M.Nachit; Options Méditerranéennes: Série A. Séminaires Méditerranéens; N22; Zaragoza, CIHEAM; pp 173-180.

Badaeva ED, Dedkova OS, Gay G, Pukhalskyi VA, Zelenin AV, Bernard S, Bernard M (2007) Chromosomal rearrangements in wheat: their types and distributions. Genome 50: 907-926, https://doi.org/10.1139/G07-072

Baker RJ & Bushuk W (1978) Inheritance of differences in gliadin electrophoregrams in the progeny of Neepawa and Pitic 62 wheats. Can J Plant Sci 58: 325-329, https://doi.org/10.4141/cjps.78-052.

Barak S, Mudgil D, Khatkar, BS (2015) Biochemical and functional properties of wheat gliadins: A review. Crit Rev Food Sci Nutr 55: 357–368, https://doi.org/10.1080/10408398.2012.654863

Barrett BA & Kidwell KK (1998) AFLP-based genetic diversity assessment among wheat cultivars from the Pacific Northwest. Crop Sci 38: 1261-1271, https://doi.org/10.2135/cropsci1998.0011183X003800050025x

Bartels D, Altosaar I, Harberd NP, Barker RF, Thompson RD (1986) Molecular analysis of γ-gliadin gene families at the complex *Gli-1* locus of bread wheat (*T.aestivum* L.). Theor Appl Genet 72: 845-853, https://doi.org/10.1007/BF00266556

Bayram ME & Korkut KZ (2020) Effect of glutenin genes on quality parameters in common wheat. J Centr Europ Agric 21: 62-76, https://doi.org/10.5513/JCEA01/21.1.2422

Bean SR & Lookhart GL (2001) Recent developments in high-performance capillary electrophoresis of cereal proteins. Electrophoresis 22: 1503-1509, https://doi.org/10.1002/1522-2683(200105)22:8<1503::AID-EL PS1503>3.0.CO;2-P

Bean SR, Bietz JA, Lookhart GL (1998) High-performance capillary electrophoresis of cereal proteins. J Chromatogr A 814 (1-2): 25-41, https://doi.org/10.1016/50021-9673(98)00437-3

Beasley HL, Uthayakumaran S, Stoddard FL, Partridge SL, Daqiq L, Chong P, Bekes F (2002) Synergistic and additive effects of three high molecular weight glutenin subunit loci. II. Effects on wheat dough functionality and end-use quality. Cereal Chem 79: 301-307, https://doi.org/10.1094/CCHEM.2002.79.2.301

Bebyakin VM & Balabolina TG (1977) Inheritance of flour gliadin component composition in spring wheat during hybridization. Doklady VASKhNIL 10: 15-16 (in Russian),

Bell GE, McDonald MB, Danneberger TK, Martin SK (1998) Sources of variation in computer imaging of electrophoretic gels. Crop Sci 38: 523-526, https://doi.org/10.2135/cropsci1998.0011183X003800020041x

Benedettelli S, Margiotta B, Porceddu E, Ciaffi M, Lafiandra D (1992) Effects of the lack of proteins controlled by genes at the Gli-D1/Glu-D3 loci on the breadmaking quality of wheat. J Cereal Sci 16(1). 69-79, https://doi.org/10.1016/50733-5210(09)80080-5

Bennett MD & Smith JB (1976) Nuclear DNA amounts in angiosperms. Philos Trans Royal Soc London B (Biol Sci) 274(933): 227-274, https://doi.org/10.1098/rstb.1976.0044

Bennett MD & Leitch IJ (1995) Nuclear DNA amounts in angiosperms. Ann Bot 76: 113-176, https://doi.org/10.1006/anbo.1995.1085

Berzonsky WA, Clements RL, Lafever HN (1991) Identification of "Amigo" and "Kavkaz" translocations in Ohio soft red winter wheats

(*Triticum aestivum* L.). Theor Appl Genet 81: 629-634, https://doi.org/10.1007/BF00226729

Bietz JA & Burnouf T (1985) Chromosomal control of wheat gliadin: Analysis by reversed-phase high-performance liquid chromatography. Theor Appl Genet 70: 599-609, https://doi.org/10.1007/BF00252285

Bietz JA & Schmalzried E (1995) Capillary electrophoresis of wheat gliadin: initial studies and application to varietal identification. LWT-Food Sci Technol 28: 174-184, https://doi.org/10.1016/50023-6438(95)91346-7

Bietz JA, Shepherd KW, Wall JS (1975) Single-kernel analysis of glutenin: Use in wheat genetics and breeding. Cereal Chem. 52: 513-532, ISSN: 0009-0352, http://pascal-francis.inist.fr/vibad/index.php?action=getRecordDetail&idt=PASCAL7638004064

Blatter RHE, Jacomet S, Schlumbaum A (2004) About the origin of European spelt (*Triticum spelta* L.): allelic differentiation of the HMW Glutenin B1-1 and A1-2 subunit genes. Theor Appl Genet 108: 360-367, https://doi.org/10.1007/s00122-003-1441-7

Blumenthal CS, Batey IL, Bekes F, Wrigley CW, Barlow EWR (1990) Gliadin genes contain heat shock elements: possible relevance to heat-induced changes in grain quality. J Cereal Sci 11: 185-187, https://doi.org/10.1016/S0733-5210(09)80162-8

Blumenthal CS, Barlow EWR, Wrigley CW (1993) Growth environment and wheat quality: the effect of heat stress on dough properties and gluten proteins. J Cereal Sci 18: 3-21, https://doi.org/10.1006/jcrs.1993.1030

Boggini G (1978) Uno sguardo all'evoluzione varietale del frumento. L'inf Agraria 34: 3503-3508 (in Italian), https://agris.fao.org/agris-search/search.do?recordID=IT19790398992

Boggini G & Pogna NE (1989) The breadmaking quality and storage protein composition of Italian durum wheat. J Cereal Sci 9: 131-138, https://doi.org/10.1016/S0733-5210(89)80013-X

Boggini G & Pogna NE (1990) Utilizzazione del frumento duro come miglioratore della panificabilitá del frumento tenero. Tecnica Molitoria (dicembre), pp 1025-1030 (in Italian)

Bonafede MD, Tranqilli G, Pflüger LA, Peña RJ, Dubcovsky J (2015) Effect of allelic variation at the Glu-3/Gli-1 loci on breadmaking quality parameters in hexaploid wheat (*Triticum aestivum* L.). J Cereal Sci 62: 143-150, https://doi.org/10.1016/j.jcs.2015.02.001

Booth M & Ewart JAD (1969) Studies on four components of wheat gliadins. Biochim Biophys Acta 181: 226-233, https://doi.org/10.1016/0005-2795(69)90245-1

Borghi B, Boggini G, Cattaneo M, Corbellini M, Pogna NE, Dalli A, Bianchi A (1985) Salmone: il primo grano di forza italiano. Sementi Elette 31: 3-6 (in Italian), http://pascal-francis.inist.fr/vibad/index.php?action=getRecordDetail&idt=8547866

Borghi B, Corbellini M, Perenzin M, Pogna NE (1991) Breeding for high quality bread wheats in Southern Europe: Results and perspectives. Vortr Pflanzenzücht 20: 278-283, https://agris.fao.org/agris-search/search.do?recordID=DE97B7256

Borghi B, Corbellini M, Gavuzzi P, Boggini G, Ouassou A (2014) Breeding for heat and drought tolerance in bread wheat by means of field and laboratory screening procedures. Bull Sociiété Bot France, Actualités Botaniques, 137: 37-46, https://doi.org/10.1080/01811789.1990.10826985

Börner A, Chebotar S, Korzun V (2000) Molecular characterization of the genetic integrity of wheat (*Triticum aestivum* L.) germplasm after long-term maintenance. Theor Appl Genet 100: 494-497, https://doi.org/10.1007/s001220050064

Brandolini A. Vaccino P, Boggini G, Özkan H, Kilian B, Salamini F (2006) Quantification of genetic relationships among A genomes of wheats. Genome 49: 297-305, https://doi.org/10.1139/g05-110

Branlard G (1983a) Study of genetic determination of 20 gliadin bands. Theor Appl Genet 64: 155-162, https://doi.org/10.1007/BF00272727

Branlard G (1983b) Correlation between gliadin bands. Theor Appl. Genet. 64: 163-168, https://doi.org/10.1007/BF00272728

Branlard G (2009) Les bases génétiques et l'utilisation des marqueurs associés à l'amélioration de la qualité des blés. Agriculture 18: 506-511 (in French), https://doi.org/10.1684/agr.2009.0344

Branlard G & Rousset M (1980) Electrophoretical characteristics of gliadins and baking value of bread wheat. Ann Amelior Plantes 30: 133-149, https://hal.inrae.fr/hal-02732656

Branlard G & Dardevet M (1985a) Diversity of grain proteins and bread wheat quality: I. Correlation between gliadin bands and flour quality characteristics. J Cereal Sci 3: 329-343, https://doi.org/10.1016/S0733-5210(85)80006-0

Branlard G & Dardevet M (1985b) Diversity of grain proteins and bread wheat quality: II. Correlation between high molecular weight subunits of glutenin and flour quality characteristics. J Cereal Sci 3: 345-354, https://doi.org/10.1016/S0733-5210(85)80007-2

Branlard G & Dardevet M (1994) A null *Gli-D1* allele with a positive effect on bread wheat quality. J Cereal Sci 20: 235-244, https://doi.org/10.1006/jcrs.1994.1063

Branlard G & Metakovsky E (2006) Some *Gli* alleles related to common wheat dough quality. In: "Gliadin and glutenin. The unique balance of wheat quality", eds C.Wrigley, F.Békés, W.Bushuk; AACC Internat; pp 115-139, https://hal.inrae.fr/hal-02821247

Branlard G, Khelifi D, Lookhart G (1992) Identification of some wheat proteins separated by two-step acid polyacrylamide gel electrophoresis and sodium dodecyl sulphate-polyacrylamide gel electrophoresis technique. Cereal Chem 69: 677-678, https://hal.inrae.fr/hal-02710831

Branlard G, Dardevet M, Nieto-Taladriz MT, Khelifi D (1994) Allelic diversity of the omega gliadin as revealed by SDS-PAGE: Their possible implication in quality variation. In: "Gluten Proteins", Proc 5th Intern Workshop Gluten Prot (Detmold, Germany, 1993); Assoc Cereal Research Publ; pp 234-243.

Branlard G, Dardevet M, Saccomano R, Lagoutte F, Gourdon J (2001) Genetic diversity of wheat storage proteins and bread wheat quality. Euphytica 119: 59-67, https://doi.org/10.1023/A:1017586220359

Branlard G, Dardevet M, Amiour N, Igrejas G (2003) Allelic diversity of HMW and LMW glutenin subunits and omega-gliadins in French bread wheat (*Triticum aestivum* L.). Genet Resour Crop Evol 50: 669-679, https://doi.org/10.1023/A:1025077005401

Branlard G, Faye A, Rhazi L, Tahir A, Lesage V, Aussenac T (2020) Genetic and environmental factors associated to gluten polymer characteristics of wheat. Foods 9: 683, https://doi.org/10.3390/foods9050683

Brezhneva TA, Upelniek VP, Pukhal'sky VA (2010) Genetic control of gliadin components in wheat *Triticum spelta* L. Russian J Genet 46: 562-565, https://doi.org/10.1134/S102279541005008X

Brönneke V, Zimmermann G, Killermann B (2000) Effect of high molecular weight glutenins and D-zone gliadins on bread making quality in German wheat varieties. Cereal Res Comm 28: 187-194, https://doi.org/10.1007/BF03543592

Brown AHD & Hodgkin T (2015) Indicators of genetic diversity, genetic erosion, and genetic vulnerability for plant genetic resources. In: Sustainable development and biodiversity (V7), Genetic diversity and erosion in plants, pp 25-53, https://doi.org/10.1007/978-3-319-25637-5_2

Bryan GJ, Stephenson P, Collins A, Kirby J, Smith JB, Gale MD (1999) Low levels of DNA sequence variation among adapted genotypes of hexaploid wheat. Theor Appl Genet 99: 192-198, https://doi.org/10.1007/s001220051224

Bunce NAC, White RP, Shewry PR (1985) Variation in estimates of molecular weights of cereal prolamins by SDS-PAGE J Cereal Sci 3: 131-142, https://doi.org/10.1016/S0733-5210(85)80023-0

Burnett CJ, Lorenz KJ, Carver BF (1995) Effects of the 1B/1R translocation in wheat on composition and properties of grain and flour. Euphytica 86: 159-166, https://doi.org/10.1007/BF00016353

Bushuk W (1980) The baking potential of Glenlea wheat. Canad J Plant Sci 60: 737-739, https://doi.org/10.4141/cjps80-104

Bushuk W & Zillman RR (1978) Wheat cultivar identification by gliadin electrophoregrams. I. Apparatus, method and nomenclature. Can J Plant Sci 58: 505-515, https://doi.org/10.4141/cjps78-076

Butow BJ, Ma W, Gale KR, Cornish GB, Rampling L, Larroque OR, Morell MK, Bekes F (2003) Molecular discrimination of Bx 7 alleles demonstrates that a highly expressed high molecular weight glutenin allele has a major impact on wheat flour dough strength. Theor Appl Genet 107: 1524-1532, https://doi.org/10.1007/S00122-003-1396-8

Butow BJ, Gale KR, Ikea J, Juhasz A, Bedö Z, Tamas L, Gianibelli MC (2004) Dissemination of the highly expressed Bx7 glutenin subunit (GluB1al allele) in wheat as revealed by novel PCR markers and RP-HPLC. Theor Appl Genet 109: 1525-1535, https://doi.org/10.1007/s00122-004-1776-8

Caballero L, Peña RJ, Martín LM, Álvarez JB (2010) Characterization of Mexican Creole wheat landraces in relation to morphological characteristics and HMW glutenin subunit composition. Genet Resour Crop Evol 57: 657-665, https://doi.org/10.1007/s10722-009-9501-8

Cakmak I, Torun A, Millet E, Feldman M, Fahima T, Korol A, Nevo E, Braun HJ, Özkan H (2004) Triticum dicoccoides: An important genetic resource for increasing zink and iron concentration in modern cultivated wheat. Soil Sci Plant Nutr 50: 1047-1054, https://doi.org/10.1080/00380768.2004.10408573

Caldwell KS, Dvorak J, Lagudah ES, Akhunov E, Luo MC, Wolters P, Powell W (2004) Sequence polymorphism in polyploid wheat and their D-genome diploid ancestor. Genetics 107: 941-947, https://doi.org/10.1534/genetics.103.016303

Campbell WP, Wrigley CW, Cressey PJ Slack CR (1987) Statistical correlations between quality attributes and grain-protein composition for 71 hexaploid wheats used as breeding parents. Cereal Chem 64: 293-299, http://pascal-francis.inist.fr/vibad/index.php?action=getRecordDetail&idt=8389150

Cao W, Hucl P, Scoles G, Chibbar RN (1998) Genetic diversity within spelta and macha wheats based on RAPD analysis. Euphytica 104: 181-189, https://doi.org/10.1023/A:101862810.2650

Cao W, Scoles G, Hucl P, Chibbar RN (1999) The use of RAPD analysis to classify *Triticum* accessions. Theor Appl Genet 98: 602-607, https://doi.org/10.1007/s001220051110

Carrillo JM, Rodriguez de Quijano M, Galiano A, Hamie B, Vazquez JF, Orellana J (1988) Composición en gluteninas de alto peso molecular de variedades de trigo blando registradas en España y su relación con la calidad panadera. An Aula Dei 19: 239-250 (in Spanish), http://pascal-francis.inist.fr/vibad/index.php?action=getRecordDetail&idt=6949314

Carrillo JM, Vazquez JF, Orellana J (1992) Identification and mapping of the *Gli-R3* locus on chromosome 1R of rye (*Secale cereale* L.). Theor Appl Genet 84: 237-241, https://doi.org/10.1007/BF00224005

Cassidy BG, Dvorak J, Anderson OD (1998) The wheat low-molecular-weight glutenin genes: characterization of six new genes and progress in understanding gene family structure. Theor Appl Genet 96: 743-750, https://doi.org/10.1007/s001220050797

Castagna R, Maga G, Perenzin M, Heun M, Salamini F (1994) RFLP-based genetic relationships of eincorn wheat. Theor Appl Genet 88: 818-823, https://doi.org/10.1007/BF01253991

Černy J, Šašek A, Zimmermann G, Seibert I (1992) An evaluation of Bavarian and lower Austrian new varieties of winter wheat by means of genetic markers of baking quality. Scientia Agric Bohem 24: 113-123, http://pascal-francis.inist.fr/vibad/index.php?action=getRecordDetail&idt=6451201

Chantret N, Salse J, Sabot F, Rahman S, Bellec A ... Chalhoub B (19 authors) (2005) Molecular basis of evolutionary events that shaped the *Hardness* locus in diploid and polyploid wheat species (Triticum and Aegiops). The Plant Cell 17: 1033-1045, https://doi.org/10.1105/tpc.104.029181

Charbonnier L (1974) Isolation and characterization of ω-Gliadin fractions. Biochim Biophys Acta 359: 142 -151, https://doi.org/10.1016/0005-2795(74)90139-1

Charles M, Belcram H, Just J, Huneau C, Violett A, Couloux A, Segurens B, Carter M, Huteau V, Coriton O, Appels R, Samain S, Chalhoub B (2008) Dynamics and differential proliferation of transposable elements during the evolution of the B and A genomes of wheat. Genetics 180: 1071-1086, https://doi.org/10.1534/genetics.108.092304

Chebotar SV, Blagodarova EM, Kurakina EA, Semeniuk IV, Polishchuk AM, Kozub NA, Sozinov IA, Khohlov AN, Rybalka AI, Sivolap YuM (2012a) Genetic polymorphism of loci that determine the bread-making quality of Ukrainian wheat varieties. Vavilov J Genet Select 16: 87-98 (in Russian), "Sivolap Yu" *M., Patent* 19828 (2012)

Chebotar GA, Motsnyy II, Chebotar SV, Sivolap YuM (2012b) Effects of dwarfing genes on the genetic background of wheat varieties in southern Ukraine. Cytol Genet (Kiev) 46: 366-372, https://doi.org/10.3103/S0095452712060023

Chen HB, Martin JM, Lavin M, Talbet LE (1994) Genetic diversity in hard red spring wheat based on sequence-tagged-site PCR markers. Crop Sci 34: 1628-1632, https://doi.org/10.2135/cropsci1994.0011183X003400060037x

Chen H, Li S, Liu Y, Liu J, Ma X, Du L, Wang K, Ye X (2021) Effects of 1Dy12 subunit silencing on seed storage protein accumulation and flour-processing quality in common wheat somatic variation line. Food Chem 335: 127663, https://doi.org/10.1016/j.foodchem.2020.127663

Chernakov VM & Metakovsky E (1993) Spontaneous mutations at the gliadin-coding loci which have been found during analysis of spikes and families of plants of spring common wheat cultivars. Russian J Genet 29: 114-124 (in Russian),

Chernakov VM & Metakovsky E (1994) Diversity of allelic variants of gliadin-coding loci and assessment of genetic similarity of common wheat cultivars bred in different breeding centres. Russian J Genet 30: 509-517 (in Russian), https://agris.fao.org/agris-search/search.do?recordID=RU9444986

Chojecki AJS, Gale MD, Holt LM, Payne PI (1983) The intrachromosomal mapping of a glucose phosphate isomerase structural gene, using allelic variation among stocks of Chinese Spring wheat. Genet Res 41: 221-226, https://doi.org/10.1017/S0016672300021297

Christiansen MJ, Andersen SB, Ortiz R (2002) Diversity changes in an intensively bred wheat germplasm during the 20th century. Molec Breed 9: 1-11, https://doi.org/10.1023/A:1019234323372

Ciaffi M, Dominici L, Lafiandra D, Porceddu E (1992) Seed storage proteins of wild wheat progenitors and their relationships with technological properties. Hereditas 116: 315-322, https://doi.org/10.1111/j.1601-5223.1992.tb00844.x

Ciaffi M, Lafiandra D, Porceddu E, Benedettelli (1993) Storage-protein variation in wild emmer wheat (*Triticum turgidum* ssp. *dicoccoides*) from Jordan and Turkey. I. Electrophoretic characterization of genotypes. Theor Appl Genet 86: 474-480, https://doi.org/10.1007/BF00838563

Ciaffi M, Lee YK, Tamas L, Gupta R, Skerritt J, Appels R (1999) The low-molecular-weight glutenin subunit proteins of primitive wheats. III. The genes from D-genome species. Theor Appl Genet 98: 135-148,

https://doi.org/10.1007/s001220051050
Clarke BC, Phongkham T, Gianibelli MC, Beasley H, Békés F (2003) The characterization and mapping of a family of LMW-gliadin genes: effects on dough properties and bread volume. Theor Appl Genet 106: 629-635, https://doi.org/10.1007/s00122-002-1091-1
Cloutier S, Lukow OM, Humphreys DG (2000) Gene-associated selection for wheat quality. Canad J Plant Sci 80: 220-221,
Colot V, Bartels D, Thompson R, Flavell R (1989) Molecular characterization of an active wheat LMW glutenin gene and its relation to wheat and barley prolamin genes. Mol Gen Genet 216; 81-90, https://doi.org/10.1007/BF00332234
Cooke RJ (1995) The reasons for variety identification. In: "Identification of Food-Grain Varieties"; ed. C.W. Wrigley; AACC; S.Paul, MN; pp 1-17.
Cooke RJ & Law JR (1998) Seed storage protein diversity in wheat varieties. Plant Var Seed 11: 159-167, https://agris.fao.org/agris-search/search.do?recordID=GB1999005697
Cooke RJ & Wrigley CW (2004) Variety registration and breeders' rights. In: Encyclopedia of Grain Science, vol. 3. Elsevier Ltd; Oxford, UK; pp 314-321.
Cooper DB, Sears RG, Lookhart GL, Jones BL (1986) Heritable somaclonal variation in gliadin proteins of wheat plants derived from immature embryo callus culture. Theor Appl Genet 71: 784-790, https://doi.org/10.1007/BF00276418
Coriton O, Faye A, Paux E, Lemoine J, Huteau V. Branlard G, Jahier J (2019) Development of 1AS.1AL-1DL durum wheat chromosome carrying *Glu-D1a* locus encoding high molecular weight glutenin subunits 2+12. Molec Breed 39, 1-9. https://doi.org/10.1007/s11032-019-0942-2
Cox TS, Murphy JP, Rodgers DM (1986) Changes in genetic diversity in the red winter wheat regions of the United States. Proc Natl Acad Sci USA 83: 5583-5586, https://doi.org/10.1073/pnas83.15.5583
Cressey PJ, Campbell WP, Wrigley CW, Griffin WB (1987) Statistical correlation between quality attributes and grain-protein composition for 60 advanced lines of crossbred wheat. Cereal Chem 64: 299-301, http://pascal-francis.inist.fr/vibad/index.php?action=getRecordDetail&idt=8389151
Crossway A & Dvořák J (1984) Distribution of nonstructural variation along three chromosome arms between wheat cultivars Chinese Spring and Cheyenne. Genetics 106: 309-324, https://doi.org/10.1093/genetics/106.2.309

Cunsolo V, Muccilli V, Saletti R, Foti S (2012) Mass spectrometry in the proteome analysis of mature cereal kernels. Mass Spectrom Rev 31: 448-465, https://doi.org/10.1002/mas20347

Curioni A, Peruffo DB, Pressi G, Pogna NE (1991) Immunological distinction between x-type and y-type high molecular weight glutenin subunits. Cereal Chem 68: 200-204, http://pascal-francis.inist.fr/vibad/index.php?action=getRecordDetail&idt=5390184

Dachkevitch T, Redaelli R, Biancardi AM, Metakovsky E, Pogna NE (1993) Genetics of gliadins coded by the group 1 chromosomes in the high-quality bread wheat cultivar Neepawa. Theor Appl Genet 86: 389-399, https://doi.org/10.1007/BF00222107

D'Amato F (1989) The progress of Italian wheat production in the first half of the 20th century: The contribution of breeders. Agric Med 119: 157-174, http://pascal-francis.inist.fr/vibad/index.php?action=getRecordDetail&idt=7248653

Damidaux R, Autran JC, Grignac P, Feillet P (1978) Mise en evidence de relations applicables en selection entre l'electrophoregramme des gliadines et les proprietes viscoelasticues du glutin de *Triticum durum* Desf. C Rend Acad Sci 287: 701-704 (in French), http://pascal-francis.inist.fr/vibad/index.php?action=getRecordDetail&idt=PASCAL7950126295

Damidaux R, Autran JC, Grignac P (1980) Determinisme genetique des constituants gliadines de Triticum durum Desf., associes a la qualite cullinaire interinseque de varietes. C Rend Acad Sci 291: 585-588 (in French), http://pascal-francis.inist.fr/vibad/index.php?action=getRecordDetail&idt=PASCAL8110120395

Davies PA, Pallotta MA, Ryan SA, Scowcroft WR, Larkin PJ (1986) Somaclonal variation in wheat: genetic and cytogenetic characterization of alcohol dehydrogenase 1 mutants. Theor Appl Genet 72: 644-653, https://doi.org/10.1007/BF00289003

Dekaprelevich LL (1964) Die Art Triticum Macha Dek. et Men. im Lichte neuester Untersuchungen über die Herkunft der hexaploiden Weizen. Z. Pflanzenzücht. 45: 17-30 (in German)

Depauw RM, Lukow OM, Adams K (2009) Origin and continuity of glutenin and gliadin subunits in Canada Western Red Spring wheat. In: "Gluten Proteins 2009"; Proc 10th Intern Gluten Workshop (7-9 Sept. 2009; Clermont-Ferrand, France); ed. G.Branlard; INRA; pp 185-189.

Devos KM, Bryan GJ, Collins AJ, Stephenson P, Gale MD (1995) Application of two microsatellite sequences in wheat storage proteins as molecular markers. Theor Appl Genet 90: 247-252, https://doi.org/10.1007/BF00222209

Devos KM, Brown JKM, Bennetzen JL (2002) Genome size reduction through illegitimate recombination counteracts genome expansion in Arabidopsis. Genome Res 12; 1075-1079, https://doi.org/10.1101/gr.132102

Dhaliwal AS, Mares DJ, Marshall DR (1987) Effect of 1B/1R chromosome translocation on milling and quality characteristics of bread wheat. Cereal Chem 64: 72-76, http://pascal-francis.inist.fr/vibad/index.php?action=getRecordDetail&idt=7503512

Dobrotvorskaya TV, Martynov SP, Pukhalsky VA (2004) Trends in the change in the genetic diversity of spring common wheat cultivars, implemented in Russia in 1928-2003. Russian J Genet 40: 1245-1257, https://doi.org/10.1023/B:RUGE.0000048667.39464.54

Dobrotvorskaya TV, Dragovich AYu, Martynov SP, Pukhalsky VA (2009) Genealogical and statistical analysis of the inheritance of alleles of gliadin-coding loci in a model set of common wheat cultivars. Russian J Genet 45: 685-695, https://doi.org/10.1134/S1022795409060088

Doekes GJ (1973) Inheritance of gliadin composition in bread wheat, *Triticum aestivum* L. Euphytica 22: 28-34, https://doi.org/10.1007/BF00021552

Dong H, Cox TS, Sears RG, Lookhart GL (1991) High molecular weight glutenin genes: Effects on quality in wheat. Crop Sci 31: 974-979, https://doi.org/10.2135/cropsci1991.0011183X003100040027x

Dong L, Zhang X, Liu D, Fan H, Sun J, Zhang Z, Quin H, Li B, Hao S, Li Z, Wang D, Zhang A, Ling HQ (2010) New insight into the organization, recombination, expression and functional mechanism of low molecular weight glutenin subunit genes in bread wheat. PLoS ONE 5: e13548, https://doi.org/10.1371/journal.pone.0013548

Dong L, Huo N, Wang Y, Deal K, Luo MC, Wang D, Anderson OD, Gu YQ (2012) Exploring the diploid wheat ancestral A genome through sequence comparison at the high-molecular-weight glutenin locus region. Mol Genet Genomics 287: 855-866, https://doi.org/10.1007/s00438-012-0721-9

Donini P, Law JR, Koebner RMD, Reeves JC, Cooke RJ (2000) Temporal trends in the diversity of UK wheats. Theor Appl Genet 100: 912-917, https://doi.org/10.1007/s001220051370

D'Ovidio R & Masci S (2004) The low-molecular-weight glutenin subunits of wheat gluten. J Cereal Sci 39: 321-339, https://doi.org/10.1016/j.jcs.2003.12.002

D'Ovidio R, Lafiandra D, Tanzarella OA, Anderson OD, Greene FC (1991) Molecular characterization of bread wheat mutants lacking the entire cluster of chromosome 6A controlled gliadin components. J

Cereal Sci 14: 125-129, https://doi.org/10.1016/S0733-5210(09)80133-1

D'Ovidio R, Tanzarella OA, Masci S, Lafiandra D, Porceddu E (1992) RFLP and PCR analyses at *Gli-1*, *Gli-2*, *Glu-1* and *Glu-3* loci in cultivated and wild wheats. Hereditas 116: 79-85, https://doi.org/10.1111/j.1601-5223.1992.tb00803.x

D'Ovidio R, Porceddu E, Lafiandra D (1994) PCR analysis of genes encoding allelic variants of high-molecular-weight glutenin subunits at the *Glu-D1* locus. Theor Appl Genet 88: 175-180, https://doi.org/10.1007/BF00225894

D'Ovidio R, Simeone M, Masci S, Porceddu E, Kasarda DD (1995) Nucleotide sequence of a γ-gliadin type gene from durum wheat: Correlation with a γ-type glutenin subunit from the same biotype. Cereal Chem 72: 443-449, http://pascal-francis.inist.fr/vibad/index.php ?action=getRecordDetail&idt=3667449

D'Ovidio R, Marchitelli C, Ercoli Cardelli L, Porceddu E (1999) Sequence similarity between allelic *Glu-B3* genes related to quality properties of durum wheat. Theor Appl Genet 98: 455-461, https://doi.org/10.1007/s001220051091

Dragovich AYu, Fisenko AV, Mitrofanova OP (2009) Genetic diversity of local varieties of common wheat (*Triticum aestivum* L.) at genes of storage proteins (gliadins). Proc Appl Bot Genet Breed 166: 75-81 (in Russian),

Dreisigacker S, Xiao Y, Srhgal D, Guzman C, He Z, Xia X, Peña RJ (2020) SNP markers for low molecular glutenin subunits (LMW-GSs) at the *Glu-A3* and *Glu-B3* loci in bread wheat. PLoS ONE 15: e0233-056, https://doi.org/10.1371/journal.pone. 0233056

Du X, Wei J, Luo X, Liu Z, Qian Y, Zhu B, Weng Q, Tang H (2020) Low-molecular-weight glutenin subunit LMW-N13 improves dough quality of transgenic wheat. Food Chem 327: 127048, https://doi.org/10.1016/j.foodchem.2020.127048

Dubcovsky J & Dvořák J (2007) Genome plasticity a key factor in the success of polyploid wheat under domestication. Science 316 (5833): 1862-1866, https://doi.org/10.1126/science.1143986

Dubcovsky J, Luo MC, Dvořák J (1995) Differentiation between homoeologous chromosomes 1A of wheat and $1A^m$ of *Triticum monococcum* and its recognition by the wheat *Ph1* locus. Proc Nat Acad Sci USA 92: 6645-6649, https://doi.org/10.1073/pnas.92.14.6645

Dubcovsky J, Echaide M, Giancola S, Rousset M, Luo MC, Joppa LR, Dvořák J (1997) Seed-storage-protein loci in RFLP maps of diploid, tetraploid and hexaploid wheat. Theor Appl Genet 95: 1169-1180, https://doi.org/10.1007/s001220050678

Du Cros DL & Wrigley CW (1979) Improved electrophoretic methods for identifying cereal varieties. J Sci Food Agric 30: 785-794, https://doi.org/10.1002/jsfa2740300807

Dumur J, Jahier J, Bancel E, Laurière M, Bernard M, Branlard G (2004) Proteomic analysis of aneuploid lines in the homeologous group 1 of the hexaploid wheat cultivar Courtot. Proteomics 4: 2685-2695, https://doi.org/10.1002/pmic200300800

Dumur J, Branlard G, Tanguy AM, Dardevet M, Coriton O, Huteau V, Lemoine J, Jahier J (2009) Development of isohomoeoallelic lines within the wheat cv. Courtot for high molecular weight glutenin subunits: transfer of the *Glu-D1* locus to chromosome 1A. Theor Appl Genet 119: 471-481, https://doi.org/10.1007/s00122-009-1053-y

Dumur J, Jahier J, Dardevet M, Coriton O, Tanguy AM, Branlard G (2010) Effects of the replacement of *Glu-A1* by *Glu-D1* locus on agronomic performance and bread-making quality of the hexaploid wheat cv. Courtot. J Cereal Sci 51: 175-181, https://doi.org/10.1016/j.jcs.2009.11.009

DuPont FM, Vensel WH, Chan R, Kasarda DD (2000) Characterization of the 1B-type ω-gliadins from *Triticum aestivum* cultivar Butte. Cereal Chem 77: 607-614, https://doi.org/10.1094/CCHEM.2000.77.5.607

DuPont FM, Vensel W, Encarnacao T, Chan R, Kasarda DD (2004) Similarities of omega gliadins from *Triticum urartu* to those encoded on chromosome 1A of hexaploid wheat and evidence for their post-translational processing. Theor Appl Genet 108: 1299-1308, https://doi.org/10.1007/s00122-003-1565-9

Dupont FM, Vensel W, Tanaka CK, Hurkman WJ, Chen R, Altenbach SB (2011) Deciphering the complexities of the wheat flour proteome using quantitative two-dimensional electrophoresis, three proteases and tandem mass spectrometry. Proteome Sci 9: 10, https://doi.org/10.1186/1477-5956-9-10

Dvořák J & Appels R (1986) Investigation of homologous crossing over and sister chromatid exchange in the wheat *Nor-B2* locus coding for rRNA and *Gli-B2* locus coding for gliadins. Genetics 113: 1037-1056, https://doi.org/10.1093/genetics/113.4.1037

Dvořák J & Chen KC (1984) Distribution of nonstructural variation between wheat cultivars along chromosome arm 6Bp: evidence from the linkage map and physical map of the arm. Genetics 106: 325-333, https://doi.org/10.1093/genetics/106.2.325

Dvořák J & McGuire PE (1981) Nonstructural chromosome differentiation among wheat cultivars, with special reference to differentiation of chromosomes in related species. Genetics 97: 391-414, https://doi.org/10.1093/genetics/97.2.391

Dvořák J, McGuire PE, Cassidy B (1988) Apparent sources of the A genomes of wheat inferred from polymorphism in abundance and restriction fragment of repeated nucleotide sequences. Genome 30: 680-689, https://doi.org/10.1139/g88-115

Dvořák J, Luo MC, Yang ZL (1998a) Restriction fragment length polymorphism and divergence in the genomic regions of high and low recombination in self-fertilizing and cross-fertilizing *Aegilops* spesies. Genetics 148: 423-434, https://doi.org/10.1093/genetics/148.1.423

Dvorak J, Luo MC, Yang ZL, Zhang HB (1998b) The structure of the *Aegilops tauschii* gene pool and the evolution of hexaploid wheat. Theor Appl Genet 97: 657-670, https://doi.org/10.1007/s001220050942

Dvořák J, Yang ZL, You FM, Luo MC (2004) Deletion polymorphism in wheat chromosome regions with contrasting recombination rates. Genetics 168: 1665-1675, https://doi.org/10.1093/genetics/113.4.1037

Dvorak J, Akhunov ED, Akhunova AR, Deal K, Luo MC (2006) Molecular characterization of a diagnostic DNA marker for domesticated tetraploid wheat provides evidence for gene flow from wild tetraploid wheat to hexaploid wheat. Molec Biol Evol 23: 1386-1396, https://doi.org/10.1093/molbev/msl004

Dvorak J, Luo MC, Akhunov ED (2011) N.I.Vavilov theory of centers of diversity in the light of current understanding of wheat diversity, domestication and evolution. Czech J Genet Plant Breed 47(special issue): S20-S27, https://doi.org/10.17221/3249-CJGPB

Dvorak J, Deal K, Luo MC, You FM, von Borstel K, Dehghani H (2012) The origin of spelt and free-threshing hexaploid wheat. J Heredity 103: 426-441, https://doi.org/10.1093/hered/esr152

Eagles HA, Hollamby GJ, Gororo NN, Eastwood RF (2002) Estimation and utilization of glutenin gene effects from the analysis of unbalanced data from wheat breeding programs. Austral J Agric Res 53: 367-377, https://doi.org/10.1071/AR01074

Edwards NM, Gianibelli MC, McCaig TN, Clarke JM, Ames NP, Larroque ORDexter JE (2007) Relationships between dough strengt, polymeric protein quantity and composition for diverse durum wheat genotypes. J Cereal Sci, 45: 140-149, https://doi.org/10.1016/jcrs.2006.07.012

Ehdaie B & Waines JG (1989) Adaptation of landraces and improved spring wheat genotypes to stress environments. J Genet Breed 43: 151-156, http://pascal-francis.inist.fr/vibad/index.php?action=getRecordDetail&idt=6613010

Ellis JRS & Beminster CH (1977) The identification of UK wheat varieties by starch gel electrophoresis of gliadin proteins. J Natl Inst Agric Botany 14: 221-231, https://agris.fao.org/agris-search/search.do?recordID=XE7902402

Endo TR & Gill BS (1984) Somatic karyotype, heterochromatin distribution, and nature of chromosome differentiation in common wheat, *Triticum aestivum* L. em Thell. Chromosome 89: 361-369, https://doi.org/10.1007/BF00331253

Espí A, Rodriguez-Quijano M, Vazquez JF, Carrillo JM, Giraldo P (2014) Molecuar characterization of *Glu-B3* locus in wheat cultivars and segregating populations. J Cereal Sci 60: 374-381, https://doi.org/10.1016/j.jcs.2014.05.008

Ewart JAD (1977) Isolation of two γ-gliadins from "Cappelle-Desprez". J Sci Food Agile 28: 843-848, https://doi.org/10.1002/jsfa.2740280911

FAOSTAT (Food and Agriculture Organization of the United Nations). Available online: http://www.fao.org/faostat/es/#data/QC (accessed on 25 January 2021).

Feeney KA, Wellner N, Gilbert S, Halford NG, Tatham AS, Shewry PR, Belton PS (2003) Molecular structures and interactions of repetitive peptides based on wheat glutenin subunits depend on chain length. Biopolymers 72: 123-131, https://doi.org/10.1002/bip.10298

Feldman M & Sears ER (1981) The wild gene resources of wheat. Sci Am 244: 102-112, https://www.jstor.org/stable/24964263

Feldman M & Kislev ME (2007) Domestication of emmer wheat and evolution of free-threshing tetraploid wheat. Israel J Plant Sci 55: 207-221, https://doi.org/10.1560/IJPS.55.3-4.207

Felix I, Martinant JP, Bernard M, Bernard S, Branlard G (1996) Genetic characterization of storage proteins in a set of F1-derived haploid lines in bread wheat. Theor Appl Genet 92: 340-346, https://doi.org/10.1007/BF00223677

Fenn D, Lukow OM, Bushuk W, Depauw RM (1994) Milling and Baking quality of 1BL/1RS translocation wheats.1. Effects of genotype and environment. Cereal Chem 71: 189-195, http://pascal-francis.inist.fr/vibad/index.php?action=getRecordDetail& idt=4215357

Ferrante P, Masci S, D'Ovidio R, Lafiandra D, Volpi C, Mattei B (2006) A proteomic approach to verify in vivo expression of a novel γ-gliadin containing an extra cysteine residue. Proteomics 6: 1908-1914,

https://doi.org/10.1002/pmic.200500236
Figliuolo G, Mazzeo M, Greco I (2007) Temporal variation of diversity in Italian durum wheat germplasm. Genet Resour Crop Evol 54: 615-626, https://doi.org/10.1007/s10722-006-0019-z
Finney KF, Jones BL, Shogren (1982) Functional (bread-making) properties of wheat protein fractions obtained by utracentrifugation. Cereal Chem 59: 449-453, http://pascal-francis.inist.fr/vibad/index.php ?action=getRecordDetail& idt=PASCAL83X0140154
Flaete NES & Uhlen AK (2003) Association between allelic variation at the combined *Gli-1*, *Glu-3* loci and protein quality in common wheat (Triticum aestivum L.). J Cereal Sci 37: 129-137, https://doi.org/10.1006/jcrs.2001.0447
Forde J, Malpica HM, Halford NG, Shewry PR, Anderson OD, Greene FC, Miflin BJ (1985) The nucleotide sequence of a HMW glutenin subunit gene located on chromosome 1A of wheat (*Triticum aestivum* L). Nucl Acid Res 13: 6817-6831, https://doi.org/10.1093/nar/13.19.6817
Francis HA, Leitch AR, Koebner RMD (1995) Conversion of a RAPD-generated PCR product, containing a novel dispersed repetitive element, into a fast and robust assay for the presence of rye chromatin in wheat. Theor Appl Genet 90: 636-642, https://doi.org/10.1007/BF00222127
Friebe B & Gill BS (1994) C-band polymorphism and structural rearrangements detected in common wheat (*Triticum aestivum*). Euphytica 78: 1-5, https://doi.org/10.1007/BF00021392
Fu TK & Sears ER (1973) The relationships between chiasmata and crossing over in *Triticum aestivum*. Genetics 75: 231-246, https://doi.org/10.1093/genetics/75.2.231
Fu YB & Somers DJ (2009) Genome-wide reduction of genetic diversity in wheat breeding. Crop Sci 49: 161-168, https://doi.org/10.2135/cropsci2008.03.0125
Fu YB, Somers DJ (2011). Allelic changes in bread wheat cultivars were associated with long-term wheat trait improvements. Euphytica 179: 209-225, https://doi.org/10.1007/s10681-010-0235-7
Fu YB; Peterson GW, Yu JK, Gao L, Jia J, Richards KW (2006) Impact of plant breeding on genetic diversity of the Canadian hard red spring wheat germplasm as revealed by EST-derived SSR markers. Theor Appl Genet 112: 1239–1247, https://doi.org/10.1007/s00122-006-0225-2
Galili G & Feldman M (1984) Mapping of glutenin and gliadin genes located on chromosome 1B of common wheat. Molec Gen Genet 193:

293-298, https://doi.org/10.1007/BF00330683
Ganeva G, Korzun V, Landjeva S, Popova Z, Christov NK (2010) Genetic diversity assessment of Bulgarian durum wheat (*Triticum durum* Desf.) landraces and modern cultivars using microsatellite markers. Genet Resour Crop Evol 57: 273-285, https://doi.org/10.1007/s10722-009-9468-5
Gao S, Gu YQ, Wu J, Coleman-Derr D, Huo N, Crossman C, Jia J, Zuo Q, Ren Z, Anderson OD, Kong X (2007) Rapid evolution and complex structural organization in genomic regions harboring multiple prolamin genes in the polyploid wheat genome. Plant Molec Biol 65: 189-203, https://doi.org/10.1007/s11103-007-9208-1
Gao L, Ma W, Chen J, Wang K, Li J, Wang S, Bekes F, Appels R, Yan Y (2010) Characterization and comparative analysis of wheat high molecular weight glutenin subunits by SDS-PAGE, RP-HPLC, HPCE, and MALDI-TOF-MS. J Agric Food Chem 58: 2777- 2786, https://doi.org/10.1021/jf903363z
Gao X, Appelbee MJ, Mekuria GT, Chalmers KJ, Mather DE (2012) A second "overexpression" allele at the *Glu-B1* high-molecular-weight glutenin locus of wheat: sequence characterization and functional effects. Theor Appl Genet 124: 333-343, https://doi.org/10.1007/s00122-011-1708-3
Gianibelli MC & Solomon RG (2003) A novel y-type High Mr glutenin subunit present in *Triticum tauschii*. J Cereal Sci 37: 253-256, https://doi.org/10.1006/jcrs.2002.0486
Gianibelli MC, Gupta RB, Lafiandra D, Margiotta B, MacRitchie F (1991) Polymorphism of high Mr glutenin subunits in *Triticum tauschii*: Characterisation by chromatography and electrophoretic methods. J Cereal Sci 33: 39-52, https://doi.org/10.1006/jcrs.2002.0328
Gianibelli MC, Larroque OR, MacRitchie F, Wrigley CW (2001) Biochemical, genetic and molecular characterization of wheat-gluten proteins. Cereal Chem 78: 635-646, https://www.researchgate.net/profile/Oscar-Larroque/publication/ 265455778
Gianibelli MC, Masci S, Larroque OR, Lafiandra D, MacRitchie F (2002a) Biochemical characterization of a novel polymeric protein subunit from bread wheat (*Triticum aestivum* L.). J Cereal Sci 35: 265-276, https://doi.org/10.1006/jcrs.2001.0401
Gianibelli MC, Lagudah ES, Wrigley CW, MacRitchie F (2002b) Biochemical and genetic characterization of a monomeric storage protein (T1) with an unusually high molecular weight in *Triticum tauschii*. Theor Appl Genet 104: 497-504, https://doi.org/10.1007/s00122-001-0778-z

Gianibelli MC, Echaide M, Larroque OR, Carrillo JM, Dubcovsky J (2002c) Biochemical and molecular characterisation of *Glu-1* loci in Argentinean wheat cultivars. Euphytica 128: 61-73, https://doi.org/10.1023/A: 1020643702867

Gianibelli MC, Wrigley CW, MacRitchie F (2002d) Polymorphism of low *Mr* glutenin subunits in *Triticum tauschii*. J Cereal Sci 35: 277-286, https://doi.org/10.1006/jcrs.2001.0424

Gil-Humanes J, Pistón F, Giménez MJ, Martín A, Barro F (2012) The introgression of RNAi silencing of γ-gliadins into commercial lines of bread wheat changes the mixing and technological properties of the dough. PLoS ONE 7(9): e45937, https://doi.org/10.1371/journal.pone.0045937

Giles RJ & Brown TA (2006) *GluDy* allele variations in *Aegilops tauschii* and *Triticum aestivum*: implications for the origins of hexaploid wheats. Theor Appl Genet 112: 1563-1572, https://doi.org/10.1007/s00122-006-0259-5

Gill BS & Friebe B (2002) Cytogenetics, phylogeny and evolution of cultivated wheats. In: "Bread Wheat. Improvement and Production"; eds B.C.Curtis, S.Rajaram, H.Gómez Macpherson; FAO Plant Prod Protec Series; 30: 71-88, https://www.fao.org/3/y4011e/y4011e07.htm

Gill BS & Kimber G (1977) Recognition of translocations and alien chromosome transfers in wheat by the Giemsa C-banding technique. Crop Sci 17: 264-266, https://doi.org/10.2135/cropsci1977.0011183X001700020008x

Gill KS, Gill BS, Endo TR, Taylor T (1996) Identification and high-density mapping of gene-rich regions in chromosome group 1 of wheat. Genetics 144: 1883-1891, https://doi.org/10.1093/genetics/144.4.1883

Gobaa S, Bancel E, Kleijer G, Stamp P, Branlard G (2007) Effect of the 1BL.1RS translocation on the wheat endosperm, as revealed by proteomic analysis. Proteomics 7: 4349-4357, https://doi.org/10.1002/pmic.200700488

Gobaa S, Bancel E, Branlard G, Kleijer G, Stamp P (2008) Proteomic analysis of wheat recombinant inbred lines: Variations in prolamin and dough rheology. J Cereal Sci 47: 610–619, https://doi.org/10.1016/j.jcs.2007.07.001

Goldsbrough AP, Bulleid NJ, Freedman RB, Flavell RB (1989) Conformational differences between two wheat *Triticum aestivum* high molecular weight glutenin subunits are due to a short region containing six amino acids differences. Biochem J 263: 837-842, https://doi.org/10.1042/bj2630837

Golenberg EM & Nevo E (1987) Multilocus differentiation and population structure in a selfer, wild emmer wheat, *Triticum dicoccoides*. Heredity 58: 451-456, https://doi.org/10.1038/hdy.1987.74

Graybosch RA (1992) High molecular weight glutenin subunit composition of cultivars, germplasm and parents of U.S. red winter wheat. Crop Sci 32: 1151-1155, https://doi.org/10.2135/cropsci1992.0011183X003200050018x

Graybosch RA (2001) Uneasy unions: Quality effects of rye chromatin transfers to wheat. J Cereal Sci 33: 3-16, https://doi.org/10.1006/jcrs.2000.0336

Graybosch RA, Peterson CJ, Hansen LE, Worrall D, Shelton DR, Lukaszewski A (1993a) Comparative flour quality and protein characteristics of 1BL/1RS and 1AL/1RS wheat-rye translocation lines. J Cereal Sci 17: 95-106, https://doi.org/10.1006/jcrs.1993.1010

Graybosch RA, Seo YW, Peterson CJ (1993b) Detection of wheat-rye chromosomal translocations using an antisecalin monoclonal antibody. Cereal Chem 70: 458-463, http://pascal-francis.inist.fr/vibad/index.php?action=getRecordDetail&idt=3812507

Graybosch RA, Lee JH, Peterson CJ, Porter DR, Chung OK (1999) Genetic, agronomic and quality comparisons of two 1AL.1BS wheat-rye chromosomal translocations. Plant Breed 118: 125-130, https://doi.org/10.1046/j.1439-0523.1999.118002125.x

Gu YQ, Coleman-Derr D, Kong X, Anderson OD (2004a) Rapid genome evolution revealed by comparative sequence analysis of orthologous regions from four *Triticeae* genomes. Plant Phys 135: 459-470, https://doi.org/10.1104/pp.103.038083

Gu YQ, Crossman C, Kong X, Luo M, You FM, Coleman-Derr D, Dubcovsky J, Anderson OD (2004b) Genomic organization of the complex α-gliadin gene loci in wheat. Theor Appl Genet 109: 648-657, https://doi.org/10.1007/s00122-004-1672-2

Gu YQ, Salse J, Coleman-Derr D, Dupin A, Crossman C, Lazo GR, Huo N, Beleram H, Ravel C, Charmet G, Charles M, Anderson OD, Chalhoub B (2006) Types and rates of sequence evolution at the high-molecular-weight glutenin locus in hexaploid wheat and its ancestral genomes. Genetics 174: 1493-1504, https://doi.org/10.1534/genetics.106.060756

Gubareva NK, Gavriliuk IP, Chernoburova AD (1975) Determination of authentity and seed purity of wheat cultivars from the electrophoregram of gliadin. Methodical indications and the catalogue of formulas of varieties. Doklady VASKhNIL: 1-36 (in Russian),

Gupta RB & MacRitchie F (1994) Allelic variation at glutenin subunit and gliadin loci, *Glu-1*, *Glu-3* and *Gli-1* of common wheats. II.Biochemical basis of the allelic effects on dough properties. J Cereal Sci 19: 19-29, https://doi.org/10.1006/jcrs.1994.1004

Gupta RB & Shepherd KW (1988) Low-molecular-weight glutenin subunits in wheat: their variation, inheritance and association with bread-making quality. Proc 7th Intern Wheat Genetics Symp; eds T.E. Miller & R.M.D. Koebner; IPSR, Cambridge, UK; v.2, pp 943-949, https://agris.fao.org/agris-search/search.do?recordID=GB9010352

Gupta RB & Shepherd KW (1990) Two-step one-dimensional SDS-PAGE analysis of LMW subunits of glutenin. I. Variation and genetic control of the subunits in hexaploid wheats. Theor Appl Genet 80: 65-74, https://doi.org/10.1007/BF00224017

Gupta RB & Shepherd KW (1992) Identification of rye chromosome 1R translocations and substitutions in hexaploid wheats using storage proteins as genetic markers. Plant Breed 109: 130-140, https://doi.org/10.1111/j.1439-0523.1992.tb00163.x

Gupta RB & Shepherd KW (1993) Production of multiple wheat-rye 1RS translocation stocks and genetic analysis of LMW subunits of glutenin and gliadins in wheats using these stocks. Theor Appl Genet 85: 719-728, https://doi.org/10.1007/BF00225011

Gupta RB, Singh NK, Shepherd KW (1989) The cumulative effect of allelic variation in LMWand HMW glutenin subunits on physical dough properties in the progeny of two bread wheats. Theor Appl Genet 77: 57-64, https://doi.org/10.1007/BF00292316

Gupta RB, Bekes F, Wrigley CW (1991) Prediction of physical dough properties from glutenin subunit composition in bread wheats: Correlation studies. Cereal Chem 68: 328-333, http://pascal-francis.inis t.fr/vibad/index.php?action=getRecordDetail&idt=5380905

Gupta RB, Metakovsky EV, Wrigley CW (1994a) The relationship between LMW-glutenin-subunit and gliadin alleles in Australian wheat cultivars. In: "Gluten Proteins", Proc 5th Intern Workshop on gluten proteins (Detmold, Germany, 1993); Assoc Cereal Research Publ, pp 589-597, http://hdl.handle.net/102.100.100/239396?index=1

Gupta RB, Paul JG, Cornish GB, Palmer CA, Bekes F, Rathjen AJ, Wrigley CW (1994b) Allelic variation at glutenin subunit and gliadin loci, Glu-1, Glu-3 and Gli-1 of common wheats. I. Its additive and interaction effects on dough properties. J Cereal Sci 19: 9-17, https://doi.org/10.1006/jcrs.1994.1003

Gupta RB, Popineau Y, Lefebvre J, Cornec M, Lawrence GJ, MacRitchie F (1995) Biochemical basis of flour properties in bread wheats. II.

Changes in polymeric protein formation and dough/gluten properties associated with the loss of low Mr and high Mr glutenin subunits. J Cereal Sci 21: 103-116, https://doi.org/10.1016/0733-5210(95) 90026-8

Guzmán C, Caballero L, Martín LM, Álvarez JB (2012) *Waxy* genes from spelt wheat: new alleles for modern wheat breeding and new phylogenetic inferences about the origin of this species. Ann Botany 110: 1161-1171, https://doi.org/10.1093/aob/mcs201

Haider N, Nabulsi I, Mir Ali N (2010) Comparison of the efficiency of A-PAGE and SDS-PAGE, ISSRs and RAPDs in resolving genetic relationships among *Triticum* and *Aegilops* species. Genet Resour Crop Evol 57: 1023-1039, https://doi.org/10.1007/s10722-010-9545-9

Hamauzu ZI, Toyomasu T, Yonezawa D (1974) Molecular weight determination of gliadin fractions in gel filtration by SDS-polyacrylamide gel electrophoresis and sedimentation equilibrium. Agric Biol Chem 38: 2445-2450, https://doi.org/10.1080/00021369.1974.10861523

Hamrick JL & Godt JW (1997) Allozyme diversity in cultivated crops. Crop Sci 37: 26-30, https://doi.org/10.2135/cropsci1997.0011183X003700010004x

Hao CY, Zhang XY, Wang LF, Dong YS, Shang XW, Jia JZ (2006) Genetic diversity and core collection evaluations in common wheat germplasm from the Northwestern spring wheat region in China. Molec Breed 17: 69-77, https://doi.org/10.1007/s11032-005-2453-6

Harberd NP, Bartles D, Thompson RD (1985) Analysis of the gliadin multigene loci in bread wheat using nullisomic-tetrasomic lines. Molec Gen Genet 198: 234-242, https://doi.org/10.1007/BF00383001

Harlan JR (1972) Genetics of disaster. J Environmtl Quality 1: 212-215, https://doi.org/10.2134/jeq1972.00472425000100030002x

Hassani ME, Gianibelli MC, Shariflou MR, Sharp PJ (2005) Molecular structure of a novel y-type HMW glutenin subunit gene present in *Triticum tauschii*. Euphytica 141: 191-198, https://doi.org/10.1007/s10681-005-6982-1

Hassani ME, Shariflou MR, Gianibelli MC, Sharp PJ (2008) Characterization of a ω-gliadin gene in *Triticum tauschii*. J Cereal Sci 47: 59-67, https://doi.org/10.1016/j.jcs.2007.01.008

He S, Ohm H, MacKenzie S (1992) Detection of DNA sequence polymorphisms among wheat varieties. Theor Appl Genet 84: 573-578, https://doi.org/10.1007/BF00224154

He ZH, Liu L, Xia XC, Liu JJ, Peña RG (2005) Composition of HMW and LMW glutenin subunits and their effects on dough properties, pan bread, and noodle quality of Chinese bread wheats. Cereal Chem 82:

345-350, https://doi.org/10.1094/CC-82-0345

Howes NK (1986) Linkage between the Lr10 gene controlling resistance to leaf rust, two endosperm proteins, and hairy glumes in hexaploid wheat. Canad J Genet Cytol 28: 595-600, https://doi.org/10.1139/g86-087

Howes NK, Lukow OM, Dawood MR, Bushuk W (1989) Rapid detection of 1BL/1RS chromosome translocation in hexaploid wheats by monoclonal antibodies. J Cereal Sci 10: 1-4, https://doi.org/10.1016/S0733-5210(89)80029-3

Hsia CC & Anderson OD (2001) Isolation and characterization of wheat ω-gliadin genes. Theor Appl Genet 103: 37-44, https://doi.org/10.1007/s00122-001-0552-2

Huang L, Milet E, Feldman M (1996) Genetic relationships among Mediterranean common wheat cultivars as determined by RFLP and parentage analysis. J Genet Breed 50: 377-382, https://weizmann.esploro.exlibrisgroup.com/993285748903596#file-0

Huang S, Sirikhachornkit A, Su X, Faris J, Gill B, Haselkorn R, Gornicki P (2002) Genes encoding plastid acetil-CoA carboxylase and 3-phosphoglycerate kinase of the *Triticum/Aegiops* complex and evolutionary history of polyploid wheat. Proc Nat Acad Sci USA 99: 8133-8138, https://doi.org/10.1073/pnas.072223799

Huang XQ & Cloutier S (2008) Molecular characterization and genomic organization of low molecular weight glutenin subunit genes at the *Glu-3* loci in hexaploid wheat (*Triticum aestivum* L.). Theor Appl Genet 116: 953-966, https://doi.org/10.1007/s00122-008-0727-1

Huang XQ, Börner A, Röder MS, Ganal MW (2002) Assessing genetic diversity of wheat (*Triticum aestivum* L.) germplasm using microsatellite markers. Theor Appl Genet 105: 699-707, https://doi.org/10.1007/s00122-002-0959-4

Huang XQ, Cloutier S, Lycar L, Radovanovic N, Humphreys DG, Noll JS, Somers DJ, Brown PD (2006) Molecular detection of QTLs for agronomic and quality traits in a doubled haploid population derived from two Canadian wheats (*Triticum aestivum* L.). Theor Appl Genet 113: 753-766, https://doi.org/10.1007/s00122-006-0346-7

Huo N, Zhang S, Zhu T, Dong L, Wang Y, Mohr T, Hu T, Liu Z, Dvorak J, Luo MC, Wang D, Lee JY, Altenbach S, Gu YQ (2018a) Gene duplication and evolution dynamics in the homoeologous regions harbouring multiple prolamin and resistance gene families in hexaploid wheat. Font Plant Sci 9: 673, https://doi.org/10.3389/fpls.2018.00673

Huo N, Zhu T, Altenbach S, Dong L, Wang Y, Mohr T, Liu Z, Dvorak J, Luo MC, Gu YQ (2018b) Dynamic evolution of α-gliadin prolamin gene family in homologous genomes of hexaploid wheat. Sci Reports 8: 5181, https://doi.org/10.1038/541598-018-23570-5

Hurka H (1994) Conservation genetics and the role of botanical gardens. In: "Conservation Genetics"; eds V.Loeschcke, J.Tomiuk, S.K.Jain; Birkhauser Verlag, Basel, Switzerland; pp 371-380, https://doi.org/10.1007/978-3-0348-8510-2_24

Hurkman WJ & Tanaka CK (2004) Improved method for separation of wheat endosperm proteins and anaysis by two-dimensional gel electrophoresis. J Cereal Sci 40: 295-299, https://www.elsevier.com/locate/jnlabr/yjers

Hussain A & Lukow OM (1994) Characterization of the 1B/1R translocation in wheat using water extractable protein concentrate. Euphytica 78: 109-113, https://doi.org/10.1007/BF00021405

Iakobashvili ZA (1989) The phylogenetic relations between wheat species by means of analysis of the polymorphism and inheritance of grain storage proteins. PhD Thesis; Vavilov Institute of General Genetics. Moscow.

Ibba MI, Kiszonas AM, Morris CF (2017a) Evidence of intralocus recombination at the Glu-3 loci in bread wheat (*Triticum aestivum* L.). Theor Appl Genet 130: 891-902, https://doi.org/10.1007/s00122-017-2858-8

Ibba MI, Kiszonas AM, Guzmán C, Morris CF (2017b) Definition of the low molecular weight glutenin subunit gene family members in a set of standard bread wheat (*Triticum aestivum* L.) varieties. J Cereal Sci 74: 263-271, https://doi.org/10.1016/j.jcs.2017.02.015

Ibba MI, Kiszonas AM, Morris CF (2017c) Influence of low molecular weight glutenin subunit haplotypes on dough rheology in elite common wheat varieties. Cereal Chem 94: 1016-1027, https://doi.org/10.1094/CCHEM-07-17-0137-R

ICC (1995) Wheat – Identification of varieties by electrophoresis. In: ICC Standard №143, Intern. Assoc. Cereal Sci. Technol.; Vienna, Austria; p.1-13.

Igrejas G & Branlard G (2020) The importance of wheat. In: Wheat quality for improving processing and human health; eds G.Igrejas, T.M.Ikeda, C.Guzmán; Springer, Cham, pp 1-7, https://doi.org/10.1007/978-3-030-34163-3_1

Ikeda TM, Nagamine T, Fukuoka H, Yano H (2002) Identification of new low-molecular-weight glutenin subunit genes in wheat. Theor Appl Genet 104: 680-687, https://doi.org/10.1007/s001220100756

Ikeda TM, Araki E, Fujitan Y, Yano H (2006) Characterization of low-molecular-weight glutenin subunit genes and their protein products in common wheat. Theor Appl Genet 112: 327-334, https://doi.org/10.1007/s00122-005-0131-z

Illichevsky NN, Metakovsky E, Sozinov AA (1989) Polymorphism and genetic control of alpha-amylase in soviet spring bread wheat varieties. Russian J Genet 25: 2176-2186 (in Russian)

Isidore E, Scherrer B, Chalhoub B, Feuillet C, Keller B (2005) Ancient haplotypes resulting from extensive molecular rearrangements in the wheat A genome have been maintained in species of three different ploidy levels. Genome Res 15: 526-536, https://doi.org/10.1101/gr.3131005

Jackson EA (1984) The genetics and biochemistry of the low molecular weight subunits of glutenin in relation to bread-making quality. PhD thesis, University of Cambridge, UK.

Jackson EA, Holt LM, Payne PI (1983) Characterisation of high molecular weight gliadin and low-molecular-weight glutenin subunits of wheat endosperm by two-dimensional electrophoresis and the chromosomal localization of their controlling genes. Theor Appl Genet 66: 29-37, https://doi.org/10.1007/BF00281844

Jackson EA, Holt LM, Payne PI (1985) *Glu-B2*, a storage protein locus controlling the D group of LMW glutenin subunits in bread wheat (*Triticum aestivum*). Genet Res 46: 11-17, https://doi.org/10.1017/S0016672300022412

Jackson EA, Morel MH, Sontag-Strohm T, Branlard G, Metakovsky E, Redaelli R (1996) Proposal for combining the classification systems of alleles of *Gli-1* and *Glu-3* loci in bread wheat (*Triticum aestivum* L.). J Genet Breed 50: 321-336, https://agris.fao.org/agris-search/search.do?recordID=IT1998061310

Javornik B (1989) Wheat cultivar identification by gliadin electrophoresis. Podravka 7: 49-62.

Javornik B, Sinkovič T, Vapa L, Koebner RMD, Rogers WJ (1991b) A comparison of methods for identifying and surveying the presence of 1BL.1RS translocations in bread wheat. Euphytica 54: 45-53, https://doi.org/10.1007/BF00145630

Jiang N, Bao Z, Zhang X, Eddy SR, Wessler SR (2004) Pack-MULE transposable elements mediate gene evolution in plants. Nature 431: 569-573, https://doi.org/10.1038/nature02953

Jin H, Zhang Y, Li G, Mu P, Fan Z, Xia X, He Z (2013) Effects of allelic variation of HMW-GS and LMW-GS on mixograph properties and Chinese noodle and streamed bread qualities in a set of Aroona near-isogenic wheat lines. J Cereal Sci 57: 146-152, https://doi.org/10.1016/j.jcs.2012.10.011

Johansson E (1996) Quality evaluation of D-zone omega gliadins in wheat. Plant Breed 115: 57-62, https://doi.org/10.1111/j.1439-0523.1996.tb00871.x

Johansson E, Branlard G, Cuniberti M, Flagelia Z, Hüsken A, Nurit E, Peña RJ, Sissons M, Vazquez D (2020) Genotypic and environmental effects on wheat technological and nutritional quality. In: Wheat quality for improving processing and human health; eds G.Igrejas, T.M.Ikeda, C.Guzman; Springer, Cham, pp 171-204, https://doi.org/10.1007/978-3-030-34163-3_8

Jones RW, Taylor NW, Senti FR (1959) Electrophoresis and fractionation of wheat gluten. Arch Biochem Biophys 84: 363-376, https://doi.org/10.1016/0003-9861(59)90599-5

Jones BL, Lookhart GL, Hall SB, Finney KF (1982) Identification of wheat cultivars by gliadin electrophoresis: electrophoregrams of the 88 wheat cultivars most commonly grown in the Unated States in 1979. Cereal Chem 59: 181-188, http://pascal-francis.inist.fr/vibad/index.php?action=getRecordDetail&idt=PASCAL82X0384620

Joppa LR, Khan K, Williams ND (1982) Chromosomal location of genes for gliadin polypeptides in durum wheat *Triticum turgidum* L. Theor Appl Genet 64: 289-293, https://doi.org/10.1007/BF00274164

Juhász A & Gianibelli MC (2006) Low-molecular-weight glutenin subunits: insights into this abundant subunit group present in glutenin polymers. In: "Gliadin and glutenin. The unique balance of wheat quality". eds C.Wrigley, F.Békés, W.Bushuk; AACC Internat.; pp 171-212, http://hdl.handle.net/102.100.100/176785?index=1

Juhász A, Larroque OR, Támas L, Hsam SLK, Zeller FJ, Békés F, Bedö Z (2003) Bánkúti 1201 - An old Hungarian wheat variety with special storage protein composition. Theor Appl Genet 107: 697-704, https://doi.org/10.1007/s00122-003-1292-2

Karp A & Maddock SE (1984) Chromosome variation in wheat plants regenerated from cultured immature embryos. Theor Appl Genet 67: 249-255, https://doi.org/10.1007/BF00317047

Kasarda DD (1980) Structure and properties of α-gliadins. Ann Technol Agric 29: 151-173, http://pascal-francis.fr/vibad/index.php?action=getRecordDetail&idt=PASCAL8110386171

Kasarda DD (1989) Glutenin structure in relation to wheat quality. In: "Wheat is Unique", ed. Y.Pomeranz; AACC, St Paul; pp 277-302.

Kasarda DD, Bernardin JE, Qualset CO (1976) Relationship of gliadin protein components to chromosomes in hexaploid wheats (*Triticum aestivum* L.) Proc Nat Acad Sci USA 73: 3646-3650, https://doi.org/10.1073/pnas.73.10.3646

Kasarda DD, Autran JC, Lew EJL, Nimmo CC, Shewry PR (1983) N-terminal amino acid sequences of ω-gliadins and ω-secalins. Implications for the evolution of prolamin genes. Biochem Biophys Acta 747: 138-150, https://doi.org/10.1016/0167-4838(83)90132-2

Kasarda DD, Lafiandra D, Morris R, Shewry PR (1984a) Genetic relationships of wheat gliadin proteins. Kulturpflanze 32: 41-60 (S33-S52)

Kasarda DD, Okita TW, Bernardin JE, Baecker PA, Nimmo CC Lew EJL, Dietler MD, Green FC (1984b) Nucleic acid (cDNA) and amino acid sequence of α-type gliadins from wheat (*Triticum aestivum*). Proc Natl Acad Sci USA. 81: 4712-4716, https://doi.org/10.1073/pnas.81.15.4712

Kasarda DD, Adelsteins AE, Laird NF (1987) γ-Gliadins with α-type structure coded on chromosome 6B of the wheat (*Triticum aestivum* L.) cultivar "Chinese Spring". In: 3rd Int. Workshop on Gluten Proteins; eds R.Làsztity & F.Békés; World Scientific, Singapore; pp 20-29, https://agris.fao.org/agris-search/search.do?recordID=US201302687144

Keck B, Köhler P, Wieser H (1995) Disulphide bonds in wheat gluten: further cystine peptides derived from gluten proteins following peptic and thermolytic digestion. Z Lebensm- Untersch Forsch 196: 239-247, https://doi.org/10.1007/BF00193253

Keller B & Feuillet C (2000) Colinearity and gene density in grass genomes. Trends Plant Sci 5: 246-251, https://doi.org/10.1016/S1360-1385(00)01629-0

Kerby K, Kuspira J, Jones BL (1988) Biochemical data bearing on the relationship between the genome of *Triticum urartu* and the A and B genomes of the polyploid wheats. Genome 30: 576-581, https://doi.org/10.1139/g88-097

Khan K, Hamada AS, Patek J (1985) Polyacrylamide gel electrophoresis for wheat variety identification: Effect of variables on gel properties. Cereal Chem 62: 310-313, http://pascal-francis.inist.fr/vibad/index.php?action=getRecordDetail&idt=8440421

Khelifi D & Branlard G (1992) The effects of HMW and LMW subunits of glutenin and of gliadins on the technological quality of progeny from four crosses between poor breadmaking quality and strong wheat cultivars. J Cereal Sci 16: 195-209, https://doi.org/10.1016/S0733-5210(09)80084-2

Khelifi D, Branlard G, Bourgoin-Greneche M (1992) Diversity of some D-zone omega gliadins of bread wheat as revealed by 2 step A-PAGE/SDS-PAGE technique. J Genet Breed 46: 351-358,

https://hal.inrae.fr/hal-02712824
Khlestkina EK, Huang XQ, Quenum FJB, Chebotar S, Röder MS, Börner A (2004a) Genetic diversity in cultivated plants – loss or stability? Theor Appl Genet 108: 1466-1472, https://doi.org/10.1007/s00122-003-1572-x
Khlestkina EK, Röder MS, Efremova TT, Börner A, Shumny VK (2004b) The genetic diversity of old and modern Siberian varieties of common spring wheat as determined by microsatellite markers. Plant Breed 123: 122-127, https://doi.org/10.1046/j.1439-0523.2003.00934.x
Khoshro HH, Bihamta MR, Hassani ME (2021) Relationship between allelic variation at the Glu-3 loci and qualitative traits in bread wheat. Cereal Res Comm, https://doi.org/10.1007/s42976-021-00203-8
Kihara H (1954) Origin of wheat. Wheat Inf Serv 1: 36-41
Koebner RMD (1995) Generation of PCR-based markers for detection of rye chromatin in a wheat background. Theor Appl Genet 90: 740-745, https://doi.org/10.1007/BF00222142
Koebner RMD & Shepherd KW (1986) Controlled introgression to wheat of genes from rye chromosome arm 1RS by induction of allosyndesis. Theor Appl Genet 73: 197-208, https://doi.org/10.1007/BF00289275
Kolyuchii VT & Sozinov AA (2000). The effect of natural selection on the frequency of alleles at gliadin loci, and their association in the artificially hybrid population of winter wheat. Cytol Genet (Kiev) 34: 32-38, https://www.cabdirect.org/cabdirect/abstract/20001617298
Konarev VG (1980) Wheat proteins. M., "Kolos", pp. 83-87 (in Russian), https://www.cabdirect.org/cabdirect/abstract/19831618324
Konarev VG, Gubareva NK, Gavrilyuk IP (1970) Protein markers of wheat and its wild relatives. II. Comparative electrophoretic analysis of the gliadins. Vestnik Selskokhoz Nauki 8: 109-114, https://www.cabdirect.org/cabdirect/abstract/19711602503
Konarev VG, Gavrilyuk IP, Gubareva NK, Peneva TI (1979) Seed proteins in genome analysis, cultivar identification, and documentation of cereal genetic resources: A review. Cereal Chem 56: 272-278, http://pascal-francis.inist.fr/vibad/index.php?action=getRecordDetail&idt=PASCAL8050195930
Kosmolak FG (1979) Gliadin composition of the bread-wheat cultivars BW20 and Sinton. Canad J Plant Sci 59: 1001-1005, https://doi.org/10.4141/cjps79-157
Kosmolak FG, Leisle D, Dexter JE, Matsuo RR, Marchylo BA (1980) A relationship between durum wheat quality and gliadin electrophoregrams. Canad J Plant Sci 60: 427-432, https://doi.org/10.4141/cips80-064

Kota RS, Gill KS, Gill BS, Endo TR (1993) A cytogenetically based physical map of chromosome 1B in common wheat. Genome 36: 548-554, https://doi.org/10.1139/g93-075

Koval SF & Metakovsky E 1985. Adaptive value of some quantitative and qualitative characters in model hybrid population of *T.aestivum*. Selskokhoz Biol (Moscow) 11: 48-52 (in Russian).

Koval SF, Metakovsky E, Kudryavtsev AM, Sozinov AA (1986) On the relationship of genes controlling the hairness and colour of spike with certain alleles of the gliadin-coding loci in wheat. Selskokhoz Biol 2: 31-36 (in Russian), https://agris.fao.org/agris-search/search.do?recordID=SU19870062310

Koval SF, Metakovsky E, Sozinov AA (1988) A series of near-isogenic spring bread wheat lines on the basis of the variety Novosibirskaya 67. Cereal Res Comm 6: 183-187, https://www.jstor.org/stable/23782844

Kozub NA, Sozinov IA, Sozinov AA (2001) Conjugation of the 1BL/1RS translocation with qualitative and quantitative traits in the common wheat *T.aestivum*. Cytol Genet (Kiev) 35: 74-80 (in Russian), http://europepmc.org/article/med/11944320

Kozub NA, Sozinov IA, Sozinov AA (2003) Recombination frequency in the locus *Gli-D1* of common wheat *T.aestivum* L. Cytol Genet (Kiev) 37: 80-83 (in Russian), https://europepmc.org/article/med/14650331

Kozub NA, Sozinov IA, Kolyuchii VT, Vlasenko VA, Sobko TA, Sozinov AA (2005) Identification of the 1AL/1RS translocation in common wheat cultivars bred in Ukraine. Cytol Genet (Kiev) 39: 16-24, https://www.researchgate.net/profile/N-Kozub/publication/7376035

Kozub NA, Sozinov IA, Sozinov AA (2008a) Dependence of cross-pollination indices on the genotypic characteristics of common wheat. Cytol Genet (Kiev) 42: 87-93 (in Russian), https://doi.org/10.3103/S0095452708030080

Kozub NA, Sozinov IA, Sozinov AA (2008b) Recombination of gliadin genes of chromosome 1D in the common wheat hybrid carrying the introgression from *Aegilops cylindrica*. Plant Breed 122: 86-88, https://doi.org/10.1046/j.1439-0523.2003.00803.x

Kozub NA, Sozinov IA, Sobko TA, Kolyuchii VT, Kuptsov SV, Sozinov AA (2009) Variation at storage protein loci in winter common wheat cultivars of the central forest-steppe of Ukraine. Cytol Genet (Kiev) 43: 55-62, https://doi.org/10.3103/S0095452709010101

Kozub NA, Sozinov IA, Sobko TA, Dedkova OS, Badaeva ED, Netsvetaev VP (2012) Rye translocations in the varieties of winter common wheat. Agric Biol 3: 68-74. https://doi.org/10.15389/agrobiology.2012.3.68eng

Kozub NA, Sozinov IA, Karelov AV, Blume YaB, Sozinov AA (2017) Diversity of Ukrainian winter common wheat varieties with respect to storage protein loci and molecular markers for disease resistance genes. Cytology and Genetics (Kiev) 51: 117-129, https://doi.org/10.3103/S0095452717020050

Kozub NA, Sozinov IA, Chaika VM, Sozinova OI, Janse LA, Blume YaB (2020) Changes in allele frequencies at storage protein loci of winter common wheat under climate change. Cytol Genet (Kiev) 54: 305-317, https://doi.org/10.3103/S0095452720040076

Krasileva KV, Vasquez-Gross HA, Howell T, Bailey P, Paraiso F, Clissold L, Simmonds J, Ramirez-Gonzalez RH, Wang X, Borrill P, Fosker C, Ayling S, Phillips AL, Uauy C, Dubcovsky J (2017) Uncovering hidden variation in polyploid wheat. PNAS 114 (6): E913-E921, https://doi.org/10.1073/pnas.1619268114

Krattiger AF, Payne PI, Law CN (1996) Effects of homoeologous group 1 and 6 chromosomes of the Capelle-Desprez (Bezostaya 1) substitution lines on aspects of bread-making quality of wheat. Euphytica 89: 17-25, https://doi.org/10.1007/BF00015714

Kuckuck H (1959) On the findings of *Triticum spelta* L. in Iran and on the arising of *Triticum aestivum*-types through crossing of different *spelta*-types. Wheat Inf Serv 9: 1-2.

Kuckuck H (1964) Experimentelle untersuchungen für enstehung der kulturweizen (Experimental investigation on the origin of cultivated wheat). Z Pflanzenzücht 51: 97-140 (in German), https://doi.org/ci.nii.ac.jp/crid/1573387450807482624

Kudryavtsev AM, Metakovsky E, Sozinov AA (1988) Polymorphism and inheritance of gliadin components controlled by chromosome 6A of spring durum wheat. Biochem Genet 26: 693-703, https://doi.org/10.1007/BF00553869

Kudryavtsev AM, Illichevskii NN, Boggini G, Benedettelli S (1996) Gliadin polymorphism and genetic diversity of modern Italian durum wheat. J Genet Breed 50: 239-248, https://agris.fao.org/agris-search/search.do?recordID=IT1998061300

Kuspira J, MacIagan J, Bhambhani RN, Sadasivaiah RS, Kim NS (1989) Genetic and cytogenetic analises of the A genome of *T.monococcum* L. V. Inheritance and linkage relationships of genes determining the expression of 12 qualitative characters. Genome 32: 869-881, https://doi.org/10.1139/g89-524

Laemmli UK (1970) Cleavage of structural proteins during the assembly of the head of bacteriophage T4. Nature 227: 680-685, https://doi.org/10.1038/227680a0

Lafiandra D, Kasarda DD, Morris R (1984) Chromosomal assignment of genes coding for the wheat gliadin protein components of the cultivars "Cheyenne" and "Chinese Spring" by two-dimensional (two-pH) electrophoresis. Theor Appl Genet 68: 531-539, https://doi.org/10.1007/BF00285007

Lafiandra D, Colaprico G, Kasarda DD, Porceddu E (1987a) Null alleles for gliadin blocks in bread and durum wheat cultivars. Theor Appl Genet 74: 610-616, https://doi.org/10.1007/BF00288860

Lafiandra D, Margiotta B, Porceddu E (1987b) A possible association between heading time and the *Gli-A2* locus in bread wheat. Plant Breed 99: 333-335, https://doi.org/10.1111/j.1439-0523.1987.tb01191.x

Lafiandra D, Benedettelli S, Margiotta B, Porceddu E (1989) Chromosomal location of gliadin coding genes in *Triticum aestivum* ssp. *spelta* and evidence of the lack of components controlled by *Gli-2* loci in wheat aneuploids. Theor Appl Genet 78: 177-183, https://doi.org/10.1007/BF00288796

Lafiandra D, D'Ovidio R, Porceddu E, Margiotta B, Colaprico G (1993) New data supporting high Mr glutenin subunit 5 as the determinant of quality differences among the pairs 5+10 vs 2+12. J Cereal Sci 18: 197-205, https://doi.org/10.1006/jcrs1993.1046

Lagudah ES & Halloran GM (1988) Phylogenetic relationships of *Triticum tauschii* the D genome donor to hexaploid wheat. 1. Variation in HMW subunits of glutenin and gliadins. Theor Appl Genet 75: 592-598, https://doi.org/10.1007/BF00289125

Lagudah ES, O'Brien LO, Halloran GM (1988) Influence of gliadin composition and high molecular weight subunits of glutenin on dough properties in an F3 population of a bread wheat cross. J Cereal Sci 7: 33-42, https://doi.org/10.1016/S0733-5210(88)80057-2

Landjeva S, Korzun V, Ganeva G (2006a) Evaluation of genetic diversity among Bulgarian winter wheat (*Traticum aestivum* L.) varieties during the period 1925-2003 using microsatellites. Genet Res Crop Evol 53: 1605-1614, https://doi.org/10.1007/s10722-005-8718-4

Landjeva S, Korzun V, Tsanev V, Vladova R, Ganeva G (2006b) Distribution of the wheat-rye tranlocation 1RS.1BL among bread wheat varieties of Bulgaria. Plant Breed 125: 102-104, https://doi.org/10.1111/j.1439-0523.2006.01142.x

Landjeva S, Korzun V, Börner A (2007) Molecular markers: actual and potential contributions to wheat genome characterization and breeding. Euphytica 156: 271-296, https://doi.org/10.1007/s10681-007-9371-0

Lange W, Linde-Laursen I, Larsen J, Ljungberg A, Ellerstrom S (1987) Cytogenetic analysis of structural rearrangements in the varieties of common wheat, *Triticum aestivum*. Theor Appl Genet 73: 635-645, https://doi.org/10.1007/BF00260770

Lapitan NLV, Sears RJ, Rayburn AL Gill BS (1986) Wheat-rye translocations: detection of chromosome break-points by *in situ* hybridization with a biotin-labeled DNA probe. J Heredity 77: 415-419, https://doi.org/10.1093/oxfordjournals.jhered.a110272

Larkin PJ & Scowcroft WR (1981) Somaclonal variation – A novel source of variability from cell vultures for plant improvement. Theor Appl Genet 60: 197-214, https://doi.org/10.1007/BF02342540

Larkin PJ, Ryan SA, Brettell RIS, Scowcroft WR (1984) Heritable somaclonal variation in wheat. Theor Appl Genet 67: 443-455, https://doi.org/10.1007/BF00263410

Laurière M, Bouchez I, Doyen C, Eynard L (1996) Identification of glycosylated forms of wheat storage proteins using two-dimensional electrophoresis and blotting. Electrophoresis 17: 497-501, https://doi.org/10.1002/elps.1150170312

Law JR, Donini P, Koebner RMD, Reeves JC, Cooke RJ (1998) DNA profiling and plant variety registration. III. The statistical assessment of distinctness in wheat using amplified fragment length polymorphisms. Euphytica 102: 335-342, https://doi.org/10.1023/A:1018318701458

Lawrence GJ (1986) The high-molecular-weight glutenin subunit composition of Australian wheat cultivars. Austral J Agric Res 37: 125-133, https://doi.org/10.1071/AR9860125

Lawrence GJ, Moss HJ, Shepherd KW, Wrigley CW (1987) Dough quality of biotypes of eleven Australian wheat cultivars that differ in high-molecular-weight glutenin subunit composition. J Cereal Sci 6: 99-101, https://doi.org/10.1016/S0733-5210(87)80045-0

Lawrence GJ, MacRitchie F, Wrigley CW (1988) Dough and baking quality of wheat lines deficient in glutenin subunits controlled by the *Glu-A1*, *Glu-B1* and *Glu-D1* loci. J Cereal Sci 7: 109-112, https://doi.org/10.1016/S0733-5210(88)80012-2

Lee JH, Graybosch RA, Peterson CJ (1995) Quality and biochemical effects of 1BL/1RS wheat rye translocation in wheat. Theor Appl Genet 90: 105-112, https://doi.org/10.1007/BF00221002

Lee SJ, Penner GA, Devos KM (1995) Characterization of loci containing microsatellite sequences among Canadian wheat cultivars. Genome 38: 1037-1040, https://doi.org/10.1139/g95-137

Lee YK, Bekes F, Gras P, Ciaffi M, Morell MK, Appels R (1999a) The low-molecular weight glutenin subunit proteins of primitive wheats. IV. Functional properties of products from individual genes. Theor

Appl Genet 98: 149-155, https://doi.org/10.1007/s001220051051
Lee YK, Ciaffi M, Appels R, Morell MK (1999b) The low-molecular weight glutenin subunit proteins of primitive wheats. II. The genes from A-genome species. Theor Appl Genet 98: 126-134, https://doi.org/10.1007/s001220051049
Lei ZS, Gale KR, He ZH, Gianibelli C, Larroque O, Xia XC, Butow BJ, Ma J (2006) Y-type gene specific markers for enhanced discrimination of high-molecular weight glutenin alleles at the *Glu-B1* locus in hexaploid wheat. J Cereal Sci 43: 94-101, https://doi.org/10.1016/j.jcs.2005.08.003
Lein A (1975) Introgression of a rye chromosome to wheat strains by Georg Riebesel – Salzmünde after 1926. In: "Triicale, Study and Selection", Proc Int Symp (July 1973, Leningrad, USSR); EUCARPIA, VASKHNIL, USSR; pp 158-167.
Leisle D, Kosmolak FG, Kovacs MI (1981) Association of glume color with gluten strength and gliadin proteins in durum wheat. Canad J Plant Sci 61: 149-151, https://doi.org/10.4141/cjps81-020
Lerner E, Kolman MA, Rogers WJ (2009) Quality and endosperm storage protein variation in Argentinean grown bread wheat. I. Allelic diversity and discrimination between cultivars. J Cereal Sci 49: 337-345, https://doi.org/10.1016/j.jcs.2008.04.003
Lew EJL, Kuzmicky DD, Kasarda DD (1992) Characterization of low molecular weight glutenin subunits by reversed-phase high-performance liquid chromatography, sodium dodecyl sulphate-polyacrylamide gel electrophoresis, and N-terminal amino acid sequencing. Cereal Chem 69: 508-515, https://www.cerealgrains.org/p ublications/cc/backissues/1992/Documents/69_508.pdf
Li YC, Korol AB, Fahima T, Beiles A, Nevo E (2002) Microsatellites: genomic distribution, putative functions and mutational mechanisms: a review. Molec Ecol 11: 2453-2465, https://doi.org/10.1046/j.1365-294X.2002.01643.x
Li YC, Korol AB, Fahima T, Nevo E (2004) Microsatellites within genes: structure, function, and evolution. Molec Biol Evol 21: 991-1007, https://doi.org/10.1093/molbev/msh073
Li W, Zhang P, Fellers JP, Friebe B, Gill BS (2004) Sequence composition, organization, and evolution of the core *Triticeae* genome. Plant J 40: 500-511, https://doi.org/10.1111/j.1365-313X.2004.02228.x
Li Y, Song Y, Zhou R, Branlard G, Jia J (2009) Detection of QTLs for bread-making quality in wheat using a recombinant inbred line population. Plant Breed 128: 235-243, https://doi.org/10.1111/j.1439-0523.2008.01578.x

Li Y, Zhou R, Branlard G, Jia J (2010) Development of introgression lines with 18 alleles of glutenin subunit and evaluation of the effects of various alleles on quality related traits in wheat (*Triticum aestivum* L.). J Cereal Sci 51: 127-133, https://doi.org/10.1016/j.jcs.2009.10.008

Li D, Jin H, Zhang K, Wang Z, Wang F, Zhao Y, Huo N, Liu X, Gu YQ, Wang D, Dong L (2018) Analysis of the *Gli-D2* locus identifies a genetic target for simultaneously improving the breadmaking and health-related traits of common wheat. The Plant J 95: 414-426, https://doi.org/10.1111/tpj.13956

Li Y, Fu J, Shen Q, Yang D (2021) High-molecular-weight glutenin subunits: genetics, structures, and relation to end use qualities. Intern J Molec Sci 22: 184. https://doi.org/10.3390/ijms2201084

Liu CY (1995) Identification of a new low Mr glutenin subunit locus on chromosome 1B of durum wheat. J Cereal Sci 21: 209-213, https://doi.org/10.1016/0733-5210(95)90037-3

Liu CY & Shepherd KW (1995) Inheritance of B subunits of glutenin and ω- and γ-gliadins in tetraploid wheats. Theor Appl Genet 90: 1149-1157, https://doi.org/10.1007/BF00222936

Liu JG, Ikeda TM, Tsunewaki K (1992) Moderately repeated, dispersed, and highly variable (MRDHV) genomic sequences of common wheat usable for cultivar identification. Theor Appl Genet 84: 535-543, https://doi.org/10.1007/BF00224149

Liu L, He Z, Yan J, Zhang Y, Xia X, Peña RJ (2005) Allelic variation at the *Glu-1* and *Glu-3* loci, presence of the 1B.1R translocation, and their effects on mixographic properties in Chinese bread wheats. Euphytica 142: 197-204, https://doi.org/10.1007/s10681-005-1682-4

Liu L, Ikeda TM, Branlard G, Peña RJ, Rogers WJ, Lerner SE, Kolma MA, Xia X, Wang L, Ma W, Appels R, Yoshida H, Wang A, Yan Y, He Z (2010) Comparison of low molecular weight glutenin subunits identified by SDS-PAGE, 2-DE, MALDI-TOF-MS and PCR in common wheat. BMC Plant Biology 10: 124, https://doi.org/10.1186/1471-2229-10-124

Lookhart GL & Bean S (1995) Separation and characterization of wheat protein fractions by high-performance capillary electrophoresis. Cereal Chem 72: 527-532, https://www.cerealgrains.org/publications/cc/backissues/1995/Documents/72_527.pdf

Lookhart GL & Finney KF (1984) Polyacrylamide gel electrophoresis of wheat gliadins: the effect of environment and germination. Cereal Chem. 61: 496-499, http://pascal-francis.inist.fr/vibad/index.php?action=getRecordDetail&idt=8938236

Lookhart GL, Jones BL, Hall SB, Finney KF (1982) An improved method for standardizing polyacrylamide gel electrophoresis of wheat gliadin proteins. Cereal Chem 59: 178-181, https://www.cerealgrains.org/publications/cc/backissues/1982/Documents/chem59_178.pdf

Lookhart GL, Cooper DB, Jones BL (1985) Effect of temperature and alternate lactate buffer systems on resolution of wheat gliadin proteins by polyacrylamide gel electrophoresis. Cereal Chem 62: 19-22, http://pascal-francis.inist.fr/vibad/index.php?action=getRecordDetail&idt=9039201

Lookhart GL, Albers LD, Bietz JA (1986) Comparison of polyacrylamide gel electrophoresis and high-performance liquid chromatography analyses of gliadin polymorphism in the wheat cultivar "Newton". Cereal Chem 63: 497-500, https://www.cerealgrains.org/publications/cc/backissues/1995/Documents/chem63_497.pdf

Lukaszewski AJ (1990) Frequency of 1RS.1AL and 1RS.1BL translocations in United States wheats. Crop Sci 30: 1151-1153, https://doi.org/10.2135/cropsci1990.0011183X003000050041x

Lukow OM, Payne PI, Tkachuk R (1989) The HMW glutenin subunit composition of Canadian wheat cultivars and their association with bread-making quality. J Sci Food Agric 46: 451-460, https://doi.org/10.1002/jsfa.2740460407

Lukow OM, Forsyth SA, Payne PI (1992) Over-production of HMW glutenin subunits coded on chromosome 1B in common wheat, *Triticum aestivum* [in Israel and Canada] J Genet Breed 46: 187-192, http://pascal-francis.inist.fr/vibad/index.php?action=getRecordDetail&idt=6127523

Luo C, Griffin WB, Branlard G, McNeil DL (2001) Comparison of low and high molecular-weight wheat glutenin allele effects on flour quality. Theor Appl Genet 102: 1088-1098, https://doi.org/10.1007/s001220000433

Luo MC, Yang ZL, You FM, Kawahara T, Waines JG, Dvorak J (2007) The structure of wild and domesticated emmer wheat populations, gene flow between them and the site of emmer domestication. Theor Appl Genet 114: 947-959, https://doi.org/10.1007/s00122-006-0474-0

Luo MC, Gu YQ, Puiu D, Wang H, Twardziok SO, ... Devos KM, Dvorak J (45 authors) (2017) Genome sequence of the progenitor of the wheat D genome *Aegilops tauschii*. Nature 551: 498-502, https://doi.org/10.1038/nature24486

Maccaferri M, Sanguineti MC, Donini P, Tuberosa R (2003) Microsatellite analysis reveals a progressive widening of the genetic basis in the elite durum wheat germplasm. Theor Appl Genet 107: 783-797, https://doi.org/10.1007/s00122-003-1319-8

Mac Key J (1960) Radiogenetics in *Triticum*. Genetica Agraria 12: 201-230, https://www.cabdirect.org/cabdirect/abstract/19611603091

Mac Key J (1968) Mutagenesis in *vulgare* wheat. Hereditas 59: 505-517, https://cyberleninka.org/article/n/484344.pdf

MacPhee DG (1991) The significance of deletions in spontaneous and induced mutations associated with movement of transposable DNA elements: possible implications for evolution and cancer. Mutant Res.250: 35-47, https://doi.org/10.1016/0027-5107(91)90160-P

MacRitchie F, duCros DL, Wrigley CW (1990) Flour polypeptides related to wheat quality. Adv Cereal Sci Technol 10: 79-145, http://pascal-francis.inist.fr/vibad/index.php?action=getRecordDetail& idt=19430180

Maddock SE, Risiott R, Parmar S, Jones MGK, Shewry PR (1985) Somaclonal variation in the gliadin patterns of grains of regenerated wheat plants. J Exp Bot 36: 1976-1984, https://doi.org/10.1093/jxb/36.12.1976

Manifesto MM, Feingold S, Hopp HE, Schlatter AR, Dubcovsky J (1998) Molecular markers associated with differences in bread-making quality in a cross between bread wheat cultivars with the same high Mr glutenins. J Cereal Sci 27: 217-227, https://doi.org/10.1006/S0733-5210(97)91000-6

Manifesto MM, Schlatter AR, Hopp HE, Suárez EY, Dubcovsky J (2001) Quantitative evaluation of genetic diversity in wheat germplasm using molecular markers. Crop Sci 41: 682-690, https://doi.org/10.2135/cropsci2001.413682x

Manley M, Randall PG, McGill AEJ (1992) The prediction of dough properties of South African wheat cultivars by SDS-PAGE analysis of HMW glutenin subunits. J. Cereal Sci 15: 39-47, https://doi.org/10.1016/S0733-5210(09)80055-6

Marchylo BA, Lukow OM, Kruger JE (1992) Quantitative variation in high molecular weight glutenin subunit 7 in some Canadian wheats. J Cereal Sci15: 29-37, https://doi.org/10.1016/S0733-5210(09)80054-4

Margiotta B, Colaprico G, D'Ovidio R, Lafiandra D (1993) Characterization of high *M*r subunits of glutenin by combined chromatographic (RP-HPLC) and electrophoretic separations and restriction fragment length polymorphism (RFLP) analyses of their encoding genes. J Cereal Sci 17: 221-236, https://doi.org/10.1006/jcrs.1993.1021

Martin P & Carrillo JM (1999) Cumulative and interaction effects of prolamin allelic variation and of 1BL/1RS translocation on flour quality in bread wheat. Euphytica 108: 29-39, https://doi.org/10.1023/A:1003648110480

Martynov SP, Dobrotvorskaya TV, Stehno Z, Dotlačil L, Faberova I, Holubec V (1992) Catalogue genealogies and gene alleles identified in 31 000 cultivars and lines of wheat. Research Inst Crop Production; Prague, Czech Rep.

Maruyama-Funatsuki W, Takata K, Nishio Z, Tabiki T, Yahata E, Kato A, Saito K, Funatsuki H, Saruyama H, YamnchiH (2004) Identification of low-molecular-weight glutenin subunits of wheat associated with bread-making quality. Plant Breed 123: 355-360, https://doi.org/10.1111/j.1439-0523.2004.00978.x

Maruyama-Funatsuki W, Takata K, Funatsuki H, Tabiki T, Ito M, Nishio Z, Kato A, Saito K, Yahata E, Saruyama H (2005) Identification and characterization of a novel LMW-s glutenin gene of a Canadian Western Extra-Strong wheat. J Cereal Sci 41: 47-57, https://doi.org/10.1016/j.jcs.2004.07.003

Maryami Z, Huertas-Garcia AB, Azimi MR, Hernández- Espinosa N, Payne T, Cervantes F, Govindan V, Ibba MI, Guzman C (2020) Variability for glutenins, gluten quality, Iron, Zinc and Phytic acid in a set of one hundred and fifty-eight common wheat landrace from Iran. Agronomy 10: 1797, https://doi.org/10.3390/agronomy10111797

Masci S, Porceddu E, Colaprico G, Lafiandra D (1991) Comparison of the B and D subunits of glutenin encoded at the *Glu-D3* locus in two biotypes of the common wheat cultivar Newton with different technological characteristics. J Cereal Sci 14: 35-46, https://doi.org/10.1016/S0733-5210(09)80016-7

Masci S, Lafiandra D, Porceddu E, Lew EJL, Tao HP, Kasarda DD (1993) D-glutenin subunits: N-terminal sequences and evidence for the presence of cysteine. Cereal Chem 70: 581-585, http://pascal-francis.inist.fr/vibad/index.php?action=getRecordDetail&idt=3834491

Masci S, Lew EJL, Lafiandra D, Kasarda DD (1995) Characterization of low-molecular-weight glutenin subunits in durum wheat by RP-HPLC and N-Terminal sequencing. Cereal Chem 72: 100-104, http://pascal-francis.inist.fr/vibad/index.php?action=getRecordDetail&idt=3432333

Masci S, D'Ovidio R, Lafiandra D, Kasarda DD (1998) Characterization of a low-molecular-weight glutenin subunit gene from bread wheat and the corresponding protein that represents a major subunit of the glutenin polymer. Plant Physiol 118: 1147- 1158, https://doi.org/10.1104/pp.118.4.1147

Masci S, Egorov TA, Ronchi C, Kuzmicky DD Kasarda DD, Lafiandra D (1999) Evidence for the presence of only one cysteine residue in the D-type low-molecular- weight subunits of wheat glutenin. J Cereal Sci

29: 17-25, https://doi.org/10.1006/jcrs.1998.0224

Masci S, D'Ovidio R, Lafiandra D, Kasarda DD (2000) A 1B-coded low-molecular weight glutenin subunit associated with quality in durum wheats shows strong similarity to a subunit present in some bread wheat cultivars. Theor Appl Genet 100: 396-400, https://doi.org/10.1007/s001220100761

Masci S, Rovelli L, Kasarda DD, Vensel WH, Lafiandra D (2002) Characterisation and chromosomal localisation of C-type low-molecular-weight glutenin subunits in the bread wheat cultivar Chinese Spring. Theor Appl Genet 104: 422-428, https://doi.org/10.1007/s001220050052

Masci S, D'Ovidio R, Scossa F, Patacchini C, Lafiandra D, Anderson OD, Blechl AE (2003) Production and characterization of a transgenic bread wheat line over-expressing a low-molecular-weight glutenin subunit gene. Molec Breed 12: 209-222, https://doi.org/10.1023/A:1026346823301

Matsuoka Y (2011) Evolution of polyploid *Triticum* wheats under cultivation: The role of domestication, natural hybridization and allopolyploid speciation in their diversification. Plant Cell Physiol 52: 750-764, https://doi.org/10.1093/pcp/pcr018

Maucher T, Figueroa JDC, Reule W, Peña RJ (2009) Influence of low molecular weight glutenins on viscoelastic properties of intact wheat kernels and their relation to functional properties of wheat dough. Cereal Chem 86: 372-375, https://doi.org/10.1094/CCHEM-86-4-0372

McFadden ES & Sears ER (1946) The origin of *Triticum spelta* and its free-threshing hexaploid relatives. J Hered 37: 81-89, https://doi.org/10.1093/oxfordjournals.jhered.a105594

McIntosh RA & Bennett FGA (1978) Telocentric mapping of genes *Pm3a* and *Hg* on chromosome 1A of hexaploid wheat. Cereal Res Comm 6: 9-14, https://www.jstor.org/stable/23779381

McIntosh RA & Cusick JE (1987) Linkage map of hexaploid wheat. In: Wheat and Wheat Improvement; ed. E.G.Heyne; Amer Soc. Agronomy, Inc; Madison, WI; pp 289-297, https://doi.org/10.2134/agronmonogr13.2ed.cl3

McIntosh RA, Yamazaki Y, Devos KM, Dubcovsky J, Rogers WJ, Appels R (2003) Catalogue of gene symbols for wheat. Proc 10th Intern Wheat Genet Symp; eds N.E.Pogna, M.Romanò, E.A.Pogna, G Galterio; Istituto Sperim Cerealicoltura, Rome; v.4.

McKendry AL, Tague DN, Finney PL, Miskin KE (1996) Effect of 1BL.1RS on milling and baking quality of soft red winter wheat. Crop Sci 36: 848-851,

https://doi.org/10.2135/cropsci1996.0011183X003600040003x
Mecham DK, Kasarda DD, Qualset CO (1978) Genetic aspects of wheat gliadin proteins. Biochem Genet 16: 831-853, https://doi.org/10.1007/BF00484739
Melnikova NV & Kudryavtsev AM (2009) Allelic variation of the gliadin-encoding genes in breeding cultivars of spring durum wheat (*Triticum durum* Desf.), bred in Russia and in the former USSR in the 20th century. Russian J Genet 45: 1208-1214, https://doi.org/10.1134/S1022795409100081
Melnikova NV, Ganeva GD, Popova ZG, Landjeva SP, Kudryavtsev AM (2010) Gliadins of Bulgarian durum wheat (*Triticum durum* Desf.) landraces: genetic diversity and geographical distribution. Genet Resour Crop Evol 57: 587-595, https://doi.org/10.1007/s10722-009-9497-0
Melnikova NV, Kudryavtseva AV, Kudryavtsev AM (2012) Catalogue of alleles of gliadin-coding loci in durum wheat (*Triticum durum* Desf.). Biochimie 94: 551-557, https://doi.org/10.1016/j.biochi.2011.09.004
Metakovsky E (1990) Organization of the family of gliadin-coding genes, genetic markers in wheat. Molecular mechanisms of genetic processes. Molecular genetics, evolution and genetic basis of selection, eds A.A.Sozinov & N.G.Shuppe; Moscow, Nauka; pp 157-168 ref 33 (in Russian), https://www.cabdirect.org/cabdirect/abstract/19911698742
Metakovsky E (1991) Gliadin allele identification in common wheat. II. Catalogue of gliadin alleles in common wheat. J Genet Breed 45: 325-344, ISSN: 0394-9257; eLIBRARY, ID: 24799235
Metakovsky E & Baboev SM (1992a) Polymorphism and inheritance of gliadin polypeptides in wheat *T.monococcum*. Theor Appl Genet 84: 971-978, https://doi.org/10.1007/BF00227412
Metakovsky E & Baboev SM (1992b) Polymorphism of gliadin and unusual gliadin alleles in *Triticum boeoticum*. Genome 35: 1007-1012, https://doi.org/10.1139/g92-154
Metakovsky E & Branlard G (1998) Genetic diversity of French common wheat germplasm studied using gliadin alleles. Theor Appl Genet 96: 209-218, https://doi.org/10.1007/s001220050729
Metakovsky E & Graybosch RA (2006) Gliadin alleles in wheat: Identification and applications. In: Gliadin and glutenin. The unique balance of wheat quality; eds C.Wrigley, F.Békés, W.Bushuk; AACC Internat.; pp 85-114; https://www.cabdirect.org/cabdirect/abstract/20073205856
Metakovsky E & Iakobashvili ZA (1990) Homology of chromosomes of *Triticum macha* Dek. et Men., and *Triticum aestivum* L. as shown with

the help of genetic markers. Genome 33: 755-757, https://doi.org/10.1139/g90-114

Metakovsky E & Kopus MM (1991) Alleles at the gliadin-coding loci in common wheat cultivars of Zernograd: analysis in starch and polyacrylamide gels. Doklady VASHNIL N3, 5-9 (in Russian).

Metakovsky E & Novoselskaya AYu (1991) Gliadin allele identification in common wheat. 1.Methodological aspects of the analysis of gliadin pattern by one-dimensional polyacrylamide gel electrophoresis. J Genet Breed 45: 317-324, http://pascal-francis.inist.fr/vibad/index.php ?action=getRecordDetail&idt=6127540

Metakovsky E & Sozinov AA (1990) Genetics of gliadin proteins and the problems of interpreting results obtained with somaclonal variation in wheat. Biotechnology in Ariculture and Forestry. v 13 (Wheat; ed. Y.P.S.Bajaj); pp 526-537, https://doi.org/10.1007/978-3-662-10933-5_29

Metakovsky E, Novoselskaya AYu, Kopus MM, Sobko TA, Sozinov AA (1984a) Blocks of gliadin components in winter wheat detected by one-dimensional polyacrylamide gel electrophoresis. Theor Appl Genet 67: 559-568, https://doi.org/10.1007/BF00264904

Metakovsky E, Novoselskaya AYu, Sozinov AA (1984b) Genetic analysis of gliadin components in winter wheat using two-dimensional polyacrylamide gel electrophoresis. Theor Appl Genet 69: 31-37, https://doi.org/10.1007/BF00262533

Metakovsky E, Akhmedov MG, Sozinov AA (1986a) The genetic analysis of gliadin-encoding genes reveals gene clusters as well as single remoted genes. Theor Appl Genet 73: 275-285, https://doi.org/10.1007/BF00289286

Metakovsky E, Seitova AM, Koval SF, Sozinov AA (1986b) Complex organization of the family of genes that control gliadin synthesis on the chromosomes of the first homoeological group in hexaploid wheat. Reports Acad Sci USSR 291: 465-468 (in Russian), http://pascal-francis.inist.fr/vibad/index.php?action=getRecordDetail&idt=8352314

Metakovsky E, Il'ina LG, Galkin AN, Koval SF, Sozinov AA (1987a) Allelic variants of blocks of gliadins in common wheat cultivars bred in Saratov. Selektsiia i Semenovod (Moscow) 1: 11-15 (in Russian).

Metakovsky E, Novoselskaya AYu, Sozinov AA (1987b) Problems of interpreting results obtained in studies of somaclonal variation in gliadin proteins in wheat. Theor Appl Genet 73: 764-766, https://doi.org/10.1007/BF00260788

Metakovsky E, Koval SF, Movchan VK, Sozinov AA (1988) Genetic formulas of gliadin in spring wheat cultivars of nothern Kazakhstan.

Selektsiia i Semenovod (Moscow) 1: 11-13 (in Russian), https://agris.fao.org/agris-search/search.do?recordID=US201302656470

Metakovsky E, Kudryavtsev AM, Iakobashvili ZA, Novoselskaya AYu (1989) Anlysis of philogenetic relations of durum, carthlicum and common wheats by means of comparison of alleles of gliadin loci. Theor Appl Genet 77: 881-887, https://doi.org/10.1007/BF00268342

Metakovsky E, Akhmedov MG, Payne PI (1990a) A case of spontaneous intergenomic transfer of genetic material containing gliadin genes in bread wheat. J Genet Breed 44: 127-132, http://pascal-francis.inist.fr/vibad/index.php?action=getRecordDetail&idt=6952806

Metakovsky E, Wrigley CW, Békés F, Gupta R (1990b) Gluten polypeptides as useful genetic markers of dough quality in Australian wheats. Austral J Agric Res 41: 289-306, https://doi.org/10.1071/AR9900289

Metakovsky E, Knezevic D, Javornik B (1991) Gliadin allele composition of Yugoslav winter wheat cultivars. Euphytica 54: 285-295, https://doi.org/10.1007/BF00023005

Metakovsky E, Davydov SD, Chernakov VM, Upelniek VP (1993a) Gliadin allele identification in common wheat. III. Frequency of occurrence and appearance of spontaneous mutations at the gliadin-coding loci. J Genet Breed 47: 221-236, http://pascal-francis.inist.fr/vibad/index.php?action=getRecordDetail&idt=6050743

Metakovsky E, Ng PKW, Chernakov VM, Pogna NE, Bushuk W (1993b) Gliadin alleles in Canada Western Red Spring wheat cultivars: Use of two different procedures of APAGE for gliadin separation. Genome 36: 743-749, https://doi.org/10.1139/g93-099

Metakovsky E, Vaccino P, Accerbi M, Redaelli R, Pogna NE (1993c) Polymorphism and spontaneous mutations at the gliadin-coding loci. Proc V Intern Workshop on gluten proteins (Detmold, Germany, 1993); ACR; pp 218-226.

Metakovsky E, Pogna NE, Biancardi AM, Redaelli R (1994) Gliadin allele composition of common wheat cultivars grown in Italy. J Genet Breed 48: 55-66, https://agris.fao.org/agris-search/search.do?recordID=IT9462350

Metakovsky E, Chernakov VM, Upelniek VP, Redaelli R, Dardavet M, Branlard G, Pogna NE (1996) Minor ω-gliadin-coding loci on chromosome 1A of common wheat: A revision. J Genet Breed 50: 277-286, https://agris.fao.org/agris-search/search.do?recordID=IT1998061304

Metakovsky E, Annicchiarico P, Boggini G, Pogna NE (1997a) Relationships between gliadin alleles and dough strength in Italian common wheats. J Cereal Sci 25: 229-236, https://doi.org/10.1006/jcrs.1996.0088

Metakovsky E, Branlard G, Chernakov VM, Upelniek VP, Redaelli R, Pogna NE (1997b) Recombiation mapping of some chromosome 1A-, 1B-, 1D- and 6B-controlled gliadins and low molecular weight glutenin subunits in common wheat. Theor Appl Genet 94: 788-795, https://doi.org/10.1007/s001220050479

Metakovsky E, Felix I, Branlard G (1997c) Association between dough quality (W value) and certain gliadin alleles in French common wheat cultivars. J Cereal Sci 26: 371-373, https://doi.org/10.1006/jcrs.1997.0130

Metakovsky E, Gòmez M, Vàzquez JF, Carrillo JM (2000) High genetic diversity of Spanish common wheats as judged from gliadin alleles. Plant Breed 119: 37-42, https://doi.org/10.1046/j.1439-0523.2000.00450.x

Metakovsky E, Branlard GP, Graybosch RA (2006) Gliadins of common wheat: Polymorphism and genetics. In: "Gliadin and glutenin. The unique balance of wheat quality". eds C.Wrigley, F.Békés, W.Bushuk; AACC Internat; pp 35-84, https://www.cabdirect.org/cabdirect/abstract/20073205855.

Metakovsky E, Melnik VA, Vaccino P, Rodriguez-Quijano M (2018a) Comparison of alleles at the *Gli-1* loci of common wheat by means of two-imensional electrophoresis of gliadin and RFLP analysis. Cytol Genet (Kiev) 52: 11-20, https://doi.org/10.3103/S0095452718010085

Metakovsky E, Melnik VA, Vaccino P, Rodriguez-Quijano M (2018b). Comparison of alleles at the *Gli-2* loci of common wheat by means of two-imensional electrophoresis of gliadin. Cytol Genet (Kiev) 52: 87-94, https://doi.org/10.3103/S0095452718020068

Metakovsky E, Melnik VA, Rodriguez-Quijano M, Upelniek VP, Carrillo JM (2018c) A catalogue of gliadin alleles: Polymorphism of 20th-century common wheat germplasm. The Crop J 6: 629-641, https://doi.org/10.1016/j.cj.2018.02.003

Metakovsky E, Melnik VA, Pascual L, Wrigley CW (2019a) Gliadin genotypes worldwide for spring wheats (*Triticum aestivum* L). 1.Genetic diversity and grain-quality gliadin alleles during the 20th century. J Cereal Sci 87: 172-177, https://doi.org/10.1016/j.jcs.2019.03.008

Metakovsky E, Melnik VA, Pascual L, Wrigley CW (2019b) Gliadin genotypes worldwide for spring wheats (*Triticum aestivum* L). 2. Strong differentiation of polymorphism between countries and regions of origin. J. Cereal Sci 87: 311-317, https://doi.org/10.1016/j.jcs.2019.04.015

Metakovsky E, Melnik VA, Pascual L, Romanov GA, Wrigley CW (2019c) Types, frequencies and value of intra-varietal genotypic non-uniformity in common wheat cultivars: Authentic biotypes and foreign seeds. J Cereal Sci 89: 102813, http://doi.org/10.1016/j.jcs.2019.102813

Metakovsky E, Melnik VA, Pascual L, Wrigley CW (2020) Over 40% of 450 registered wheat cultivars (*Triticum aestivum*) worldwide are composed of multiple biotypes. J Cereal Sci 96: 103088, http://doi.org/10.1016/j.jcs.2020.103088

Metakovsky E, Melnik VA, Pascual L, Wrigley CW (2021a) Some intra-varietal mon-uniformities in *Triticum aestivum* can be explained as repeated spontaneous mutations at the Gli loci. J Cereal Sci 100: 103243, http://doi.org/10.1016/j.jcs.2021.103243.

Metakovsky E, Pascual L, Vaccino P, Melnik VA, Rodriguez-Quijano M, Popovych Yu, Chebotar S, Rogers WJ (2021b) Heteroalleles in common wheat: Multiple differences between allelic variants of the *Gli-B1* locus. Intern J Molec Sci 22: 1832, https://doi.org/10.3390/ijms22041832

Miller TE (1984) The homoeologous relationship between the chromosome of rye and wheat: Carrent status. Canad J Genet Cytol 26: 578-579, https://doi.org/10.1139/g84-091

Mirali N (2000) Heterogeneity within old and modern durum and bread wheat grown in Syria using the A-PAGE and SDS-PAGE electrophoretic techniques. Plant Var & Seeds 13: 149-157.

MirAli N (2002) Cluster analysis of Syrian grown bread wheat genotypes based on gliadin composition (*Triticum aestivum* L.). J Genet Breed 56: 177-183, https://agris.fao.org/agris-search/search.do?recordID=IT2003063358

Mitrofanova OP (1976) Genetic control of the wheat gliadin of *T.aestivum* (L.) variety Chinese Spring. Cytol Genet (Kiev) 10: 244-247 (in Russian), PMID: 1014091

Mitrofanova OP (1979) The analysis of inheritance of gliadin electrophoretic components in common wheat. Trudy Prikl Bot Genet Sel 63: 9-23 (in Russian).

Morgunov AI, Rogers WJ, Sayers EJ, Metakovsky E (1990) The-high-molecular weight subunits composition of Soviet wheat varieties. Euphitica 51: 41-52, https://doi.org/10.1007/BF00022891

Morgunov AI, Peña RJ, Crossa J, Rajaram S (1993) Worldwide distribution of *Glu-1* alleles in bread wheat. J Genet Breed 47: 53-60, http://pascal-francis.inist.fr/vibad/index.php?action=getRecordDetail&

idt=6093161
Mori N, Ishii T, Ishido T, Hirosawa S, Watatani H, Kawahara T, Nesbitt M, Belay G, Takumi S, Ogihara Y, Nakamura C (2003) Origin of domesticated emmer and common wheat inferred from chloroplast DNA fingerprinting. In: 10th Intern Wheat Genet Symp (1-6 Sept 2003) Paestum, Italy; pp 25-28.
Morrell PL, Toleno DM, Lundy KE, Clegg MT (2006) Estimating the contribution of mutation, recombination and gene conversion in the generation of haplotypic diversity. Genetics 173: 1705-1723, https://doi.org/10.1534/genetics.105.054502
Morris R & Sears ER (1967) The cytogenetics of wheat and its relatives. In: "Wheat and Wheat Improvement", eds O.Quisenberry & L.P.Reitz; Amer. Soc. Agronomy; Madison, WI; pp 19-98.
Morris CF, Pasha I, Beecher BS, Bettge AD, See D (2009) Glutenin alleles in U.S. Pacific Northwest wheat. In: “Gluten Proteins 2009“; ed. G.Branlard; Proc 10th Intern Gluten Workshop (7-9 Sept 2009), Clermont-Ferrand, France; INRA; pp 190-194.
Motsnyi II, Fayt VI, Blagodarova EM (2009) Identification and characteristics of the 1R (1B) substitution lines of bread wheat. Cytol Genet (Kiev) 43: 169-176, https://doi.org/10.3103/S0095452709030050.
Motsny II, Blagodarova EM, Rybalka AI (2017) Cytogenetic characteristics of wheat lines with modified 1RS.1BL rye-wheat translocation. Cytol Genet (Kiev) 51: 331-338, https://doi.org/10.3103/S0095452717050073
Movchan VK (1988) Development of spring common wheat breeding in northern Kazakhstan. Selektia i Semenovod 1: 6-10 (in Russian).
Moxon ER, Rainey PB, Nowak MA, Lenski RE (1994) Adaptive evolution of highly mutable loci in pathogenic bacteria. Current Biol 4: 24-33, https://doi.org/10.1016/S0960-9822(00)00005-1
Müller H & Wieser H (1995) The location of disulphide bonds in α-type gliadins. J Cereal Sci 22: 21-27, https://doi.org/10.1016/S0733-5210(95)80004-9
Müller H & Wieser H (1997) The location of disulphide bonds in monomeric γ-type gliadins. J Cereal Sci 26: 169-176, https://doi.org/10.1006/S0733-5210(97)90100-4
Naeem HA & MacRitchie F (2005) Polymerization of glutenin during grain development in near-isogenic wheat lines differing at *Glu-D1* and *Glu-B1* in greenhouse and field. J Cereal Sci 41: 7-12, https://doi.org/10.1016/j.jcs.2004.04.009

Nagy I.J., Takács I., Juhász A., Tamás L., Bedö Z. 2005 Identification of a new class of recombinant prolamin genes in wheat. Genome 48: 840-847, https://doi.org/10.1139/g05-042

Neel JV (1983) Frequency of spontaneous and induced "point" mutations in higher eukaryotes. J Hered 74, 2-15, https://doi.org/10.1093/oxfordjournals.jhered.a109711

Nei M (1972) Genetic distance between populations. Amer Nat 106: 283-292, https://www.jstor.org/stable/2459777

Nei M (1973) Analysis of gene diversity in subdivided populations. Proc Natl Acad Sci USA 70: 3321-3323, https://doi.org/10.1073/pnas.70.12.3321

Nesbitt M & Samuel D (1996) From staple crop to extinction? The archaeology and history of hulled wheats. In: "Hulled wheats. Promoting the conservation and use of underutilizied and neglected crops". Proc First Intern Worcshop Hulled Wheats, eds S.Padulosi, K.Hummer, J.Heller; Intern Plant Genet Resourse Inst; Rome, Castelvecchio Pacoli, Tuscany, Italy; pp 41-100.

Netsvetaev VP, Kopus MM, Ryzhkova TA (2013) Variants of gliadin and the number of disulfide bonds in protein complex of common wheat. Sovrem Problemy Nauki i Obrazov N5 (in Russian).

Nettevich ED, Berkutova NS, Pogorelova LG (1981) Intravarietal polymorphism of spring common wheat cultivar Moskovskaya-35 and grain quality. Doklady VASKhNIL 2: 4-6 (in Russian), http://pascal-francis.inist.fr/vibad/index.php?ction=getRecordDetail &idt=PASCALAGROLINEINRA82X01734776093161

Nettevich ED, Morgunov AI, Rogers WJ, Bespalov AM, Metakovsky E (1991) Characteristic of domestic wheat cultivars for high-molecular-weight glutenins of grain. Doklady VASKhNIL 7: 2-5 (in Russian), https://agris.fao.org/agris-search/search.do?recordID=US9190892

Nevo E (1988) Genetic diversity in nature: Patterns and theory. Evol Biol 23: 217-247, https://doi.org/10.1007/978-1-4613-1043-3_6

Nevo E (1994) Evolutionary significance of genetic diversity in nature: environmental stress, pattern and theory. In: "Isozymes. Organization and Roles in Evolution, Genetics and Physiology", Proc 7th Intern Congr Isozymes (Novosibirsk, Rusia 6-13.09.1992); eds C.L.Markert, J.G. Scandalios, H.A.Lim, O.L.Serov; World Scientific, New Jersey; pp 267-296.

Nevo E (2011) Evolution under environmental stress at macro- and microscales. Genome Biol Evol 3: 1039-1052, https://doi.org/10.1093/gbe/evr052

Nevo E & Beiles A (1989) Genetic diversity of wild emmer wheat in Israel and Turkey. Theor Appl Genet 77: 421-455, https://doi.org/10.1007/BF00305839

Nevo E, Golenberg E, Beiles A, Brown AHD, Zohary D (1982) Genetic diversity and environmental associations of wild wheat, *Triticum dicoccoides*, in Israel. Theor Appl Genet 62: 241-254, https://doi.org/10.1007/BF00276247

Nevo E, Beiles A, Krugman T (1988) Natural selection of allozyme polymorphisms: A microgeographical differentiation by edaphic, topographical, and temporal factors in wild emmer wheat (*Triticum dicoccoides*). Theor Appl Genet 76: 737-752, https://doi.org/10.1007/BF00276761

Nevo E. Gerechter-Amitai Z, Beiles A (1991) Resistance of wild emmer wheat to stem rust: Ecological, pathological and allozyme associations. Euphytica 53: 121-130, https://doi.org/10.1007/BF00023792

Nevo E, Pagnotta MA, Beiles A, Porceddu E (1995) Wheat storage proteins: glutenin DNA diversity in wild emmer wheat, *Triticum dicoccoides*, in Israel and Turkey. 3. Environmental correlates and allozyme associations. Theor Appl Genet 91: 415-420, https://doi.org/10.1007/BF00222968

Nevo E, Fu YB, Pavlicek T, Khalifa S, Tavasi M, Beiles A (2012) Evolution of wild cereals during 28 years of global warming in Israel. Proc Nat Acad Sci USA 109: 3412-3415, https://doi.org/10.1073/pnas.1121411109

Ng PKW & Bushuk W (1989) Concerning the nomenclature of high molecular weight glutenin subunits. J Cereal Sci 9: 53-60, https://doi.org/10.1016/S0733-5210(89)80021-9

Ng PKW & Bushuk W (1990) Genetic origin of a unique gliadin doublet of Thatcher-type bread wheat. Canad J Plant Sci 70: 303-306, https://doi.org/10.4141/cjps90-035

Ng PKW, Scanlon MG, Bushuk W (1988) A catalogue of biochemical fingerprints of registered Canadian wheat cultivars by electrophoresis and high-performance liquid chromatography. Publication №139 of the Food Sci. Dep. (Oct.1988); Univ. of Manitoba, Winnipeg, Canada; pp 1-83.

Ng PKW, Pogna NE, Mellini F, Bushuk W (1989) *Glu-1* allele composition of the wheat cultivars registered in Canada. J Genet Breed 43: 53-59, https://pascal-francis.inist.fr/vibad/index.php?ction=getRecordDetail&idt=7225769

Nieto-Taladriz MT & Carrillo JM (1996) Complexity of the *Gli-A3* locus in bread wheat. Plant Breed 115: 192-194, https://doi.org/10.1111/j.1439-0523.1996.tb00900.x

Nieto-Taladriz MT, Perretant MR, Rousset M (1994) Effects of gliadins and HMW and LMW subunits of glutenin on dough properties in the F6 recombinant inbred lines from a bread wheat cross. Theor Appl Genet 88: 81-88, https://doi.org/10.1007/BF00222398

Nieto-Taladriz MT, Ruiz M, Martinez MC, Vázquez JF, Carrillo JM (1997) Variation and classification of B low-molecular-weight glutenin subunit alleles in durum wheat. Theor Appl Genet 95: 1155-1160, https://doi.org/10.1007/s001220050676

Nieto-Taladriz MT, Rodriguez-Quijano M, Carrillo JM. (1998) Biochemical and genetic characterization of a D glutenin subunit encoded at the *Glu-B3* locus. Genome 41: 215-220, https://doi.org/10.1139/g98-010

Nikolaev AA, Pukhalsky VA, Upelniek VP (2009) Genetic diversity of local common wheat (*Triticum aestivum* L.) cultivars grown in Western and Eastern Siberia by gliadin alleles. Russian J Genet 45: 189-197, https://doi.org/10.1134/S1022795409020094

Nishikawa K, Furuta Y, Hina Y, Yaada T (1981) Genetic studies of α-amylase isozymes in wheat. IV. Genetic analysis in hexaploid wheat. Japan J Genet 56: 385-395, https://doi.org/10.1266/jjg.56.385

Noma S, Kawaura K, Hayakawa K, Abe C, Tsuge N, Ogihara J (2016) Comprehensive molecular characterization of the α/β-gliadin multigen family in hexaploid wheat. Molec Genet Genom 291: 65-77, https://doi.org/10.1007/s00438-015-1086-7

Noma S, Hayakawa K, Abe C, Suzuki S, Kawaura K, (2019) Contribution of α-gliadin alleles to the extensibility of flour dough in Japanese wheat cultivars. J Cereal Sci 86: 15-21, https://doi.org/10.1016/j.jcs2018.12.017

Novoselskaya AYu, Upelniek VP, Sutka I, Galiba TG, Metakovsky E, Sozinov AA (1987) Studying of somaclonal variation in wheat with the help of biochemical markers. Proc III Inter Workshop on Gluten Proteins; eds R.Lasztity & F.Bekes; Budapest; World Sci, pp 57-70, https://agris.fao.org/agris-search/search.do?recordID=US20130268714 7

Novoselskaya-Dragovic AYu, Fisenko AV, Imasheva AG, Pukhalsky VA. (2007) Comparative analysis of the dynamics of genetic diversity for gliadin-coding loci among winter common wheat cultivars created over a 40-year period of scientific breeding in Serbia and Italy. Russian J Genet 43: 1236-1242, https://doi.org/10.1134/S1022795407110051

Nunes-Miranda J, Bancel E, Viala D, Chambon C, Capelo JL, Branlard G, Ravel C, Igrejas G (2017) Wheat glutenin: the "tail" of the 1By protein subunits. J Proteomics 169: 136-142, https://doi.org/10.1016/j.jprot.2017.05.019

O'Brien L, Morell M, Wrigley C, Appels R (2001) Genetic pool of Australian wheats. In: The World Wheat Book (Chapter 23), eds A.P.Bonjean & W.J.Angus; Lavoisier Publishing, Paris, pp 603-640, http://hdl.handle.net/102.100.100/202221?index=1

Okita TW, Cheesbrough V, Reeves CD (1985) Evolution and heterogeneity of the α/β-type and γ-type gliadin DNA sequences. J Biol Chem 260: 8203-8213, https://doi.org/10.1016/S0021-9258(17)39582-0

Orford SE, Law JR, Ganal M, Koebner RMD (2006) Has modern breeding led to a genetic narrowing in European winter wheat? Proc 13th Intern EWAC Conf, eds A.Börner, K.Pankova, J.W.Snape; Prague (27.06.-01.07.2005), pp 98-100.

Oury FX, Chiron H, Faye A, Gardet O, Giraud A, Heumez E, Rolland B, Rousset M, Trottet M, Charmet G, Branard G (2010) The prediction of bread wheat quality: joint use of the phenotypic information brought by technological tests and the genetic information brought by HMW and LMW glutenin subunits. Euphytica 171: 87-109, https://doi.org/10.1007/s10681-009-9997-1

Ozkan H, Brandolini A, Pozzi C, Effgen S, Wunder J, Salamini F (2005) A reconsideration of the domestication geography of tetraploid wheat. Theor Appl Genet 110: 1052-1060, https://doi.org/10.1007/s00122-005-1925-8

Pagnotta MA, Nevo E, Beiles A, Porceddu E (1995) Wheat storage proteins: glutenin diversity in wild emmer, *Triticum dicoccoides*, in Israel and Turkey. 2. DNA diversity detected by PCR. Theor Appl Genet 91: 409-414, https://doi.org/10.1007/BF00222967

Panchy N, Lehti-Shiu M, Shiu SH (2016) Evolution of gene duplication in plants. Plant Physiol 171: 2204-2316, https://doi.org/10.1104/pp.16.00523

Panin VM & Netsvetaev VP (1986) Genetic control of gliadin and some morphological traits of a spike of durum winter wheat. Nauchno-Techn Bull VSGI (Odessa) 60: 31-36 (in Russian).

Paris R, Petruzzino G, Savino M, DeSimone V, Ficco DBM, Trono D (2021) Genome-wide identification, characterization and expression pattern analysis of the γ-gliadin gene family in the durum wheat (*Triticum durum* Desf) cultivar Svevo. Genes 12: 1743, https://doi.org/10.3390/genes12111743

Pasha I, Anjum FM, Morris CF (2010) Grain hardness: A major determinant of wheat quality. Food Sci Technol Intern 16: 511-512, https://doi.org/10.1177/1082013210379691

Patey AL & Waldron NM (1976) Gliadin proteins from Maris Widgeon wheat. J Sci Food Agric 27: 838-842, https://doi.org/10.1002/jsfa.2740270908

Payne PI (1987) Genetics of wheat storage proteins and the effect of allelic variation on bread-making quality. Annual Reviews Plant Physiol 38: 141-153, https://doi.org/10.1146/annurev.pp.38.060187.00104

Payne PI & Corfield KG (1979) Subunit composition of wheat glutenin proteins, isolated by gel filtration in a dissociating medium. Planta 145: 83-88, https://doi.org/10.1007/BF00379931

Payne PI & Lawrence GJ (1883) Catalogue of alleles for the complex gene loci, *Glu-A1*, *Glu-B1* and *Glu-D1* which code for high-molecular-weight subunits of glutenin in hexaploid wheat. Cereal Res Comm 11: 29-35, https://www.jstor.org/stable/23781365

Payne PI, Corfield KG, Holt LM, Blackman JA (1981) Correlation between the inheritance of certain high-molecular weight subunits of glutenin and bread-making quality in progenies of six crosses of bread wheat. J Sci Food Agrc 32: 51-60, https://doi.org/10.1002/jsfa.2740320109

Payne PI, Holt LM, Lawrence GJ, Law CN (1982) The genetics of gliadin and glutenin, the major storage proteins in the wheat endosperm. Qual Plant Plant Foods Human Nutr 31: 229-241, https://doi.org/10.1007/BF01108632

Payne PI, Jackson EA, Holt LM (1984a) The association between γ-gliadin 45 and gluten strength in durum wheat varieties: a direct causal effect or the result of genetic linkage? J Cereal Sci 2: 73-81, https://doi.org/10.1016/S0733-5210(84)80020-X

Payne PI, Jackson EA, Holt LM, Law CN (1984b) Genetic linkage between endosperm storage protein genes on each of the short arms of chromosomes 1A and 1B in wheat. Theor Appl Genet 67: 235-243, https://doi.org/10.1007/BF00317044

Payne PI, Holt LM, Hutchinson J, Bennett MD (1984c) Development and characterization of a line of bread wheat, *Triticum aestivum*, which lacks the short-arm satellite of chromosome 1B and the *Gli-B1* locus. Theor Appl Genet 68: 327-334, https://doi.org/10.1007/BF00267886

Payne PI, Holt LM, Johnson R, Snape JW (1986a) Linkage mapping of four gene loci, *Glu-B1*, *Gli-B1*, *Rg-1* and *Yr10* on chromosome 1B of bread wheat. Genet Agraria 40: 231-242, http://pascal-francis.inist.fr/vibad/index.php?ction=getRecordDetail& idt=8808943

Payne PI, Roberts MS, Holt LM (1986b) Location of genes controlling the D group of LMW glutenin subunits on chromosome 1D of bread wheat. Genet Res 47: 175-179,

https://doi.org/10.1017/S0016672300023090
Payne PI, Nightingale MA, Krattiger AF, Holt LM (1987a) The relationship between HMW glutenin subunit composition and bread-making quality of British-grown wheat varieties. J Sci Food Agric 40: 51-65, https://doi.org/10.1002/jsfa.2740400108
Payne PI, Seekings JA, Worand AJ, Jarvis MG, Holt LM (1987b) Allelic variation of glutenin subunits and gliadins and its effect on breadmaking quality in wheat: Analysis of F5 progeny from Chinese Spring x Chinese Spring (Hope 1A). J Cereal Sci 6: 103-118, https://doi.org/10.1016/S0733-5210(87)80047-4
Payne PI, Holt LM, Krattiger AF, Carrillo JM (1988a) Relationship between seed quality characteristics and HMW glutenin subunit composition determined using wheats grown in Spain. J Cereal Sci 7: 229-235, https://doi.org/10.1016/S0733-5210(88)80004-3
Payne PI, Holt LM, Lister PG (1988b) *Gli-A3* and *Gli-B3*, two newly designated loci coding for omega-type gliadins and D subunits of glutenin. Proc 7th Int Wheat Genet Symp; eds T.E.Miller & R.M.D.Koebner; IPSR, Cambridge; v. 2, pp 999-1002, https://agris.fao.org/agris-search/search.do?recordID=GB9010360
Payne PI, Seekings JA, Kaur H, Krattiger AF, Rogers WJ (1990) Contribution of allelic variation of glutenin subunits and gliadins to bread-making quality. Proc 4th Intern Workshop Gluten Prot; eds W.Bushuk & R.Tkachuk; AACC, St Paul, MN; pp 504-511.
Pecetti L, Damania AB, Jana S (1992) Practical problems in large-scale germplasm evaluation: A case study in durum wheat. FAO/IBPGR Plant Genet Res Newsl 88/89: 5-10,
Peng JH, Sun D, Nevo E (2011) Domestication evolution, genetics and genomics in wheat. Molec Breeding 28: 281, https://doi.org/10.1007/s11032-011-9608-4
Perera E, Lacadena JR, Sendino AM (1983) Cytogenetic structure of durum wheat cultivars from, or introduced into Spain. Z Pflanz 91: 36-45, https://agris.fao.org/agris-arch/search.do?recordID=US201302165782
Perronne R, Makowski D, Goffaux R, Montalent P, Goldringer L (2017) Temporal evolution of varietal, spatial and genetic diversity of bread wheat between 1980 and 2006 strongly depends upon agricultural regions in France. Agric Ecosyst Environ 236: 12-20, https://doi.org/10.1016/j.agee.2016.11.003
Peruffo ADB (1982) Variations in the biotype composition of common wheat cultivars examined by gliadin electrophoresis. Cereal Res Comm 10: 111-113, https://www.jstor.org/stable/23781394

Peruffo ADB, Pallavicini C, Varanini Z, Pogna NE (1981) Analysis of wheat varieties by gliadin electrophoregrams. I. Catalog of electrophoregram formulas of 29 common wheat cultivars grown in Italy. Genet Agraria 35: 195-200, https://agris.fao.org/agris-search/search.do?recordID=XE8164707

Peruffo ADB, Pogna NE, Pallavicini C, Pegoraro E, Mellini F, Bianchi A (1984) Diagrammi elettroforetici delle gliadine e chiave di identificazione delle varietá di grano tenero iscritte nel registro delle varietá. Sementi Elette 30: IV1-IV29 (in Italian), http://pascal-francis.inist.fr/vibad/index.php?ction=getRecordDetail&idt=8808943

Peruffo ADB, Pogna NE, Tealdo E, Tutta C, Albuzio A (1985) Isolation and partial characterization of gamma-gliadins 40 and 43.5 associated with quality in common wheat. J Cereal Sci 3: 355-363, https://doi.org/10.1016/S0733-5210(85)80008-4

Petersen G, Seberg O, Yde M, Berthelsen K (2006) Phylogenetic relationships of *Triticum* and *Aegilops* and evidence for the origin of the A, B, and D genomes of common wheat (*Triticum aestivum*). Molec Phylogen Evol 39: 70-82, https://doi.org/10.1016/j.agee.ympev.2006.01.003

Picard B, Branlard G, Oury FX, Rousset M (1992) Etude de la diversité génétique du blé tendre. I. Comparaison de distances biochimiques, agromorphologiques et généalogiques. Agronomie 12: 611-622 (in French), https://hal.archives-ouvertes.fr/hal-00885504

Pistón F, Gil-Humanes J, Rodriguez-Quijano M, Barro F (2011) Down-regulating γ-gliadins in bread wheat leads to non-specific increases in other gluten proteins and has no major effect on dough gluten strength. PLoS ONE 6: e24754, https://doi.org/10.1371/journal.pone.0024754

Plaschke J, Ganal MW, Röder MS (1995) Detection of genetic diversity in closely related bread wheat using microsatellite markers. Theor Appl Genet 91: 1001-1007, https://doi.org/10.1007/BF00223912

Pogna NE & Mellini F (1986) Alla ricerca delle basi biochimiche e genetiche della qualitá del glutine. L'Inform. Agraria 42: 65-66 (in Italian).

Pogna NE & Mellini F (1988) Wheat storage protein genes and their use for improvement of pasta-making quality. Proc Third Intern Symp Durum Wheat "The Future of Cereales for Human Feeding and Development of Biotechnological Research", ed. G.Wittmer; Chamber of Commerce of Foggia, Italy; pp 135-146.

Pogna NE, Boggini G, Corbellini M, Cattaneo M, Peruffo ADB (1982a) Association between gliadin electrophoretic bands and quality in common wheat. Canad J Plant Sci 62: 913-918, https://doi.org/10.4141/cjps82-135

Pogna NE, Peruffo ADB, Boggini G, Corbellini M (1982b) Analysis of wheat varieties by gliadin electrophoregrams. II. Nature, origin and quality of biotypes present in six Italian common wheat varieties. Genet Agraria 6, 143-154.

Pogna NE, Dal Bellin Peruffo A, Mellini F (1985) Genetic aspects of gliadin bands 40 and 43,5 associated with gluten strength. Genet Agraria 39: 101-108, http://pascal-francis.inist.fr/vibad/index.php?ction=getRecordDetail& idt=9090667

Pogna NE, Borghi B, Mellini F, Peruffo ADB, Nash RJ (1986) Electrophoresis of gliadins for estimating the genetic purity in hybrid wheat seed production. Genet Agraria 40: 205-212, http://pascal-francis.inist.fr/vibad/index.php?ction=getRecordDetail&idt=8800830

Pogna NE, Lafiandra D, Peillet P, Autran JC (1988) Evidence for a direct causal effect of low molecular weight subunits of glutenins on gluten viscoelasticity in durum wheats. J. Cereal Sci. 7: 211-214, https://doi.org/10.1016/S0733-5210(88)80001-8

Pogna NE, Mellini F, Baretta A, Peruffo ADB (1989) The high-molecular-weight glutenin subunits of common wheat cultivars grown in Italy. J Genet Breed 43: 17-24, http://pascal-francis.inist.fr/vibad/index.php?action=getRecordDetail&idt=7225765

Pogna NE, Metakovsky E, Redaelli R, Raineri F, Dachkevitch T (1993) Recombination mapping of *Gli-5*, a new gliadin-coding locus on chromosome 1A and 1B in common wheat. Theor Appl Genet 87: 113-121, https://doi.org/10.1007/BF00223754

Pogna NE, Metakovsky E, Redaelli R, Dachkevitch T, Chernakov VM (1994) The group 1 chromosomes of wheat contain several loci coding for gliadins. In: Proc 8th Intern Wheat Genet Symp (Beijing, China, July 1993); eds Z.S.Li & Z.Y.Xin; CASP; pp 137-140.

Pogna NE, Redaelli R, Vaccino P, Biancardi AM, Peruffo ADB, Curioni A, Metakovsky E, Pagliaricci S (1995) Production and genetic characterization of near-isogenic lines in the bread-wheat cultivar Alpe. Theor Appl Genet 90: 650-658, https://doi.org/10.1007/BF00222129

Pogna NE, Monari AM, Caciatori P, Redaelli R, Ng PKW (1998) Development and characterization of bread wheat lines lacking chromosome 1B-, 1D-, 6A- and 6D-encoded prolamins. Proc 9th Intern Wheat Genetics Symp: Saskatoon, Saskatchewan (Canada); pp 265-268.

Polishchuk AM, Chebotar S, Blagodarova EM, Kozub NA, Sozinov IA, Syvolap YuM (2010) PCR analysis of the wheat varieties and near-isogenic wheat lines with the use of allele-specific primers for the *Gli-1* and *Glu-3* loci. Cytol Genet (Kiev) 44: 345-353. https://doi.org/10.3103/S0095452710060046

Poperelya FA (1996) Three principal genetic systems of grain quality of winter common wheat. Nauchno-Techn Bull VSGI (Odessa) 117-132 (in Ukrainian).

Poperelya FA & Babayants LT (1978) The Gld 1B3 block as a marker for the gene responsible for the resistance of wheat plants to stem rust. Doklady VASKhNIL 6: 8-9 (in Russian), https://agris.fao.org/agris-search/search.do?recordID=US19810626604

Poperelya FA & Gasanova GM (1980) Gliadin component composition and endosperm hardness as indicators of wheat grain quality. Nauchno-Techn Bull VSGI (Odessa) 3/37: 21-25 (in Russian), https://www.cabdirect.org/cabdirect/abstract/19811605517

Poperelya FA & Blagodarova AM (1998) Genetics of grain quality of the first super-strong wheat genotypes in Ukraine. Cytol Genet (Kiev) 32: 11-19 (in Ukrainian).

Poperelya FA, Sozinov AA, Omorbekova ZA (1978) Study of the nature of differences in the flour quality in cultivars Odesskaya 51 and Kavkaz by methods of biochemical genetics. Nauchno-Techn Bull VSGI (Odessa) 31: 3-9 (in Russian).

Poperelya FA, Bito M, Sozinov AA (1980) Relationship between blocks of gliadin components with pants survival, their productivity, spike colour, and quality of F2 hybrids from crossing the cultivars Bezostaya 1 and Crvena Zvezda. Doklady VASKhNIL 4: 4-7 (in Russian), https: //agris.fao.org/agris-search/search.do?recordID=US201301309644

Popineau Y & Pineau F (1988) Changes of conformation and surface hydrophobicity of gliadins. Food Sci Technol 21: 113-117, http://pascal-francis.inist.fr/vibad/index.php?ction=getRecordDetail& idt=7735529

Popineau Y, Cornec M, Lefebvre J, Marchilo B (1994) Influence of high Mr glutenin subunits on glutenin polymers and theological properties of gluten and gluten subfractions of near isogenic lines of wheat Sicco. J Cereal Sci 19: 231-242, https://doi.org/10.1006/jcrs.1994.1030

Popovych Yu, Chebotar S, Melnik VA, Rodriguez-Quijano M, Pascual L, Rogers WJ, Metakovsky E (2020) Congruity of the polymorphisms in the expressed and noncoding parts of the *Gli-B1* locus in common wheat. Agronomy 10 (10): 1510, https://doi.org/10.3390/agronomy10101510

Popovych Yu, Blagodarova OM, Chebotar S (2021) Genetic variation of *Gli-B1* locus in Ukrainian bread wheat varieties and lines. Biopolimers

& Cell 37: 379-388, https://dx.doi.org/10.7124/bc.000A63

Porceddu E, Ceoloi C, Lafiandra D, Tanzarella OA, Scarascia Munozza GT (1988) Geneic resources and plant breeding: Problems and pospects. In: Proc 7th Internat Wheat Genetics Symp (v.1); eds T.E.Miller, R.M.D.Koebner; IPSR, Cambridge, UK, pp 7-22, https://agris.fao.org/agris-search/search.do?recordID=US20130269 9090

Prasad M, Varshney RK, Roy JK, Balyan HS, Gupta PK (2000) The use of microsatellites for detecting DNA polymorphism, genotype identification and genetic diversity in wheat. Theor Appl Genet 100: 584-592, https://doi.org/10.1007/s001220050077

Pronin D, Geisslitz S, Börner A, Scherf KA (2020) Fingerprinting of wheat protein profile for improved distinction between wheat cultivars and species. Cereal Chem 97: 999-1009, https://doi.org/10.1002/cche.10323

Qi PF, Wei YM, Ouellet T, Chen Q, Tan X, Zheng YL (2009) The γ-gliadin multigene family in common wheat (*Triticum aestivum*) and its closely related species. BMC Genom 10: 168–181, https://doi.org/10.1186/1471-2164-10-168

Rabinovich SV (1998) Importance of wheat-rye translocations for breeding modern cultivas of *Triticum aestivum* L. Euphytica 100: 323-340, https://doi.org/10.1023/A:1018361819215

Rafalski JA (1986) Structure of wheat gamma-gliadin genes. Gene 43: 221-229, https://doi.org/10.1016/0378-1119(86)90210-6

Rakszegi M, Bekes F, Lang L, Tamas L, Shewry PR, Bedo Z (2005) Technological quality of transgenic wheat expressing an increased amount of a HMW glutenin subunit. J Cereal Sci 42 (1): 15-23, https://doi.org/10.1016/j.jcs.2005.02.006

Ram S, Sharma S, Verma A, Tyagi BS, Peña RJ (2011) Comparative analyses of LMW glutenin alleles in bread wheat using allele-specific PCR and SDS-PAGE. J Cereal Sci 54: 488-493, https://doi.org/10.1016/j.jcs.2011.09.004

Raquin AL, Depaulis F, Lambert A, Galic N, Brabant P, Goldringer I (2008). Experimental estimation of mutation rates in a wheat population with a gene genealogy approach. Genetics 179: 2195-2211, https://doi.org/10.1534/genetics.107.071332

Rathan ND, Mahendru-Singh, Govindan V, Ibba MI (2020) Impact of high and low-molecular-weight glutenins on the processing quality of a set of biofortified common wheat (*Triticum aestivum* L) lines. Front Sustain Food Syst 4, 175, https://10.3389/fsufs.2020.583367

Ravel C, Praud S, Murigneux A, Canaguier A, Sapet F, Samson D, Balfourier F, Dufour P, Chalhoub B, Brunel D, Beckert M, Charmet G

(2006) Single-nucleotide polymorphism frequency in a set of selected lines of bread wheat (*Triticum aestivum* L.). Genome 49: 1131-1139, https://doi.org/10.1139/g06-067

Ravel C, Praud S, Canaguier A, Dufour P, Giancola S, Balfourier F, Chalhoub B, Brunel D, Linossier L, Dardevet M, Beckert M, Rousset M, Murigneux A, Charmet G (2007) DNA sequence polymorphisms and their application to bread wheat quality. Euphytica 158: 331-336, https://doi.org/10.1007/s10681-006-9288-z

Ravel C, Faye A, Ben-Sadoun S, Ranoux M, Dardevet M, Dupuits C, Exbrayat F. Poncet C, Sourdille P, Branlard G (2020) SNP markers for early identification of high molecular weight glutenin subunits (HMW-GSs) in bread wheat. Theor Appl Genet 133: 751-770, https://doi.org/10.1007/s00122-019-03505-y

Rayburn AL & Carver BF (1988) Cytological identification of 1B/1R wheat-rye translocation in winter wheat breeding lines. Euphytica 38: 237-240, https://doi.org/10.1007/BF00023526

Redaelli R, Pogna NE, Dachkevitch T, Cacciatori P, Biancardi AM, Metakovsky EV (1992) Inheritance studies of the 1AS/1DS chromosome translocation in the bread wheat variety Perzivan-1. J Genet Breed 46: 253-262, http://www.sidalc.net/cgi-bin/wxis.exe/?IsisScript=CIMMYTxis&method=post&formato=2&cantidad=1&expression=mfn=015952

Redaelli R, Metakovsky E, Davydov SD, Pogna NE (1994) Two-dimensional mapping of gliadins using biotypes and null mutants of common wheat cultivar Saratovskaya 29. Hereditas 121*:* 131–137, https://doi.org/10.1111/j.1601-5223.1994.00131.x

Redaelli R, Morel MH, Autran JC, Pogna NE (1995) Genetic analysis of low *Mr* glutenin subunits fractionated by two-dimensional electrophoresis (A-PAGE x SDS-PAGE). J Cereal Sci. 21: 5-13, https://doi.org/10.1016/S0733-5210(95)80003-4

Redaelli R, Pogna NE, Ng PKW (1997) Effect of prolamins encoded by chromosomes 1B and 1D on the rheological properties of dough in near-isogenic lines of bread wheat. Cereal Chem 74: 102-107, https://doi.org/10.1094/CCHEM.1997.74.2.102

Reddy P & Appels R (1993) Analysis of a genomic DNA segment carrying the wheat high-molecular-weight (HMW) glutenin Bx17 subunit and its use as an RFLP marker. Theor Appl Genet 85: 616-624, https://doi.org/10.1007/BF00220921

Reeves CD & Okita TW (1987) Analyses of α/β gliadin genes from diploid and hexaploid wheats. Gene 52: 257-266, https://doi.org/10.1016/0378-1119(87)90052-7

Reif JC, Zhang P, Dreisigacker S, Warburton ML, van Ginkel M, Hoisington D, Bohn M, Melchinger AE (2005), Wheat genetic diversity trends during domestication and breeding. Theor Appl Genet 110: 859-864, https://doi.org/10.1007/s00122-004-1881-8

Ribeiro M, Carvalho C, Carnide V, Guedes-Pinto H, Igrejas G (2011) Towards allelic diversity in the storage proteins of old and currently growing tetraploid and hexaploid wheats in Portugal. Genet Resour Crop Evol 58: 1051-1073, https://doi.org/10.1007/s10722-010-9642-9

Ribeiro M, Nunes-Miranda JD, Branlard G, Carrillo JM, Rodriguez-Quijano M, Igrejas G (2013) One hundred years of grain omics: identifying the glutens that feed the world. J Proteome Res 12: 4702-4716, https://doi.org/10.1021/pr400663t

Ribeiro-Carvalho C, Guedes-Pinto H, Harrison G, Heslop-Harrison JS (1997) Wheat-rye chromosome translocations involving small terminal and intercalary rye chromosome segments in the Portuguese wheat landrace Barbela. Heredity 78: 539-546, https://doi.org/10.1038/hdy.1997.84

Ribeiro-Carvalho C, Guedes-Pinto H, Heslop-Harrison JS, Schwarzacher T (2001) Introgression of rye chromatine on chromosome 2D in the Portuguese wheat landrace "Barbela". Genome 44: 1122-1128, https://doi.org/10.1139/g01-088

Ribeiro-Carvalho C, Guedes-Pinto H, Igrejas G, Stephenson P, Schwarzacher T, Heslop-Harrison JS (2004) High levels of genetic diversity throughout the range of the Portuguese wheat landrace "Barbela". Ann Bot 94: 699-705, https://doi.org/10.1093/aob/mch194

Röder MS, Plaschke J, König SU, Börner A, Sorrells ME, Tanksley SD, Ganal MW (1995) Abundance, variability and chromosomal location of microsatellites. Mol Gen Genet 246: 327-337, https://doi.org/10.1007/BF00288605

Röder MS, Korzun V, Gill BS, Ganal MW (1998a) The physical mapping of microsatellite markers in wheat. Genome 41: 278-283, https://doi.org/10.1139/g98-009

Röder MS, Korzun V, Wendehake K, Plaschke J, Tixier MH, Leroy P, Ganal MW (1998b) A microsatellite map of wheat. Genetics 149: 2007-2023, https://doi.org/10.1093/genetics/149.4.2007

Rodriguez-Quijano M & Carrillo JM (1994) Relación entre subunidades de gluteninas de alto peso molecular y fuerza del gluten en cultivares indígenas españoles de Triticum aestivum ssp. vulgare. Invest Agraria 9: 327-339 (in Spanish), http://www.sidalc.net/cgi-bin/wxis.exe/IsisScript=cibago,xis&method=post&formato=2&cantidad=1&expression=mfn=035468

Rodriguez-Quijano M & Carrillo JM (1996a) Linkage map of prolamin loci *Gli-D4* and *Gli-D5* in hexaploid wheat. Pant Breed 115: 189-191, https://doi.org/10.1111/j.1439-0523.1996.tb00899.x

Rodriguez-Quijano M & Carrillo JM (1996b) Relationship between allelic variation of *Glu-1* and *Gli-1/Glu-3* prolamin loci and gluten strength in hexaploid wheat. Euphytica 91: 141 148, https://doi.org/10.1007/BF00021063

Rodriguez-Quijano M, Vázquez JF, Moita-Brites C, Carrillo JM (1998) Allelic variation of HMW glutenin subunits in Portuguese landraces of *Triticum aestivum* ssp *vulgare*. J Genet Breed 52: 95-98.

Rogers WJ, Payne PI, Harinder K (1989) The HMW glutenin subunit and gliadin compositions of German-grown wheat varieties and their relationship with bread-making quality. Plant Breed 103: 89-100, https://doi.org/10.1111/j.1439-0523.1989.tb00356.x

Rogers WJ, Rickatson JM, Sayers EJ, Law CN (1990) Dosage effects of chromosomes of homoeologous groups 1 and 6 upon bread-making quality in hexaploid wheat. Theor Appl Genet 80: 281-287, https://doi.org/10.1007/BF00224399

Rooke L, Bekes F, Fido R, Barro F, Gras P, Tatham AS, Barcelo P, Lazzeri P, Shewry PR (1999) Over-expression of a gluten protein in transgenic wheat results in greatly increased dough strength. J Cereal Sci 30: 115-120, https://doi.org/10.1006/JCRS.1999.0265

Roussel V, Koenig J, Beckert M, Balfourier F (2004) Molecular diversity in French bread wheat accessions related to temporal trends and breeding programs. Theor Appl Genet 108: 920-930, https://doi.org/10.1007/s00122-003-1502-y

Roussel V, Leisova L, Exbrayat F, Stehno Z, Balfourier F (2005) SSR allelic diversity changes in 480 European bread wheat varieties released from 1840 to 2000. Theor Appl Genet 111: 162-170, https://doi.org/10.1007/s00122-005-2014-8

Rousset M, Carrillo JM, Qualset CO, Kasarda DD (1992) Use of recombinant inbred lines of wheat for study of associations of high-molecular-weight glutenin subunits alleles to quantitative traits. Theor Appl Genet 83: 403-412, https://doi.org/10.1007/BF00226527

Roy N, Islam S, Yu Z, Lu M, Lafiandra D, Zhao Y, Anwar M, Mayer JE, Ma W (2020) Introgression of an expressed HMW 1Ay glutenin subunit allele into bread wheat cv. Lincoln increases grain protein content and breadmaking quality without yield penalty. Theor Appl Genet 133: 517-528, https://doi.org/10.1007/s00122-019-03483-1

Ruanet VV, Kudryavtsev AM, Dadashev SY (2001) The use of artificial neural networks for autonomic analysis and genetic identification of gliadin electrophoretic spectra in durum wheat. Russian J of Genetics 37: 1207-1209, https://doi.org/10.1023/A:1012321109086

Ruiz M & Carrillo JM (1993) Linkage relationships between prolamin genes on chromosomes 1A and 1B of durum wheat. Theor Appl Genet 87: 353-360, https://doi.org/10.1007/BF01184923

Ruiz M & Carrillo JM (1995) Separate effects on gluten strength of *Gli-1* and *Glu-3* prolamin genes on chromosomes 1A and 1B in durum wheat. J Cereal Sci 21: 137-144, https://doi.org/10.1016/0733-5210(95)90029-2

Ruiz M & Carrillo JM (1996) *Gli-B3*/*Glu-B2*-encoded prolamins do not affect selected quality properties in the durum wheat cross "Abadia" x "Mexicali 75". Plant Breed 115: 410-412, https://doi.org/10.1111/j.1439-0523.1996.tb00943.x

Ruiz M, Metakovsky EV, Rodriguez-Quijano M, Vazquez JF, Carrillo JM (2002a) Assessment of storage protein variation in relation to some morphological characters in a sample of Spanish landraces of common wheat (*Triticum aestivum* L. ssp. *aestivum*). Genet Resour Crop Evol 49: 371-382, https://doi.org/10.1023/A:1020626430815

Ruiz M, Rodriguez-Quijano M, Metakovsky E, Vazquez JF, Carrillo JM (2002b) Polymorphism, variation and genetic identity of Spanish common wheat germplasm based on gliadin alleles. Field Crops Res 79: 185-196, https://doi.org/10.1016/S0378-4290(02)00139-9

Ruiz M, Aguiriano E, Giraldo P, Carrillo JM (2011) Genetic redundancy among durum wheat accessions as assessed by SSRs and endosperm proteins. Spanish J Agric Res 9: 156-165, https://doi.org/10.5424/sjar/20110901-160-10

Rybalka AI & Sozinov AA (1979) Mapping of the Gld 1B locus controlling biosynthesis of common wheat storage proteins. Cytol Genet (Kiev) 13: 276-282 (in Russian), http://pascal-francis.inist.fr/vibad/index.php?ction=getRecordDetail&idt= PASCALAGROLINEINRA8110070764

Sabelli PA & Shewry PR (1991) Characterization and organization of gene families at the *Gli-1* loci of bread and durum wheats by restriction fragment analysis. Theor Appl Genet 83: 209-216, https://doi.org/10.1007/BF00226253

Sabelli PA, Lafiandra D, Shewry PR (1992) Restriction fragment analysis of "null" forms at the *Gli-1* loci of bread and durum wheats. Theor Appl Genet 83, 428-434, https://doi.org/10.1007/BF00554430

Sandhu D, Champoux JA, Bondareva N, Gill KS (2001) Identification and physical localization of useful genes and markers to a major gene-rich region on wheat group 1S chromosomes. Genetics 157: 1735-1747,

https://doi.org/10.1093/genetics/157.4.1735
Sapirstein HD & Bushuk W (1985a) Computer-aided analysis of gliadin electrophoregrams. I. Improvement of precision of relative mobility determination by using a three reference band standardization. Cereal Chem. 62: 372-377, http://pascal-francis.inist.fr/vibad/index.php?action=getRecordDetail&idt=9270804
Sapirstein HD & Bushuk W (1985b) Computer-aided analysis of gliadin electrophoregrams. II. Wheat cultivar identification and class comparisons. Cereal Chem. 62: 377-392, http://pascal-francis.inist.fr/vibad/index.php?action=getRecordDetail&idt=8459650
Sapirstein HD & Bushuk W (1986) Computer-aided wheat cultivar identification and analysis of densitometric scanning profiles of gliadin electrophoregrams. Seed Sci Tchnol 14: 489-517, http://pascal-francis.inist.fr/vibad/index.php?ction=getRecordDetail&idt= 8277220
Scheets K & Hedgcoth C (1988) Nucleotide sequence of a γ gliadin gene: Comparisons with other γ gliadin sequences show the structure of γ gliadin genes and the general primary structure of γ gliadins. Plant Sci. 57: 141-150, https://doi.org/10.1016/0168-9452(88)90080-5
Schlegel R & Korzun V (1997) About the origin of 1RS.1BL wheat-rye chromosomal translocations from Germany. Plant Breed 116: 537-540, https://doi.org/10.1111/j.1439-0523.1997.tb02186.x
Sears ER (1954) The aneuploids of common wheat. Research Bull Mo Coll Agr Exp Sta (Univ. Missoury) 572: 3-59.
Sears ER (1974) The wheats and their relatives. In: "Handbook of Genetics", v.2 ("Plants, Plant Viruses and Protists"), ed. R.C.King; Plenum Press, London; pp 59-91,
https://doi.org/10.1007/978-1-4684-2994-7_3
Seitova AM, Metakovsky E, Sozinov AA (1986) The biotype composition and blocks of gliadin components in the common wheat Bogarnaya-56. Cytol Genet (Kiev) 20: 196-201 (in Russian), http://pascal-francis.inist.fr/vibad/index.php?ction=getRecordDetail&idt=8121048
Senerchia N, Wicker T, Felber F, Parisod C (2013) Evolutionary dynamics of retrotransposons assessed by high-throughput sequencing in wild relatives of wheat. Genome Biol Evol 5: 1010-1020,
https://doi.org/10.1093/gbe/evt064
Sexson KR, Wu YV, Huebner FR, Wall JS (1978) Molecular weights of wheat γ2-, β5-, α7-, α8-, and α9-gliadins. Biochim Biophys Acta 532:279-285, https://doi.org/10.1016/0005-2795(78)90582-2
Shan X, Clayshulte SR, Haley SD, Byrne PF (2007) Variation for glutenin and waxy alleles in the US hard winter wheat germplasm. J Cereal Sci 45: 199-208, https://doi.org/10.1016/j.jcs.2006.09.007

Sharma A, Garg S, Sheikh I, Vyas P, Dhaliwal HS (2020) Effect of wheat grin protein composition on end-use quality. J Food Sci Technol 57:2771-2785, https://doi.org/10.1007/s13197-019-04222-6

Sharp PJ, Chao S, Desai S, Kilian A, Gale MD (1989) Use of RFLP markers in wheat and related species. In: "Development and Application of Molecular Markers to Problems in Plant Genetics"; eds T.Helentjaris & B.Burr; Cold Spr Hurb Lab; NY; pp 29-34.

Shepherd KW (1977) Utilization of a rye chromosome arm in wheat breeding. Plant Breeding Papers; Int Congr Society Advancement Breeding Researches Asia & Oceania, https://agris.fao.org/agris-search /search.do?recordID=US201302441244

Shewry PR & Halford NG (2002) Cereal seed storage proteins: structures, properties and role in grain utilization. J Exp Bot 53: 947-958, https://doi.org/10.1093/jexbot/53.370.947

Shewry PR & Tatham AS (1997) Disulphide bonds in wheat gluten proteins. J Cereal Sci 25: 207-227, https://doi.org/10.1006/jcrs.1996.0100

Shewry PR, Faulks AJ, Pratt H.M, Miflin BJ (1978) The varietal identification of single seeds of wheat by sodium dodecyl-sul**ph**ate polyacrylamide gel electrophoresis of gliadin. J Sci Food Agric 29: 841-849, https://doi.org/10.1002/jsfa.2740291004

Shewry PR, Halford NG, Tatham AS (1992) The high molecular weight subunits of wheat glutenin. J Cereal Sci 15: 105-120, https://doi.org/10.1016/S0733-5210(09)80062-3

Shewry PR, Halford NG, Lafiandra D (2003) Genetics of wheat gluten proteins. Adv Genet 49: 111-184, https://doi.org/10.1016/S0065-2660(03)01003-4

Shewry PR, Halford NG, Lafiandra D (2006) The high-molecular-weight subunits of glutenin. In: "Gliadin and glutenin. The unique balance of wheat quality", eds C.Wrigley, F.Békés, W.Bushuk; AACC Internat; pp 143-169, https://www.cabdirect.org/cabdirect/abstract/20073205858

Siedler H, Messmer MM, Schachermayer GM, Winzeler H, Winzeler M, Keller B (1994) Genetic diversity in European wheat and spelt breeding material based on RFLP data. Theor Appl Genet 88: 994-1003, https://doi.org/10.1007/BF00220807

Singh NK & Shepherd KW (1988a) Linkage mapping of genes controlling endosperm storage proteins in wheat. 1. Genes on the short arms of group 1 chromosomes. Theor Appl Genet 75: 628-641, https://doi.org/10.1007/BF00289132

Singh NK & Shepherd KW (1988b) Linkage mapping of genes controlling endosperm storage proteins in wheat. 2. Genes on the long arms of group 1 chromosomes. Theor Appl Genet 75: 642-650, https://doi.org/10.1007/BF00289132

Singh NK, Shepherd KW, McIntosh RA (1990) Linkage mapping of genes for resistance to leaf, stem and stripe rusts and ω-secalins on the short arm of rye chromosome 1R. Theor Appl Genet 80: 609-616, https://doi.org/10.1007/BF00224219

Skerritt JH (1985) A sensitive monoclonal-antibody-based test for gluten detection: Quantitative immunoassay. J Sci Food Agric 36 (10) 987-994, https://doi.org/10.1002/jsfa.2740361014

Skerritt JH, Martinuzzi O, Metakovsky E (1991) Chromosomal control of wheat gliadin epitopes: Analysis with specific monoclonal antibodies. Theor Appl Genet 82 (1): 44-53, https://doi.org/10.1007/BF00231276

Snape JW, Flavell RB, O'Dell M, Hughes WG, Payne PI (1985) Intrachromosomal mapping of the nucleolar organiser region relative to three marker loci on chromosome 1B of wheat (*Triticum aestivum*). Theor Appl Genet 69: 263-270, https://doi.org/10.1007/BF00662440

Sobko TA (1984) Identification of a locus that controls the synthesis of alkohol-soluble proteins of the endosperm of winter common wheat. Vestnik Silskogosp Nauki 7: 78-80 (in Ukrainian).

Sobko TA & Poperelya FA (1982) The linkage between gliadin-coding locus Gld 1A and the gene *Hg* for hairness of spike scales in wheat. Nauchno-Techn Bull VSGI (Odessa) 2: 28-33 (in Russian).

Sobko TA & Poperelya FA (1986) The frequency of alleles at the gliadin-encoding loci in common wheat winter cultivars. Visnik Silskogosp Nauki 5: 84-87 (in Ukranian).

Sobko TA & Sozinov AA (1999) Analysis, using genetic markers, of the genotypic structure of winter bread wheat cultivars grown in Ukraine. Cytol Genet (Kiev) 33: 30-41 (in Russian).

Sobko TA, Poperelya FA, Rybalka AI, Sozinov AA (1986) Inheritance and mapping of genes encoding the synthesis of storage proteins on the chromosome 1A of common wheat. Cytol Genet (Kiev) 20: 57-62 (in Russian), https://agris.fao.org/agris-search/search.do?recordID=US874689988

Sobko TO, Lisova GM, Sirant LV (2019) Genetic variability of the storage protein loci in the European winter common wheat varieties licensed for growing in Ukraine. Factors Experim Evol Organ 24: 284-289 (in Ukrainian), https://doi.org/10.7124/FEEO.v24.1116

Somers DJ, Kirkpatrick R, Moniwa M, Walsh A (2003) Mining single-nucleotide polymorphisms from hexaploid wheat ESTs. Genome 46: 431-437, https://doi.org/10.1139/g03-027

Sontag-Strohm T, Payne PI, Salovaara H (1996) Effect of allelic variation of glutenin subunits and gliadins on baking quality in the progeny of two biotypes of bread wheat cv. Ulla. J Cereal Sci 24: 115-124, https://doi.org/10.1006/jcrs,1996.0044

Sourdille P, Singh S, Cadalen T, Brown-Guedira GL, Gay G, Qi L, Gill BS, Dufour P, Murigneux A, Bernard M (2004) Microsatellite-based deletion bin system for the establishment of genetic-physical map relationships in wheat (*Triticum aestivum L.*). Funct Integr Genomics 4: 12-25, https://doi.org/10.1007/s10142-004-0106-1

Souza E, Fox PN, Byerlee D, Skovmand B (1994) Spring wheat diversity in irrigated areas of two developing countries. Crop Sci. 34: 774-783, https://doi.org/10.2135/cropsci1994.0011183X003400030031x

Sozinov AA & Kopus MM (1983) Mutation of the gliadin-coding locus of the chromosome 1D. Cytol Genet (Kiev) 17: 19-24 (in Russian), http://pascal-francis.inist.fr/vibad/index.php?ction=getRecordDetail&idt=PASCALAGROLINEINRA83X0347922

Sozinov AA & Poperelya FA (1977) Electrophoretic method for the identification of common wheat genotypes carrying translocation or substitution of the chromosome 1B for the chromosome 1R of rye. Doklady VASKHNIL 2: 2-4 (in Russian), https://agris.fao.org/agris-search/search.do?recordID=US201302417640

Sozinov AA & Poperelya FA (1980) Genetic classification of prolamins and its use for plant breeding. Ann. Technol. Agric. 29: 229-245, http://pascal-rancis.inist.fr/vibad/index.php?ction=getRecordDetail&idt= PASCALAGROLINEINRA8110070764

Sozinov AA & Poperelya FA (1985) Relationship of gliadin components with grain quality of winter wheat. Selektsiia i Semenovod 4: 34-35 (in Russian).

Sozinov AA, Poperelya FA, Stakanova AI (1974) The use of gliadin electrophoresis in wheat selection for quality. Vestnik Selskokhoz Nauki 7: 99-108 (in Russian).

Sozinov AA, Poperelya FA, Kopus MM (1975a) Genetically controlled differences in the gliadin components of the wheat varieties Bezostaya 1 and Dneprovskaya 521 and their role in determining flour quality. Doklady VASKhNIL 11: 10-14 (in Russian).

Sozinov AA, Poperelya FA, Stakanova AI (1975б) Hybridological analysis as a method of study of genetic regularities of gliadin biosynthesis. Naucno-Tekhn Bull VSGI (Odessa): 10-15 (in Russian).

Sozinov AA, Metakovsky E, Koval SF (1986) Regularities of genotype formation in wheat breeding. Vestnik Selskokhoz Nauki 3: 60-71 (in Russian).

Sozinov AA, Novoselskaya AYu, Lushnikova AA, Bogdanov YuF (1987б) Cytological and biochemical analysis of bread wheat varieties with 1B/1R substitutions and translocations in the caryotype. Cytol Genet (Kiev) 21: 17-24, ISSN: 0095-4527 (in Russian), http://pascal-francis.inist.fr/vibad/index.php?ction=getRecordDetail&idt=7622929

Sozinov AA, Sozinov IA, Kozub NA, Sobko TA (1999) Stable gene associations in breeding and evolution of grasses. In: "Evolutionary Theory and Processes: Modern Perspectives. Papers in Honour of Eviatar Nevo". ed. S.P.Wasser; Kluver Acad. Pabl., Netherlands, pp 97-113, https://doi.org/10.1007/978-94-011-4830-6_7

Spielmeyer W, Moullet O, Laroche A, Lagudah ES (2000) Highly recombinogenic region at seed storage protein loci on chromosome 1DS of *Aegilops tauschii*, the D-genome donor of wheat. Genetics 155: 361-367, https://doi.org/10.1093/genetics/155.1.361

Sreeramulu G & Singh NK (1997) Genetic and biochemical characterization of novel low molecular weight glutenin subunits in wheat (Triticum aestivum L.). Genome 40: 41-48, https://doi.org/10.1139/g97-006

Stachel M, Lelley T, Grausgruber H, Vollmann J (2000) Application of microsatellites in wheat (Triticum aestivum L.) for stadying genetic differentiation caused by selection for adaptation and use. Theor Appl Genet 100: 242-248, https://doi.org/10.1007/s001220050032

Storlie EW & Talbot LE (1993) Cause of tall off-types in a semidwarf spring wheat. Crop Sci 33: 1131-1135, https://doi.org/10.2135/cropsci1993.0011183X003300060003x

Storlie EW, Xie H, Talbot LE (1996) Tall off-types in semidwarf spring wheat with high-reducting genes *Rht 1* and *Rht 2*. Crop Sci 36: 1521-1522, https://doi.org/10.2135/cropsci1996.0011183X003600060019x

Sumner-Smith M, Rafalski JA, Sugiyama T, Stoll M, Söll D (1985) Conservation and variability of wheat α/β-gliadin genes. Nucleic Acid Res 13: 3905-3916, https://doi.org/10.1093/nar/13.11.3905

Sutka J (2001) Genes for frost resistance in wheat. Euphytica 119: 169-177, https://doi.org/10.1023/A:1017520720183

Sutton KH (1991) Qualitative and quantitative variation among high molecular weight subunits of glutenin detected by reversed-phase high-performance liquid chromatography. J Cereal Sci 14 (1): 25-34, https://doi.org/10.1016/S0733-5210(09)80015-5

Talbert LE, Blake NK, Chee PW, Blake TK, Magyar GM (1994) Evaluation of "sequence-tagged-site" PCR products as molecular markers in wheat. Theor Appl Genet 87: 789-794, https://doi.org/10.1007/BF00221130

Talbert LE, Smith LY, Blake NK (1998) More than one origin of hexaploid wheat is indicated by sequence comparison of low-copy DNA. Genome 41: 402-407, https://doi.org/10.1139/g98-037

Tanaka H, Shimizu R, Tsujimoto H (2005) Genetical analysis of contribution of low-molecular-weight glutenin subunits to dough strength in common wheat (*Triticum aestivum* L.). Euphytica 141: 157-162, https://doi.org/10.1007/s10681-005-6714

Tanaka M, Nagano T, Yano H, Matsuda T, Ikeda TM, Haruma K, Kato Y (2011) Impact of ω-5 gliadin on wheat-dependent exercise-induced anaphylaxis in mice. Bioscience Biotechnol. Biochem 75: 313-317, https://doi.org/10.1271/bbb.100695

Tao HP & Kasarda DD (1989) Two-dimensional gel mapping and N-terminal sequencing of LMW glutenin subunits. J Exp Bot 40: 1015-1020, https://www.jstor.org/stable/23692454

Tao HP, Adalsteins AE, Kasarda DD (1992) Intermoecuar disulfide bonds link specific high-molecular-weight glutenin subunitsin wheat endosperm. Biochem Biophys Acta 1159: 13-21, https://doi.org/10.1016/0167-4838(92)90069-P

Tatham AS & Shewry PR (1995) The S-poor prolamins of wheat, barley and rye. J Cereal Sci 22: 1-16, https://doi.org/10.1016/S0733-5210(05)80002-5

Tatham AS, Miflin BJ, Shewry PR (1985) The beta-turn conformation in wheat gluten proteins: relationship to gluten elasticity. Cereal Chem 62: 404-442, http://pascal-francis.inist.fr/vibad/index.php?ction=getRecordDetail&idt=9270808

Thomas JB & Schaalje GB (1997) Winter survival and competition in a mixture of winter wheat cultivars. Crop Sci 37: 732-738, https://doi.org/10.2135/cropsci1997.0011183X003700030007x

Thormann I & Engels JMM (2015). Genetic diversity and erosion – a global perspective. In: Genetic diversity and erosion in plants (V.7); Sustainable development and biodiversity; pp 263-298, https://doi.org/10.1007/978-3-319-25637-5_10

Thuillet AC, Bru D, David J, Roumet P, Santoni S, Sourdille P, Bataillon T (2002) Direct estimation of mutation rate for 10 microsatellite loci in durum wheat, *Triticum turgidum* (L.) Thell. ssp *durum* desf. Mol Biol Evol 19: 122-125, https://doi.org/10.1093/oxfordjournals.molbev.a003977

Thuillet AC, Bataillon T, Sourdille P, David J (2004) Factors affecting polymorphism of microsatellite loci in bread wheat [*Triticum aestivum* (L.) Thell.]: effects of mutation processes and physical distance from the centromere. Theor Appl Genet 108: 368-377, https://doi.org/10.1007/s00122-003-1443-5

Tkachuk R & Metlish VJ (1980) Wheat cultivar identification by high voltage gel electrophoresis. Ann Technol Agric 29: 207-212, http://pascal-francis.inist.fr/vibad/index.php?ction=getRecordDetail&idt=PASCAL8110404559

Tsenov N, Atanasova D, Todorov I, Ivanova I, Stoeva I (2010) Quality of winter common wheat advanced lines depending on allelic variation of *Glu-A3*. Cereal Res Comm 38: 250-258, https://doi.org/10.1556/crc.38.2010.2.11

Upelniek VP, Novoselskaya AYu, Sutka J, Galiba G, Metakovsky EV (1995) Genetic variation at storage protein-coding loci of common wheat (cv "Chinese Spring") induced by nitrosoethylurea and by the cultivation of immature embryos *in vitro*. Theor Appl Genet 90: 372-379, https://doi.org/10.1007/BF00221979

Upelniek VP, Novoselskaya-Dragovich AYu, Shishkina AA, Melnik VA, Dedova LV, Kudryavtsev AM (2013) Laboratory analysis of wheat seed proteins (technological instruction and toolkit). Diagnostics of varietal compliance and wheat seed purity. Vavilov Institute of General Genetics, Moscou, Russia (in Russian).

Urade R, Sato N, Sugiyama M (2018) Gliadins from wheat grain: an overview, from primary structure to nanostructures of aggregates. Biophys Rev 10: 435-443, https://doi.org/10.1007/s12551-017-0367-2

Uthayakumaran S, Batey IL, Wrigley CW (2005) On-the-spot identification of grain variety and weat quality type by Lab-on-a-cip capillary electrophoresis. J Cereal Sci 41: 371-374, https://doi.org/10.1016/j.jcs.2004.12.001

Uthayakumaran S, Wrigley CW, Batey IL, Bushuk W, Lookhart GL (2006) Genotype identification. In: "Gliadin and glutenin. The unique balance of wheat quality"; eds C.W.Wrigley, F.Békés, W.Bushuk; AACC International; pp 307-331, https://www.cabdirect.org/cabdirect/abstract/20073205863

Vaccino P, Accerbi M, Corbellini M (1993) Cultivar identification in *T.aestivum* using highly polymorphic RFLP probes. Theor Appl Genet 86: 833-836, https://doi.org/10.1007/BF00212609

Vaccino P & Metakovsky E (1995) Gliadin alleles in DNA RFLP patterns of common wheat: implication for analysis of organization and evolution of complex loci. Theor Appl Genet 90: 173-181, https://doi.org/10.1007/BF00222199

Vaccino P, Redaelli R, Metakovsky E, Borghi B, Corbellini M, Pogna NE (2002) Identification of novel low Mr glutenin subunits in the high quality bread wheat cv Salmone and their effects on gluten quality.

Theor Appl Genet 105: 43-49, https://doi.org/10.1007/s00122-001-0826-8

van Beuningen LT & Busch RH (1997a) Genetic diversity among North American spring wheat cultivars. 2. Ancestor contributions to gene poos of different eras and regions. Crop Sci 37: 580-585, https://doi.org/10.2135/cropsci1997.0011183X003700020044x

van Beuningen LT & Busch RH (1997b) Genetic diversity among North American spring wheat cultivars. 3. Cluster analysis based on quantitative morphological traits. Crop Sci 37: 981-988, https://doi.org/10.2135/cropsci1997.0011183X003700030046x

Van Deynze AE, Dubcovsky J, Gill KS, Nelson JC, Sorrells ME, Dvořák J, Gill BS, Lagudah ES, McCouch SR, Appels R (1995) Molecular-genetic maps for group 1 chromosomes of Triticeae species and their relation to chromosomes in rice and oat. Genome 38: 45-59, https://doi.org/10.1139/g95-006

Van de Wouw M, van Hintum T, Kik C, van Treuren R, Visser B (2010) Genetic diversity trends in twentieth century crop cultivars: a meta analysis. Theor Appl Genet 120, 1241-1252, https://doi.org/10.1007/s00122-009-1252-6

Van Herpen TWJM, Goryunova SV, van der Schoot J, Mitreva M, Salentijn E, Vorst O, Schenk MF, van Veelen PA, Koning F, van Soest LJM, Vosman B, Bosch D, Hamer RJ, Gilissen LJWJ, Smulders MJM (2006) Alpha-gliadin genes from the A, B, and D genomes of wheat contain different sets of celiac disease epitopes. BMC Genomics 7: 1-12, https://doi.org/10.1186/1471-2164-7-1

Vavilov NI (1926) Studies on the origin of cultivated plants. Bureau Appl Botany Plant Breed 14: 1-245 (in Russian).

Vavilov NI (1992) Origin and geography of cultivated plants. Translated from Russian by D Löve; Cambridge University Press.

Vawser MJ & Cornish GB (2004) Over-expression of HMW glutenin subunit *Glu-B1* 7x in hexaploid wheat varieties (*Triticum aestivum*). Austral J Agric Res 55: 577-588, https://doi.org/10.1071/AR03227

von Büren M (2001) Polymorphisms in two homeologous gamma-gliadin genes and the evolution of cultivated wheat. Genet Res Crop Evol 48: 205-220, https://doi.org/10.1023/A:1011213228222

von Büren M, Lüthy J, Hübner P (2000) A spelt-specific γ-gliadin gene: discovery and detection. Theor Appl Genet 100: 271-279, https://doi.org/10.1007/s001220050036

Walkowiak S, Gao L, Monat C, Haberer G, Kassa MT, ..., Pozniak CJ (2020) Multiple wheat genomes reveal global variation in modern breeding. Nature 588: 277-283,

https://doi.org/10.1038/s41586-020-2961-x
Wang G, Snape JW, Hu H, Rogers WJ (1993) The high-molecular-weight glutenin subunit compositions of Chinese bread wheat varieties and their relationship with bread-making quality. Euphytica 68: 205-212, https://doi.org/10.1007/BF00029874
Wang J, Luo MC, Chen Z, You FM, Wei Y, Zheng Y, Dvořák J (2013) *Aegilops tauschii* single nucleotide polymorphisms shed light on the origins of wheat D-genome genetic diversity and pinpoint the geographic origin of hexaploid wheat. New Phytologist 198: 925-937, https://doi.org/10.1111/nph.12164
Wang X, Zhang Y, Zhang B, Florides CG, Gao Z, Wang Z, Zhang X, Wei Y (2018) Comparison of quality properties between high-molecular-weight glutenin subunits 5+10 and 2+12 near isogenic lines under three common wheat genetic backgrounds. Cereal Chem 95: 575-583, https://doi.org/10.1002/cche.10061
Wang Y, Chen Q, Li Y, Guo Z, Liu C, ... Wei Y, Qi P (25 authors) (2021) Post-translational cleavage of HMW-GS Dy10 allele improves the cookie-making quality in common wheat (*Triticum aestivum*). Molec Breed 41: 49, https://doi.org/10.1007/s11032-021-01238-9
Watry H, Zerkle A, Laudencia-Chingcuanco D (2020) Modified acid-Page method for rapid screening and phenotyping of wheat gliadin mutant lines. MethodsX 7: 100858, https://doi.org/10.1016/j.mex.2020.100858
Weegels PL, Hamer RJ, Schofield JD (1996) Functional properties of wheat glutenin. J Cereal Sci 23: 1-18, https://doi.org/10.1006/jcrs.1996.0001
Werner WE, Wiktorowicz JE, Kasarda DD (1994) Wheat varietal identification by capillary electrophoresis of gliadins and high-molecular-weight glutenin subunits. Cereal Chem 71: 397-402, http://pascal-francis.inist.fr/vibad/index.php?action=getRecordDetail&idt=3346336
Wicker T, Stein N, Albar L, Feuillet C, Schlagenhauf E, Keller B (2001) Analysis of a contiguous 211 Kb sequence in diploid wheat (*Triticum monococcum*) reveals multiple mechanisms of genome evolution. The Plant J 26: 307-316, https://doi.org/10.1046/j.1365-313X.2001.01028.x
Wicker T, Yahiaoui N, Guyot R, Schlagenhauf E, Liu ZD, Dubcovsky J, Keller B (2003) Rapid genome divergence at orthologous low molecular weight glutenin loci of the A and A(m) genomes of wheat. The Plant Cell 15: 1186-1197, https://doi.org/10.1105/tpc.011023
Wicker T, Buchmann JP, Keller B (2010) Patching gaps in plant genomes results in gene movement and erosion of colinearity. Genome Res 20:

1229-1237, https://www.genome.org/cgi/doi/10.1101/gr.107284.110

Wicker T, Mayer KFX, Gundlach H, Martis M, Steuernagel B, Scholz U, Šimková H, Kubaláková M, Choulet F, Taudien S, Platzer M, Feuillet C, Fahima T, Budak H, Doležel J, Keller B, Stein N (2011) Frequent gene movement and pseudogene evolution is common to the large and complex genomes of wheat, barley, and their relatives. The Plant Cell 23: 1706-1718, https://doi.org/10.1105/tpc.111.086629

Wieser H & Zimmerman G (2000) Importance of amounts and proportions of high molecular weight subunits of glutenin for wheat quality. Europ Food Res Thecnol 210: 324-330, https://doi.org/10.1007/s002170050558

William MDHM, Riera-Lizarazu O, Mujeeb-Kazi A (1992) A combination of protein electrophoretic techniques for the detection of 1B, 1B/1R heterozygotes in *Triticum aestivum* L. J Genet Breed 46: 137-142, http://pascal-francis.inist.fr/vibad/index.php?action=getRecordDetail&idt=6127515

Witkowski E, Waga J, Witkowska K, Rapacz M, Gut M, Bielawska A, Luber H, Lukaszewski AJ (2008) Association between frost tolerance and the alleles of high molecular weight glutenin subunits present in Polish winter wheats. Euphytica 159: 377-384, https://doi.org/10.1007/s10681-007-9537-9

Worland AJ (1999) The importance of Italian wheats to worldwide varietal improvement. J Genet Breed 53: 165-173.

Worland AJ, Appendino ML, Sayers EJ (1994) The distribution, in European winter wheats, of genes that influence ecoclimatic adaptability whilst determining photoperiodic insensitivity and plant height. Euphytica 80: 219-228, https://doi.org/10.1007/BF00039653

Worland AJ, Borner A, Korzun V, Li WM, Petrovic S, Sayers EJ (1998) The influence of photoperiod genes to the adaptability of European winter wheats. Euphytica 100: 385-394, https://doi.org/10.1023/A:1018327700985

Woychik JH, Boundy JA, Dimler RJ (1961) Starch gel electrophoresis of wheat gluten proteins with concentrated urea. Arch Biochem Biophys 94: 477-482, https://doi.org/10.1016/0003-9861(61)90075-3

Wrigley CW (1970) Protein mapping by combined gel electrofocusing and electrophoresis: Application to the study of the genotypic variations in wheat gliadins. Biochem Genet 4: 509-516, https://doi.org/10.1007/BF00486601

Wrigley CW (1996) Giant proteins with flour power. Nature 381: 738-739, https://doi.org/10.1038/381738a0

Wrigley CW & Baxter RI (1974) Identification of Australian wheat cultivars by laboratory procedures: grain samples containing a mixture of cultivars. Austral J Exp Agric Anim Husb 14: 805-810, https://doi.org/10.1071/EA9740796

Wrigley CW. & Shepherd kW (1973) Electrofocusing of grain proteins from wheat genotypes. Ann NY Acad Sci 209: 154-162, https://doi.org/10.1111/j.1749-6632.1973.tb47526.x

Wrigley CW, Shepherd KW (1974) Identification of Australian wheat cultivars by laboratory procedures: Examination of pure samples of grain. Austral J Exp Agric Anim Husb 14: 796-804, https://doi.org/10.1071/EA9740805

Wrigley CW. & Shepherd KW (1977) Pedigree investigation using biochemical markers: the wheat cultivar Gabo. Austral J Exp Agric Anim Husb 17: 1028-1031, https://doi.org/10.1071/EA9771028

Wrigley CW, Autran JC, Bushuk W (1982a) Identification of cereal varieties by gel electrophoresis of the grain proteins. Adv Cereal Sci Thecnol. 5: 211-259.

Wrigley CW, Robinson PJ, Williams WT (1982b) Association between individual gliadin proteins and quality, agronomic and morphological attributes of wheat cultivars. Austral J Agric Res 33: 409-418, https://doi.org/10.1071/AR9820409

Wrigley CW, Bekes F, Gupta R, Metakovsky EV (1990) Correlations relating gluten-polypeptide composition to dough properties. Proc 39th Austral Cereal Chem Conf; Royal Austr Chem Inst, Perth (1989); pp 101-106.

Wrigley CW, Bekes F, Bushuk W (2006a) Preface. In: "Gliadin and glutenin. The unique balance of wheat quality", eds C.Wrigley, F.Békés, W.Bushuk; AACC International; pp v-vii.

Wrigley CW, Bekes F, Bushuk W (2006b) Gluten: A balance of gliadin and glutenin. In: "Gliadin and glutenin. The unique balance of wheat quality", eds C.Wrigley, F.Békés, W.Bushuk; AACC International; pp 3-32, https://hdl.handle.net/102.100.100/130774?index=1

Wrigley CW, Bekes F, Cavanagh CR, Bushuk W (2006c) Gliadin and glutenin-subunit composition data provided on the AACC international website. In: "Gliadin and glutenin. The unique balance of wheat quality", eds C.Wrigley, F.Békés, W.Bushuk; AACC International; pp 453-459.

Xu H, Wang RJ, Shen X, Zhao YL, Sun GL, Zhao HX, Guo AG (2006) Functional properties of a new low-molecular-weight glutenin subunit gene from a bread wheat cultivar. Theor Appl Genet 113: 1295-1303, https://doi.org/10.1007/s00122-006-0383-2

Xynias IN, Kozub NO, Sozinov IA (2006) Seed storage protein composition of Hellenic bread wheat cultivars. Plant Breed 125: 408-410, https://doi.org/10.1111/j.1439-0523.2006.01242.x

Yamada NA, Smith GA, Castro A, Roques CN, Boyer JC, Farber RA (2002) Relative rates of insertion and deletion mutations in dinucleotide repeats of various lengths in mismatch repair proficient mouse and mismatch repair deficient human cells. Mutation Res 499: 213-225, https://doi.org/10.1016/S0027-5107(01)00282-2

Yamauchi H, Nishio Z, Takata K, Oda Y, Yamaki K, Ishida N, Miura H (2001) The brad-making quality of a domestic flour blended with an extra-strong flour, and staling of the bread made from the blended flour. Food Sci Technol Res 7: 120-125, https://doi.org/10.3136/fstr.7.120

Yen Y & Baenziger PS (1992) A better way to construct recombinant chromosome lines and their controls. Genome 35: 827-830, https://doi.org/10.1139/g92-125

You FM, Huo N, Deal KR, Gu YQ, Luo MC, McGuire PE, Dvorak J, Anderson OD (2011) Annotation-based genome-wide SNP discovery in the large and complex *Aegilops tauschii* genome using next-generation sequencing without a reference genome sequence. BMC Genomics 12:59, https://doi.org/10.1186/1471-2164-12-59

Zemetra RS, Morris R, Schmidt JW (1986) Gene locations for heading date using reciprocal chromosome substitutions in winter wheat. Crop Sci 26: 531-533, https://doi.org/10.2135/cropsci1986.0011183X002600030020x

Zhang W, Gianibelli MC, Ma W, Rampling LR, Gale KR (2003) Identification of SNPs and development of allele-specific PCR markers for γ-gliadin alleles in *Triticum aestivum.* Theor Appl Genet 107: 130-138, https://doi.org/10.1007/s00122-003-1223-2

Zhang W, Gianibelli MC, Rampling LR, Gale KR (2004) Characterization and marker development for low moecular weight glutenin genes from *Glu-A3* alleles of bread wheat (*Triticum aestivum.* L). Theor Appl Genet 108: 1409-1419, https://doi.org/10.1007/s00122-003-1558-8

Zhang Y, Liu Q, Yang W, Liu K, Sun J, Guo X, Li Y, Wang D, Ling H, Zhang A (2011) Development of a new marker system for identifying the complex members of the low-molecuar-weight glutenin subunit family in bread wheat (*Triticum aestivum* L). Theor Appl Genet 122: 1503-1516, https://doi.org/10.1007/s00122-011-1550-7

Zhang X, Liu D, Zhang J, Jiang W, Luo G, Yang W, Sun J, Tong Y, Cui D, Zhang A (2013) Novel insights into the composition, variation, organization and expression of the low-molecular-weight glutenin

subunit gene family in common wheat. J Exp Bot 64: 2027-2040, https://doi.org/10.1093/jxb/ert070
Zhang Y, Hu M, Liu Q, Sun L, Chen X, Lv L, Liu Y, Jia X, Li H (2018) Deletion of high-molecular-weight glutenin subunits in wheat significantly reduced dough strength and bread-making quality. BMC Plant Biol 18: 319, https://doi.org/10.1186/s12870-018-1530-2
Zhao HX, Hu SW, Guo AG, Mares D (2002) Study on relationship between the size distribution of polymeric protein and flour mixing properties. Agric Sci China 1:138-142.
Zhao HX, Wang R, Guo AG, Hu SW, Sun GL (2004) Development of primers specific for LMW-GS genes located on chromosome 1D and molecular characterization of a gene from *Glu-D3* complex locus in bread wheat. Hereditas 141: 193-198, https://doi.org/10.1111/j.1601-5223.2004.01852.x
Zillman RR & Bushuk W (1979a) Wheat cultivar identification by gliadin electrophoregrams. II. Effects of environmental and experimental factors on the gliadin electrophoregrams. Can J Plant Sc. 59: 281-286, https://doi.org/10.4141/cjps79-047
Zillman RR & Bushuk W (1979b) Wheat cultivar identification by gliadin electrophoregrams. III. Catalogue of electrophoregram formulas of Canadian wheat cultivars. Can J Plant Sci 59: 287-298, https://doi.org/10.4141/cjps79-048
Zimin AV, Puiu D, Hall R, Kingan S, Clavijo B, Salzberg SL (2017) The first near-complete assembly of the hexaploid bread wheat genome, *Triticum aestivum*. GigaSci 6 (11): gix097, https://doi.org/10.1093/gigasciencej.gix097
Zlatska A, Shytikova Yu, Sherepitko D (2009) Toward understanding the genetic basis of high dough strength of two common winter wheat varieties Panna and Skrbnytsya with extra strong charactristics. In: “Gluten Proteins 2009“; Proc 10th Intern Gluten Workshop; Clermont-Ferrand, France (7-9 Sept 2009); ed. G.Branlard; INRA; pp 213-216.
Zurabishvili TG, Iordansky AB, Badaev NS (1978) Linear differentiation of cereal chromosomes. II. Polyploid wheats. Theor Appl Genet 51: 201-210, https://doi.org/10.1007/BF00273766